Fossil Fuel and the Environment

Fossil Fuel and the Environment

Edited by **Robbie Larkin**

C LANRYE
INTERNATIONAL

New Jersey

Published by Clanrye International,
55 Van Reypen Street,
Jersey City, NJ 07306, USA
www.clanryeinternational.com

Fossil Fuel and the Environment
Edited by Robbie Larkin

International Standard Book Number: 978-1-63240-240-0 (Hardback)

Printed in the United States of America.

Contents

Preface

Information and problems related to fossil fuels and the environment are described in this all-inclusive book. Geopolitics has been continuously shaped by fossil fuel and even today, the world is facing an energy crisis. As a result, the world is witnessing a shift towards alternate, profitable methods of energy production like wind-turbines which are rapidly increasing in number in US and Europe, while solar power and bio-fuels are getting world recognition. This book encompasses topics ranging from specific ones like fuel combustion and co-existence with renewable energy to basics like environment, economics of energy and food security. The book is not an elaborative scientific work, but the research and analysis of the subject with a general view point make it a comfortable read for anyone who wants to gain insight on the subject.

This book is a comprehensive compilation of works of different researchers from varied parts of the world. It includes valuable experiences of the researchers with the sole objective of providing the readers (learners) with a proper knowledge of the concerned field. This book will be beneficial in evoking inspiration and enhancing the knowledge of the interested readers.

In the end, I would like to extend my heartiest thanks to the authors who worked with great determination on their chapters. I also appreciate the publisher's support in the course of the book. I would also like to deeply acknowledge my family who stood by me as a source of inspiration during the project.

Editor

A Review of Hydrogen-Natural Gas Blend Fuels in Internal Combustion Engines

Antonio Mariani, Biagio Morrone and Andrea Unich
Dept. of Aerospace and Mechanical Engineering - Seconda Universitá degli Studi di Napoli
Italy

1. Introduction

In the last ten years, the number of natural gas (NG) vehicles worldwide has rapidly grown with the biggest contribution coming from the Asia-Pacific and Latin America regions (IANGV, 2011). As natural gas is the cleanest fossil fuel, the exhaust emissions from natural gas spark ignition vehicles are lower than those of gasoline-powered vehicles. Moreover, natural gas is less affected by price fluctuations and its reserves are more evenly widespread over the globe than oil. In order to increase the efficiency of natural gas engines and to stimulate hydrogen technology and market, hydrogen can be added to natural gas, obtaining Hydrogen - Natural Gas blends, usually named as HCNG.

This chapter gives an overview of the use of HCNG fuels in internal combustion engines. The chemical and physical properties of hydrogen and natural gas relevant for use in internal combustion engines are described. Then a survey on the impact of hydrogen on natural gas engine performance and emissions is presented with reference to research activities performed on this field.

2. Data reduction

In this section the main physical quantities used in this chapter are presented and discussed.

The stoichiometric air-fuel ratio on mass basis (AFR_{stoich}), defined in equation 1, is the mass of air needed to fully oxidize 1 kg of fuel, while AFR is the ratio between air and fuel mass flow rates, equation 2. The ratio between the actual AFR and the AFR_{stoich}, is the relative air-fuel ratio, equation 3. If $\lambda > 1$ the mixture is *lean* and the oxidation takes place with excess of air respect to the stoichiometric amount; for λ values lower than 1 the mixture is *rich*, and the fuel oxidation is not complete. The ratio $1/\lambda$ is defined as the equivalence ratio ϕ, equation 4.

$$AFR_{stoich} = \left(\frac{m_a}{m_f} \right)_{stoich} \tag{1}$$

$$AFR = \frac{m_a}{m_f} \tag{2}$$

$$\lambda = \frac{AFR}{AFR_{stoich}} \tag{3}$$

$$\phi = \frac{1}{\lambda} \tag{4}$$

Equation 5 defines the indicated mean effective pressure (imep), an engine parameter which evaluates the work obtained by an engine cycle, $\oint p\, dV$, divided by the engine displacement. The Coefficient of Variation of imep, COV_{imep}, is the ratio of the standard deviation of the indicated mean effective pressure and the average imep over a representative number of cycles, equation 6.

$$imep = \frac{1}{V_d} \oint p\, dV \tag{5}$$

$$COV_{imep} = \frac{\sigma_{imep}}{imep_{avg}} \tag{6}$$

In case the effect of mechanical efficiency has to be taken into account, the brake mean effective pressure (bmep) is considered. In 4-stroke engines, the bmep is calculated from the torque measured at the engine shaft, according to equation 7:

$$bmep = \frac{T \cdot 4\pi}{V_d} \tag{7}$$

The stoichiometric reaction equation of a methane-hydrogen blend reads as:

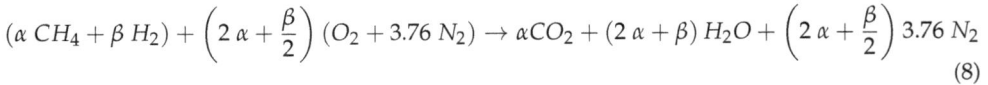

$$(\alpha\, CH_4 + \beta\, H_2) + \left(2\,\alpha + \frac{\beta}{2}\right)(O_2 + 3.76\, N_2) \rightarrow \alpha CO_2 + (2\,\alpha + \beta)\, H_2O + \left(2\,\alpha + \frac{\beta}{2}\right) 3.76\, N_2 \tag{8}$$

where $\alpha + \beta = 1$. The quantities α and β represent the mole per each species in the blend, and it is immediate to observe that the reduction of the C/H ratio, compared to pure methane, brings about a theoretical reduction of the CO_2.

The burning velocity represents a main property for the combustion characteristics of the fuels and is defined as the velocity at which unburned gases move through the combustion wave in the direction normal to the wave surface (Glassman & Yetter, 2008). The laminar burning velocities can be obtained using the following equation 9 (Mandilas et al., 2007) being S_s the unstretched flame speed, ρ_b and ρ_u the burned and unburned gas densities. Equation 10 relates the unstretched flame speed, the stretched flame speed S_n, the stretch rate κ and the Markstein length L_b.

$$u_l = S_s \frac{\rho_b}{\rho_u} \tag{9}$$

$$S_s - S_n = \kappa L_b \tag{10}$$

The stretch rate κ is calculated from the position of the flame front, $R = R(t)$, with the following equation 11 (Chen, 2009):

$$\kappa = \frac{1}{R}\frac{dR}{dt} \tag{11}$$

The Markstein length characterizes the variation in the local flame speed due to the influence of external stretching and determines the flame instability with respect to preferential diffusion (Markstein, 1964).

3. Natural gas

The main natural gas constituent is methane and the composition is strictly dependent on the origin gas field. Table 1 shows the composition of a natural gas sample obtained by the Italian distribution network, determined by means of gas chromatographic analysis.

Natural gas has been widely investigated as fuel for road vehicles because of its lower impact on the environment than gasoline and more widespread resources.

Constituent	Composition [% vol.]
Methane	88.98
Ethane	6.85
Propane	1.27
Butane	0.24
Pentane	0.04
Hexane	0.003
Nitrogen	0.96
Carbon dioxide	1.61

Table 1. Example of natural gas composition.

Ristovski et al. (2004) performed an experimental activity on a passenger car converted to operate either on gasoline or on compressed natural gas (CNG). Fuelling the engine by CNG, both regulated (CO, NOx and HC) and unregulated emissions (PAHs and formaldehyde) were lower than gasoline.

Prati, Mariani, Torbati, Unich, Costagliola & Morrone (2011) tested a bifuel passenger car fuelled alternatively by gasoline and natural gas on a chassis dynamometer over different driving cycles, in order to evaluate the effects of fuel properties on combustion, exhaust emissions and engine efficiency. The results showed that gasoline produced CO emissions higher than NG over the real world Artemis driving cycles, as a consequence of mixture enrichment during load transients. A detailed description of the driving cycles is reported in Barlow et al. (2009). Over the type approval New European Driving Cycle (NEDC), NG involved higher HC emissions compared to gasoline as a consequence of the higher light-off temperature for the catalytic oxidation of CH_4, which is the major constituent of HC when the vehicle is fuelled by NG, while there were no differences over the Artemis driving cycles which were performed after a warming up conditioning of the vehicle. NOx emissions were higher for gasoline over all the test cycles. CO_2 emissions for CNG showed a reduction between 21% and 29% over the tested driving cycles as a consequence of the reduced carbon content of the fuel and the lower fuel consumption on mass basis. A 5% fuel consumption reduction, expressed in MJ/km, is observed over the NEDC for the CNG respect to gasoline, while for the Artemis the reduction ranges between 10% and 22%. The higher gasoline consumption is the consequence of the mixture enrichment during transients. Particulate emissions referred to gasoline were higher than NG ones over the NEDC and comparable over the Artemis. Particle number observed was also higher for gasoline, with the exception of the Artemis Motorway.

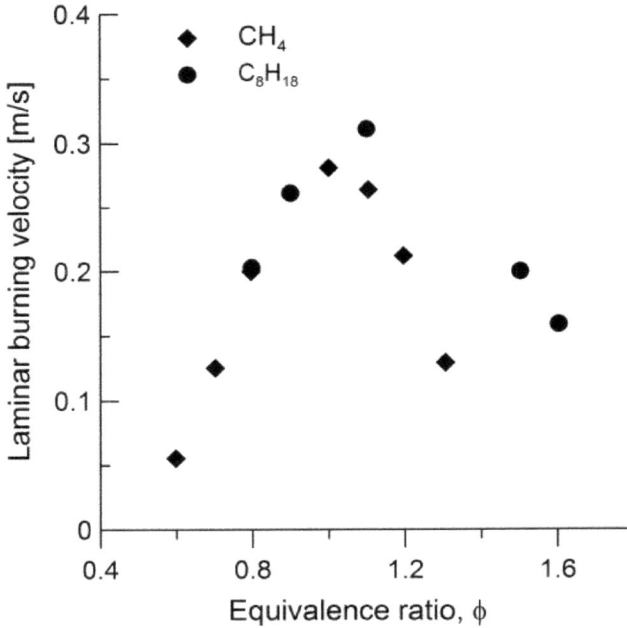

Fig. 1. Flame speeds of methane and iso-octane versus equivalence ratio ϕ (Mandilas et al., 2007).

One of the drawbacks of the NG fuel is the laminar burning velocity lower than gasoline, as shown in Figure 1 (Mandilas et al., 2007) requiring, as a consequence, a higher spark advance.

4. Hydrogen production and storage

4.1 Hydrogen production

The production of hydrogen is an important aspect since it is not present as a free chemical species in nature. Hydrogen can be produced in several ways, but reforming from fossil fuels or partial oxidation and electrolysis are the most employed from an industrial point of view.

The electrolysis consists in splitting the water molecule in hydrogen and oxygen as indicated in the next reaction equation:

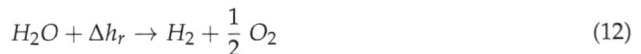

$$H_2O + \Delta h_r \rightarrow H_2 + \frac{1}{2} O_2 \tag{12}$$

If the energy for water electrolysis is provided by renewable energy sources, hydrogen production is an environmental friendly process, without green-house gas emissions. Two main types of industrial electrolysis units are used today, which differ in the type of electrolyte adopted. The first type of electrolysers is characterised by an alkaline aqueous solution of $25 - 35\%$ in weight of potassium hydroxide (KOH) to maximise the ionic conductivity, in which the hydroxide ions (OH^-) are the charge carriers (Ulleberg, 2003). In the second type of electrolysers the electrolyte is an ion conducting membrane that allows H^+ ions to be transported from the anode to the cathode side to recombine forming hydrogen. They are known as Proton Exchange Membrane (PEM) electrolysers (Barbir, 2005). However, water electrolysis powered by renewable energy sources is not competitive considering the current energy costs but it may become more economical in the future (Bartels et al., 2010).

The nowadays most economical sources of hydrogen are coal and natural gas, with significant experience in the operation of these types of plants, which will continue to be built and operated. The fuel reforming is a process in which hydrocarbon fuels, such as natural gas, are converted into a hydrogen-rich reformate gas. A reformer accomplishes the task by thermo-chemically processing hydrocarbon feedstock in high temperature reactors with steam and/or oxygen. Effective reformers should efficiently produce pure hydrogen with low pollutants emission. The methane steam reforming global reaction is reported as an example in the following reaction 13.

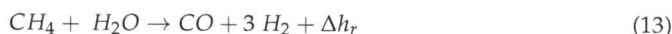

$$CH_4 + H_2O \rightarrow CO + 3\,H_2 + \Delta h_r \qquad (13)$$

The reformate gas is composed of 40% − 70% hydrogen by volume and carbon monoxide, carbon dioxide, water, nitrogen and traces of other compounds. The water-gas shift conversion removes CO and increases hydrogen content. Shift step takes place at high temperatures of about 350 − 480°C, followed by a low-temperature shift (180 − 250°C).

4.2 Hydrogen storage

Hydrogen has been recognized as an ideal energy carrier but it has not yet been widely employed in the transportation sector. The lack of an efficient storage prevents its application, in particular as fuel for transportation. Because of the low density of hydrogen at ambient conditions, it is a challenge to store enough energy on-board to allow for an acceptable vehicle range. The density can be increased by pressurizing or liquefying hydrogen. High-pressure gaseous hydrogen, up to 700 bars, is considered a potential safety hazard due to problems of material resistance. For vehicle application, cylinders are made of composite fibre due to weight considerations. Indeed, tanks add a relevant weight to the vehicle, much greater than the stored fuel, which is the 3% of the total weight (cylinder plus fuel) for a 700 bars approved system (Sørensen, 2005).

Liquid hydrogen storage requires refrigeration to a temperature of about 20 K, and the liquefaction process requires at least 15.1 MJ/kg. The on-board storage pressures for the liquid hydrogen are only slightly above the atmospheric, with typical values around 6 bars. The vessel for storing liquid hydrogen consists of several metal layers separated by highly insulating materials. The main drawback is the hydrogen boil-off from the storage caused by the need to control tank pressures by venting valves. Boil-off usually starts after a dormancy period and then proceeds at a level of 3% − 5% per day (Sørensen, 2005).

As an alternative, even more challenging options have been proposed and investigated. Most attention is paid to storage in solid materials and especially metal hydrides. Here, hydrogen gas is fed to a tank containing a metal powder and is absorbed as hydrogen atoms in the metals crystal lattice to form a metal hydride. In metal hydrides, hydrogen can be stored with energy densities up tp 15000 MJ/m^3, higher than that of liquid hydrogen, which is 8700

MJ/m^3 (Sørensen, 2005). The main disadvantage, however, is the weight of the storage alloys. Furthermore refuelling times are affected by absorption rates.

Other storage options are under investigation but still at prototypal stage (Bakker, 2010).

5. HCNG blends

Table 2 compares the main physical properties for pure fuels, methane and hydrogen. In the same table, LHV represents the Lower Heating Value of the fuel, AFR is the air-to-fuel ratio and $LHV_{stoich,\ mix}$ $[MJ/Nm^3]$ is the volumetric lower heating value for a stoichiometric air-fuel mixture.

	CH_4	H_2
Adiabatic flame temperature of stoichiometric mixtures [K]	2210	2400
Flammability limits in air at 25°C and 1 bar [% vol.]	5.0-15	4.0-75
Minimum ignition energy in air at $\phi = 1$ and 1 bar [mJ]	0.47	0.02
LHV [MJ/kg]	50.0	120.3
LHV_{vol} $[MJ/Nm^3]$	35.3	10.6
AFR_{stoich}	17.2	34.3
$LHV_{stoich,\ mix}$ $[MJ/Nm^3]$	3.351	3.143

Table 2. CH_4 and H_2 properties (Glassman & Yetter, 2008).

Table 3 shows the main fuel characteristics of natural gas and hydrogen-natural gas blends with 10% (HCNG10), 20% (HCNG20) and 30% (HCNG30) of hydrogen in volume. The volumetric hydrogen content is calculated according to equation 14.

$$H_2[\%vol.] = \frac{V_{H_2}}{V_{NG} + V_{H_2}} \tag{14}$$

The volumetric Lower Heating Value is the fuel energy per unit volume, so it is a measure of the energy that can be stored in the fuel tank. It is 7% lower than NG for HCNG10, 14% for HCNG20 and 21% for HCNG30. $LHV_{stoich,\ mix}$, which is proportional to the engine power output, is negligibly affected by hydrogen addition.

	Natural Gas	HCNG10	HCNG20	HCNG30
H_2 [% vol.]	-	10	20	30
H_2 [% energy]	-	3.2	7.0	14.4
LHV [MJ/kg]	45.3	46.2	46.7	48.5
LHV_{vol} $[MJ/Nm^3]$	36.9	34.3	31.7	29.2
AFR_{stoich}	15.6	15.8	16.1	16.4
$LHV_{stoich,\ mix}$ $[MJ/Nm^3]$	3.375	3.367	3.358	3.349

Table 3. NG and HCNG fuel properties.

5.1 Combustion characteristics

Since hydrogen laminar combustion speed is about eight times greater than methane, it provides a reduction of combustion duration when mixed with natural gas in small concentrations. Many studies have been carried out to measure the flame speed of hydrogen-methane air mixtures at different hydrogen concentrations and equivalence ratios. Ilbas et al. (2006) performed the measurements at ambient temperatures with hydrogen-methane blends up to 100% hydrogen.

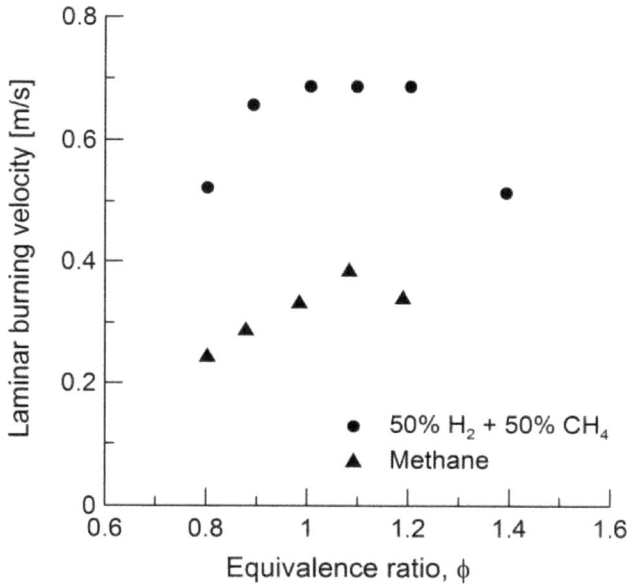

Fig. 2. Flame speed of different fuels versus equivalence ratio ϕ (Ilbas et al., 2006).

Figure 2 shows the flame speed for methane and a 50% hydrogen-methane blend plotted versus the equivalence ratio. The maximum flame speed for the blend is 0.69 m/s while the maximum for methane is 0.39 m/s for an equivalence ratio $\phi = 1.1$. The flammable regions were also widened as the hydrogen content increased in the mixtures.

Figure 3, where the flame speed is plotted versus hydrogen content, shows the non-linear dependence of this property on hydrogen percentage.

Mandilas et al. (2007) performed experiments in a spherical stainless steel vessel at initial temperatures up to 600 K and initial pressures up to 1.5 MPa to study the effects of hydrogen addition on laminar and turbulent premixed methane-air flames. The burning velocity, u_l, was found using equation 9. Methane can be ignited for $0.6 \leq \phi \leq 1.3$, with the peak burning velocity occurring at $\phi = 1.0$. The addition of H_2 extends the ignition limits to the range $0.5 \leq \phi \leq 1.4$ and increases the values of u_l at lean equivalence ratios, while u_l does not increase for rich equivalence ratios. The authors also compared the turbulent velocity u_{tr} for methane and a blend with 30% of hydrogen. As in the laminar case, the addition of hydrogen

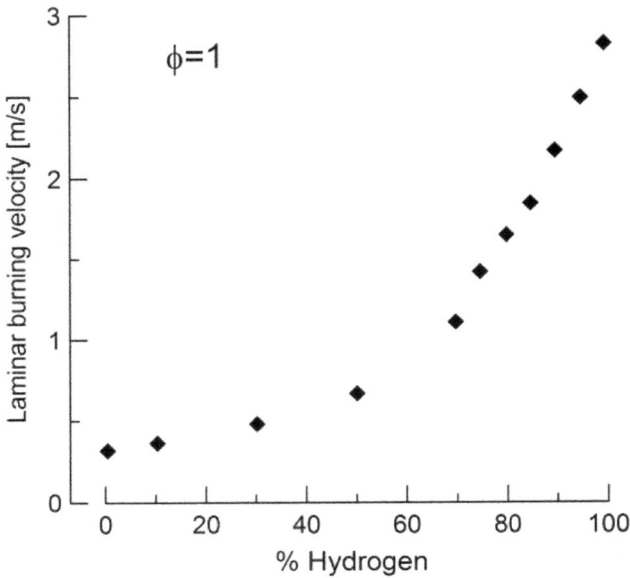

Fig. 3. Flame speeds versus hydrogen content in methane-hydrogen blends at $\phi = 1$ (Ilbas et al., 2006).

extends the ignition limits and higher u_{tr} values, in particular at lean air-fuel mixtures, are attained compared to methane.

A comparison of results obtained by several authors for the unstretched laminar burning velocity versus the equivalence ratio, for HCNG20, is shown in Figure 4 (Miao et al., 2009). It is observed in any case that the maximum flame speed is attained at $\phi \cong 1.1$ with values around 0.5 m/s.

5.2 The impact of HCNG blends on engine efficiency and exhaust emissions

The reduction of combustion duration promoted by hydrogen addition results in increased engine efficiency respect to natural gas and enhances combustion stability, reducing cycle-by-cycle variation. Nagalingam et al. (1983) proved that the high burning rate of HCNG blends requires an ignition timing lower than natural gas to obtain the Maximum Brake Torque (MBT).

The MBT spark advance versus the hydrogen content, shown in Figure 5 (Karim et al., 1996), is noticeably affected by hydrogen addition, in particular for very lean air-fuel mixtures. The plot shows that for blends containing significant amount of hydrogen, small adjustments to the ignition timing are needed when the equivalence ratio is changed.

The engine efficiency can be increased fuelling the engine by HCNG blends. Sierens & Rosseel (2000) developed a fuel system which supplies hydrogen-natural gas mixtures in variable proportion to the engine. For low brake mean effective pressures high efficiency can be achieved by increasing the hydrogen content reducing throttling losses. The authors

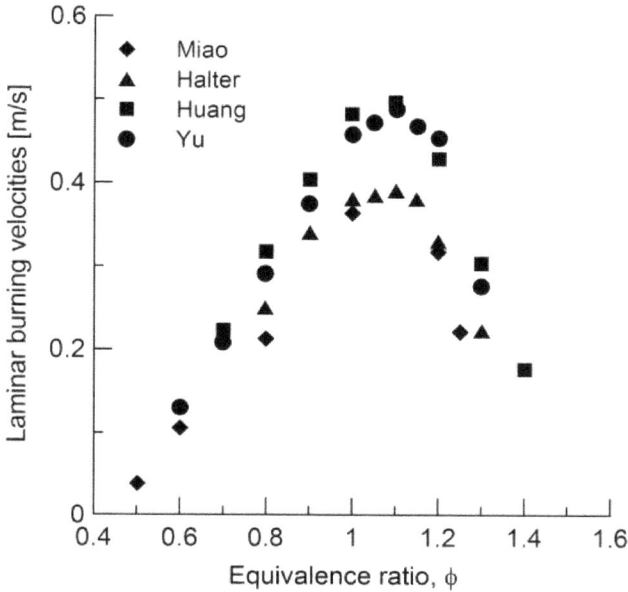

Fig. 4. Unstretched laminar burning velocity u_l versus the equivalence ratio ϕ for HCNG20 (Miao et al., 2009).

found that 10% hydrogen increases engine efficiency moderatly whereas 20% hydrogen gives negligible extra benefit, as shown in Figure 6.

Recently, Ma et al. (2010) investigated the effect of high hydrogen volumetric content, up to 55%, on the performance of a turbocharged lean burn natural gas engine. The authors found that the addition of hydrogen significantly extends the lean limit, decreases burn duration and yields higher thermal efficiency. The plot of the engine efficiency versus λ, Figure 7, shows a negative trend in engine efficiency for natural gas for λ values greater than 1.3, while the blend with the higher hydrogen content shows positive trend up to $\lambda = 1.6$.

The increased hydrogen/carbon ratio and engine efficiency bring a reduction of CO_2 emissions. By the way, as a consequence of a faster combustion, higher temperature are attained in the combustion chamber, increasing NOx emissions in HCNG fuelled engines compared to natural gas, for a given equivalence ratio ϕ. NOx can be kept down and engine efficiency further improved if the engine is run with lean mixtures or adopting EGR at stoichiometric air-fuel ratio.

Sierens & Rosseel (2000) found the maximum NOx emissions at a relative air-fuel ratio $\lambda = 1.1$. For higher λ values, the reduction in heat of combustion available for the charge mixture reduces the temperature and NOx as a consequence, as shown in Figure 8. However, such conditions cause an increase in THC emissions, as shown in Figure 9.

Hoekstra et al. (1995) obtained very low NOx emissions operating with HCNG blends close the lean limit, significantly extended compared with natural gas. Besides, the excellent anti

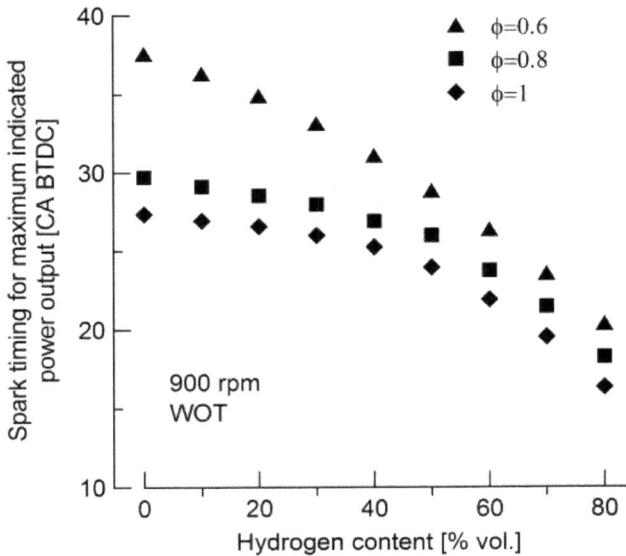

Fig. 5. Spark timing for maximum indicated power output versus hydrogen content (Karim et al., 1996).

knock qualities of natural gas are not undermined by the presence of relatively small amounts of hydrogen in the blend (Karim et al., 1996).

The effect of hydrogen on the lean limit, here defined as the λ value at which the COV_{imep} attains 10%, is shown in Figure 10 (Ma et al., 2010), with values of 1.2 for NG, 2.1 for HCNG30 and 2.5 for HCNG55.

The impact of hydrogen addition to natural gas on cycle-by-cycle variations have been investigated in many scientific activities and the results showed that the coefficient of variation in maximum pressure and in indicated mean effective pressure are reduced with increasing hydrogen content, both with lean air-to-fuel ratio as well described by Ma et al. (2008) in Figure 11 and Wang et al. (2008) and with large exhaust gas recirculation ratio values, Figure 12 (Huang et al., 2009).

Numerical simulations have also been used to predict performance and emissions of internal combustion engines fuelled by HCNG blends.

Figure 13 shows the predicted fuel consumption in terms of energy per kilometer [MJ/km] over the NEDC versus the hydrogen content (Mariani et al., 2011). Stoichiometric air-to-fuel ratio was considered for each fuel in order to assure an efficient exhaust after-treatment adopting a three-way catalyst. Exhaust gas recirculation was investigated (instead of ultra lean mixture) with the aim at improving engine efficiency and reducing NOx emissions respect to undiluted charge. In fact, HCNG blends combustion properties are particularly suitable for EGR, assuring a stable combustion even if the charge is diluted (Hu et al., 2009).

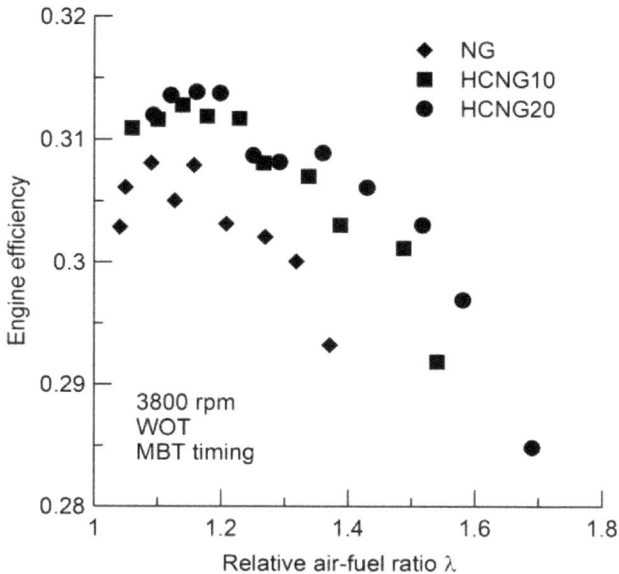

Fig. 6. Engine efficiency versus relative air-fuel ratio λ for different fuels (Sierens & Rosseel, 2000).

MBT ignition timing has been adopted for all fuels and operating conditions investigated. Fuel consumption is reduced as the hydrogen content increases due to the positive effect on average engine efficiency over the driving cycle, with values 2.5%, 4.7% and 5.7% lower than NG for HCNG10, 20 and 30 respectively. Fuel consumption is further reduced adopting 10% EGR for HCNG blends, with values 5.4%, 6.6% and 7.7% lower than NG for HCNG10, 20 and 30 respectively. NOx emissions, expressed in g/km over the driving cycles, are reported in Figure 14. Adding hydrogen higher in-cylinder temperatures are attained as a consequence of a faster combustion, resulting in increased NOx emissions with values 3.6%, 10.7% and 19.7% higher than NG for HCNG10, HCNG20 and HCNG30 respectively. The use of EGR results in lower NOx emissions with respect to the case without EGR, with values about 85% lower than CNG for each HCNG fuel.

6. Real-life cases of HCNG use

HCNG blends can be distributed by the present natural gas refuelling stations, providing them with a mixing equipment in order to obtain blends with the selected hydrogen content. The system must operate to assure a high accuracy of hydrogen percentage because the fuel composition influences engine performances hence requiring customized engine calibration. In particular, the increased combustion velocity requires a reduction of the ignition advance as the hydrogen concentration increases to obtain the maximum engine torque. Furthermore, the fuel supply system should be calibrated to compensate the variation of fuel properties caused by hydrogen addition. In fact, present natural gas vehicles requires stoichiometric air-fuel ratio to obtain a high conversion efficiency of HC, CO and NOx emissions in the three-way

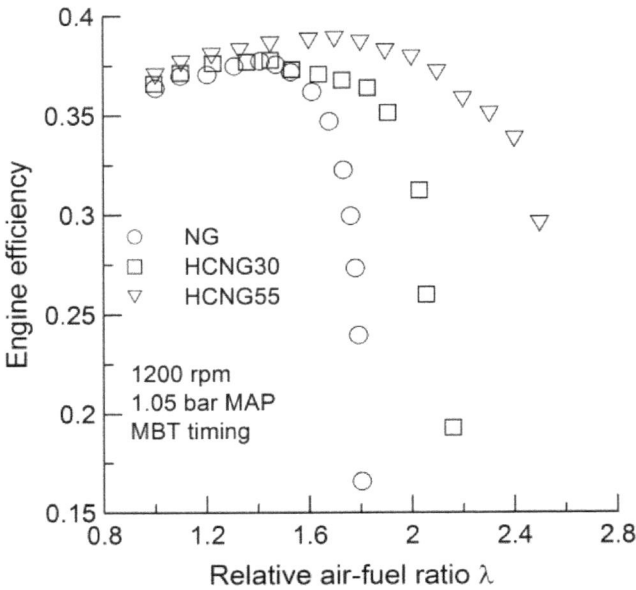

Fig. 7. Engine efficiency versus relative air-fuel ratio λ for different fuels (Ma et al., 2010).

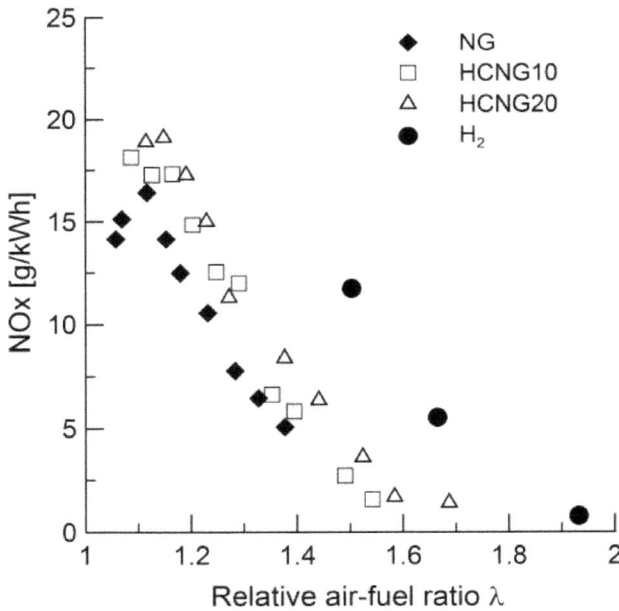

Fig. 8. NOx emissions versus relative air-fuel ratio λ (Sierens & Rosseel, 2000).

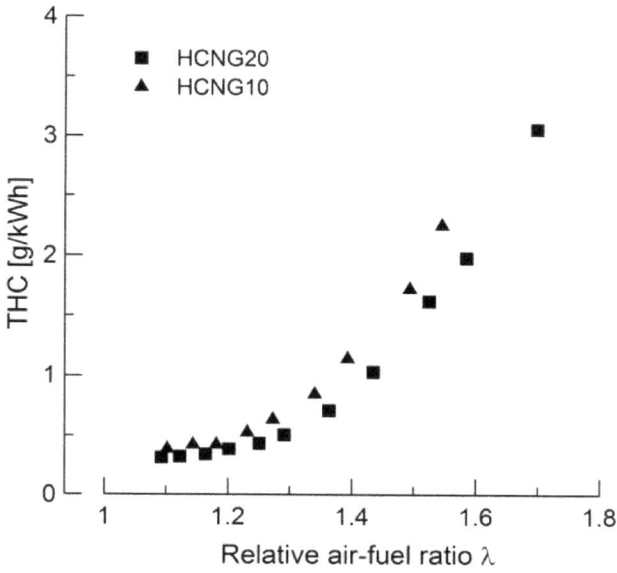

Fig. 9. Hydrocarbon emission versus relative air-fuel ratio λ (Sierens & Rosseel, 2000).

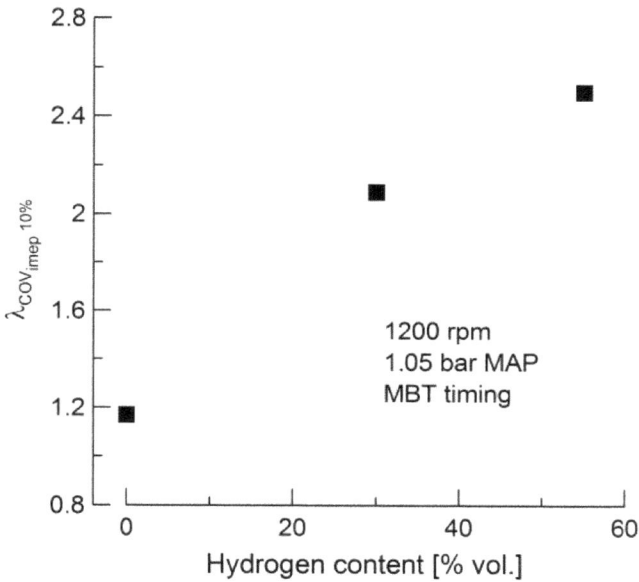

Fig. 10. Lean limit versus hydrogen content in the blend (Ma et al., 2010).

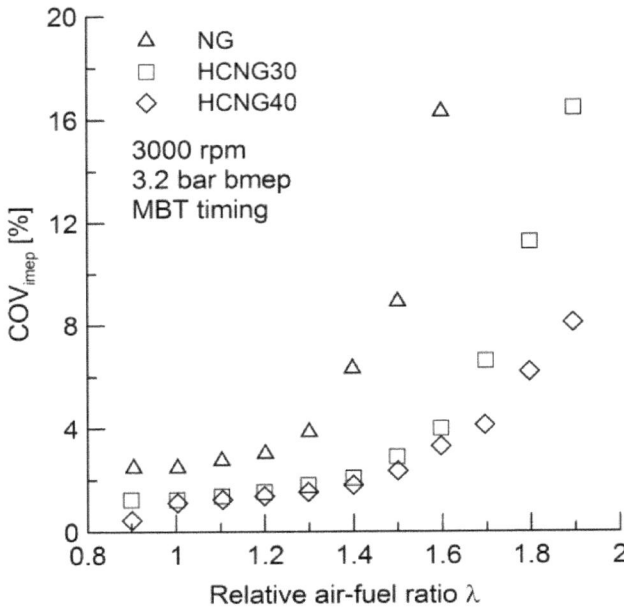

Fig. 11. COV_{imep} versus relative air-fuel ratio λ for NG and HCNG blends (Wang et al., 2008).

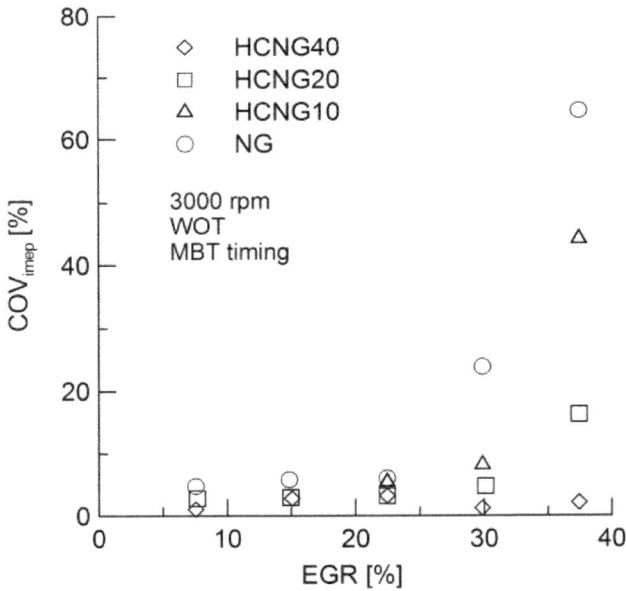

Fig. 12. COV_{imep} versus EGR for NG and HCNG blends (Huang et al., 2009).

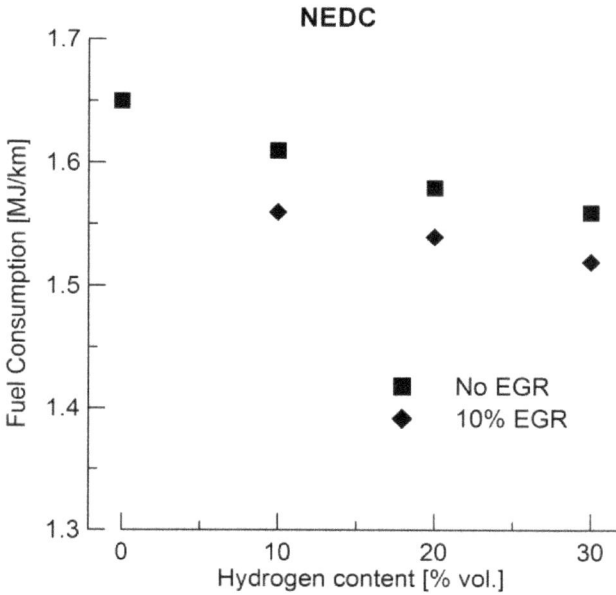

Fig. 13. Predicted fuel consumption versus hydrogen content over the NEDC (Mariani et al., 2011).

catalytic converter. HCNG fuels can be used in lean burn engines or with high EGR rates at stoichiometric conditions, exploiting their excellent combustion properties, with positive impact on engine efficiency and low exhaust emissions.

Finally, the use of HCNG fuel can stimulate the development of the hydrogen technologies and market which are, nowadays, the main practical problems preventing it to be implemented.

Many research projects have been performed in the past and others are still going on to assess the potential benefits coming by using HCNG fuels in real-life applications. The U.S. Department of Energy Advanced Vehicle Testing Activity (AVTA) teamed with Electric Transportation Applications (ETA) and Arizona Public Service (APS) to develop a hydrogen pilot plant, where hydrogen is produced by means of PEM electrolyzer and is dispensed to vehicles that operate with different HCNG blends with hydrogen ranging from 0% to 100%. The project demonstrated the safety of operating vehicles on hydrogen and the reduction of exhaust emissions attainable with hydrogen and HCNG fuelled vehicles compared to gasoline (Francfort & Karner, 2006).

A hydrogen production plant with HCNG dispenser have been built in Malmö, Sweden, for project to improve engine efficiency and reduce emissions of a bus fleet (Ridell, 2006).

In Italy, public transportation companies of Regione Emilia Romagna and the ENEA research center are involved in experimental tests to evaluate fuel consumption and exhaust emissions of buses for urban transport service, Figure 15 (Genovese et al., 2011).

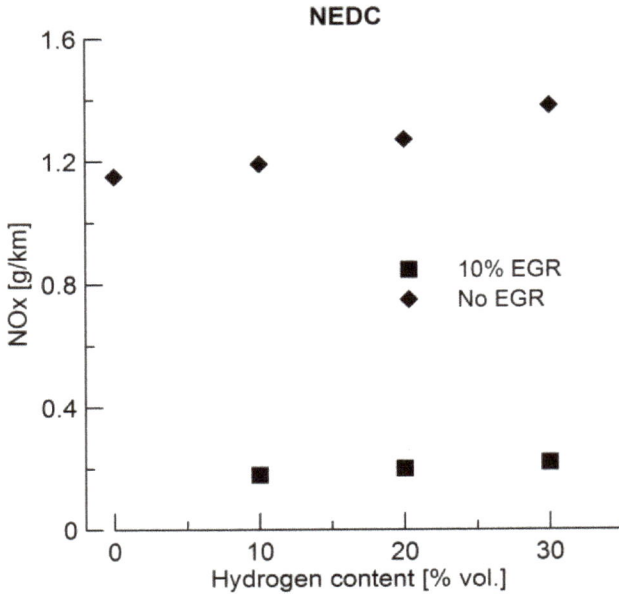

Fig. 14. Predicted NOx emissions versus hydrogen content over the NEDC (Mariani et al., 2011).

Fig. 15. Urban bus tested with HCNG blends (Genovese et al., 2011).

Regione Lombardia, Fiat Research Center, Sapio, CNR-Istituto Motori and Seconda Universitá degli studi di Napoli are involved in a project to test a passenger car fuelled by HCNG blends, varying the hydrogen content, in order to assess the impact of hydrogen addition to natural gas on combustion, exhaust emissions and fuel consumption, over different driving cycles, Figure 16 (Prati, Costagliola, Torbati, Unich, Mariani, Morrone & Gerini, 2011).

The authors of this review have designed and built an high accuracy mixing equipment to produce HCNG blends with imposed hydrogen content. The device is developed on the occasion of a project which involves the research group of the Seconda Universitá degli Studi di Napoli, the Neapolitan Transportation Company (CTP), NA-MET, the company managing the NG bus fleet and ECOS srl, an enterprise which develops CNG fuelling stations.

Fig. 16. Fiat Panda HCNG tested in the laboratory of Istituto Motori-CNR (Prati, Costagliola, Torbati, Unich, Mariani, Morrone & Gerini, 2011).

7. Conclusion

Natural gas is employed as fuel since it is the cleanest fossil fuel with exhaust emissions from natural gas vehicles lower than those of gasoline-powered vehicles. Some of its drawbacks can be mitigated by enriching it with hydrogen to produce the so called hydrogen-natural gas blends.

The laminar flame speed of methane is lower than the gasoline one and the addition of hydrogen, which presents a laminar flame speed about eight times that of methane, significantly improves this main combustion property.

In the past years, many authors have proved both experimentally and numerically that the HCNG blends improve engine efficiency and reduce CO_2 emissions because of the reduced C/H ratio and fuel consumption. NOx emissions are, instead, larger than NG because of the higher in-cylinder temperature attained, for a given equivalence ratio. Anyway, the use of lean AFR or the EGR definitely reduces NOx emissions and bring about an extra increase in engine efficiency. The good combustion patterns of HCNG blends help to keep low HC emissions.

8. Acknowledgements

This work has been supported by a PRIST 2008 grant by the Seconda Universitá degli studi di Napoli, together with a 2011 research grant funded by the Seconda Universitá degli studi di Napoli.

9. Nomenclature

AFR	Air-fuel ratio [kg_{air}/kg_{fuel}]
avg	Average
CA	Crank angle [$°$]
COV	Coefficient of variation
BTDC	Before top dead center
EGR	Exhaust gas recirculation
HCNG	Hydrogen-natural gas blend
imep	Indicated mean effective pressure [Pa]
LHV	Lower heating value [MJ/kg or MJ/Nm^3]
MAP	Manifold absolute pressure
MBT	Maximum brake torque
NG	Natural gas
NEDC	New European driving cycle
NOx	Nitrogen oxides
PAH	Polycyclic aromatic hydrocarbons
R	Flame front position [m]
rpm	Revolutions per minute
S	Flame speed [m/s]
t	Time [s]
T	Torque [N m]
THC	Total unburned hydrocarbon
u_l	Unstretched laminar burning velocity [m/s]
V	Volume [m^3]
WOT	Wide open throttle
Greek symbols	
α	Mole number of NG [mol]
β	Mole number of hydrogen [mol]
Δh_r	Enthalpy of reaction [kJ/mol]
κ	Stretch rate [1/s]
λ	Relative air-fuel ratio [-]
ϕ	Equivalence ratio [-]
σ	Standard deviation
Subscripts	
a	Air
b	Burned
d	Displacement
f	Fuel
l	Laminar
mix	Mixture
n	Stretched
s	Unstretched
stoich	Stoichiometric
tr	Turbolent
u	Unburned
vol	Volumetric

10. References

Bakker, S. (2010). Hydrogen patent portfolios in the automotive industry - the search for promising storage methods, *Int. J. of Hydrogen Energy* 35: 6784–6793.

Barbir, F. (2005). Pem electrolysis for production of hydrogen from renewable energy sources, *Solar Energy* 78: 661–669.

Barlow, T., Latham, S., McCrae, I. & Boulter, P. (2009). *A reference book of driving cycles for use in the measurement of road vehicle emissions.*
URL: *http://www.dft.gov.uk/pgr/roads/environment/emissions/ppr-354.pdf*

Bartels, J. R., Pate, M. B. & Olson, N. K. (2010). An economic survey of hydrogen production from conventional and alternative energy sources, *Int. J. of Hydrogen Energy* 35: 8371–8384.

Chen, Z. (2009). Effects of hydrogen addition on the propagation of spherical methane/air fames: A computational study, *Int. Journal Hydrogen Energy* 34: 6558–6567.

Francfort, J. & Karner, D. (2006). Hydrogen ice vehicle testing activities, *SAE paper* (2006-01-0433).

Genovese, A., Contrisciani, N., Ortenzi, F. & Cazzola, V. (2011). On road experimental tests of hydrogen/natural gas blends on transit buses, *Int. J. of Hydrogen Energy* 36: 1775–1783.

Glassman, I. & Yetter, R. A. (2008). *Combustion*, fourth edn, Academic press, San Diego.

Hoekstra, R., Collier, K., Mulligan, N. & Chew, L. (1995). Experimental study of a clean burning vehicle fuel, *Int. J. Hydrogen Energy* 20: 737–745.

Hu, E., Huang, Z., Liu, B., Zheng, J. & Gu, X. (2009). Experimental study on combustion characteristics of a spark-ignition engine fueled with natural gas hydrogen blends combining with egr, *Int. J. Hydrogen Energy* 34: 1035–1044.

Huang, B., Hu, E., Huang, Z., Zheng, J., Liu, B. & Jiang, D. (2009). Cycle-by-cycle variations in a spark ignition engine fuelled with natural gas-hydrogen blends combined with egr, *Int. J. Hydrogen Energy* 34: 8405–8414.

IANGV (2011). Natural gas vehicle statistics.
URL: *www.iangv.org*

Ilbas, M., Crayford, A., Yilmaz, I., Bowen, P. & Syred, N. (2006). Laminar-burning velocities of hydrogen-air and hydrogen-methane-air mixtures: An experimental study, *Int. J. Hydrogen Energy* 31: 1768–1779.

Karim, G. A., Wierzba, I. & Al-Alousi, Y. (1996). Methane-hydrogen mixtures as fuels, *Int. J. Hydrogen Energy* 21: 625–631.

Ma, F., Wang, M., Jiang, L., Chen, R., Deng, J., Naeve, N. & Zhao, S. (2010). Performance and emission characteristics of a turbo charged cng engine fueled by hydrogen-enriched compressed natural gas with high hydrogen ratio, *Int. J. Hydrogen Energy* 35: 6438–6447.

Ma, F., Wang, Y., Liu, H., Li, Y., Wang, J. & Ding, S. (2008). Effects of hydrogen addition on cycle-by-cycle variations in a lean burn natural gas spark-ignition engine, *Int. J. Hydrogen Energy* 33: 823–831.

Mandilas, C., Ormsby, M., Sheppard, C. & Woolley, R. (2007). Effects of hydrogen addition on laminar and turbulent premixed methane and iso-octane air flames, *Proceedings of the Combustion Institute* 31: 1443–1450.

Mariani, A., Morrone, B. & Unich, A. (2011). Numerical evaluation of internal combustion spark ignition engines performance fuelled with hydrogen - natural gas blends, *Int. J. Hydrogen Energy* p. doi:10.1016/j.ijhydene.2011.10.082.

Markstein, G. (1964). *Nonsteady Flame Propagation*, Pergamon Press.

Miao, H., Jiao, Q., Huang, Z. & Jiang, D. (2009). Measurement of laminar burning velocities and markstein lengths of diluted hydrogen-enriched natural gas, *Int. J. Hydrogen Energy* 34: 507–518.

Nagalingam, B., Duebel, F. & Schmillen, K. (1983). Performance study using natural gas, hydrogen-supplemented natural gas and hydrogen in avl research engine, *Int. J. Hydrogen Energy* 8: 715–720.

Prati, M. V., Costagliola, M. A., Torbati, R., Unich, A., Mariani, A., Morrone, B. & Gerini, A. (2011). Combustion analysis of a sparl ignition engine fuelled with natural gas-hydrogen blends, *WHTC*.

Prati, M. V., Mariani, A., Torbati, R., Unich, A., Costagliola, M. A. & Morrone, B. (2011). Emissions and combustion behavior of a bi-fuel gasoline and natural gas spark ignition engine, *SAE Int. Journal of Fuels and Lubricants* 4: 328–338.

Ridell, B. (2006). Malmö hydrogen and cng/hydrogen filling station and hythane bus project, *WHEC*.

Ristovski, Z., Morawska, L., Ayoko, G., Johnson, G., Gilbert, D. & Greenaway, C. (2004). Emissions from a vehicle fitted to operate on either petrol or compressed natural gas, *Science of the Total Environment* 323: 179–194.

Sierens, R. & Rosseel, E. (2000). Variable composition hydrogen/natural gas mixtures for increased engine efficiency and decreased emissions, *Journal of Engineering for Gas Turbines and Power* 122: 135–140.

Sørensen, B. (2005). *Hydrogen and Fuel Cells*.

Ulleberg (2003). Modeling of advanced alkaline electrolyzers: a system simulation approach, *Int. J. of Hydrogen Energy* 28: 21–33.

Wang, J., Chen, H., Liu, B. & Huang, Z. (2008). Study of cycle-by-cycle variations of a spark ignition engine fuelled with natural gas-hydrogen blends, *Int. J. Hydrogen Energy* 33: 4876–4883.

Effects of Fuel Properties on Diffusion Combustion and Deposit Accumulation

Kazuhiro Hayashida[1] and Katsuhiko Haji[2]
[1]Department of Mechanical Engineering, Kitami Institute of Technology
[2]Advanced Technology and Research Institute, Petroleum Energy Center (PEC)
(Present affiliation: Research & Development Division
JX Nippon Oil & Energy Corporation)
Japan

1. Introduction

Petroleum is still the major source of energy in the world; petroleum-based fuels are used to various combustion devices, such as automotive engines, gas turbines and industrial furnaces. Combustion of petroleum-based fuels generates undesirable exhaust emissions (e.g. unburnt hydrocarbons, NOx, soot particles), and exhaust emissions from combustion devices cause serious problems to the environment and human health. Adjustment of the properties of the fuels is an effective way to improve the combustion characteristics so that minimize pollutant emissions. Basic knowledge of relationship between the fuel properties and practical combustion performances is necessary to make effective adjustments corresponding to the ongoing diversification of fuels, such as newly-introduced crude oil, sulfur-free fuel, and synthetic fuel (Iwama, 2005).

It is well known that combustion characteristic of liquid fossil fuels vary by fuel properties, such as distillation characteristics and hydrocarbon components (Kök & Pamir, 1995). Especially in the case of diffusion combustion, soot emission is strongly affected by fuel properties (Kidoguchi et al., 2000). Diffusion combustion is widely applied to various combustion devices, but the influence of fuel properties on diffusion combustion is not fully understood. Moreover, when combustion devices are used for a long time, deposits gradually accumulate on the parts of the device that are exposed to high temperatures, such as the fuel nozzle and the combustion chamber wall (Zerda, 1999). Accumulation characteristics of deposits are also strongly affected by changes in the properties of the fuel used. Since excessive deposit accumulation can cause malfunctions of combustion devices, such as decreased output and degradation of exhaust emissions, understanding of the relationship of fuel properties to deposit accumulation is important.

The effects of fuel properties on diffusion combustion and deposit accumulation are described in this article. Several types of kerosene fraction, which have different fuel properties, were used as the test fuels. A wick combustion burner was used to form a stable laminar diffusion flame of liquid fuel; the difference of diffusion combustion characteristics was investigated. Moreover, the effects of fuel properties on deposit accumulation were investigated through deposit accumulation on wick during wick combustion.

2. Experimental apparatus

2.1 Wick combustion burner

A wick flame was formed with a wick combustion burner, as shown in Fig. 1. The burner was equipped with a pool filled with fuel, and a wick was put in the pool. Fuel was supplied from a tank to the pool through a float chamber under the fuel tank. To form a steady flame, the fuel level within the pool was kept constant by the float. The fuel flow rate was derived from the weight loss of the burner, measured by an electronic balance. The pool was made of aluminium. The pool had an outer diameter of 20 mm, an inner diameter of 16 mm, and a depth of 6 mm. The wick, made of sintered bronze metal (39 % porosity), was placed in the centre of the pool. The wick was cylindrical (8 mm diameter, 18 mm length, 6 mm wall thickness) with a flat bottom (8 mm diameter, 2 mm thickness). The wick was put in the pool so that the bottom protruded 7 mm from the pool rim. The distance from the fuel surface to top of the wick was 10 mm.

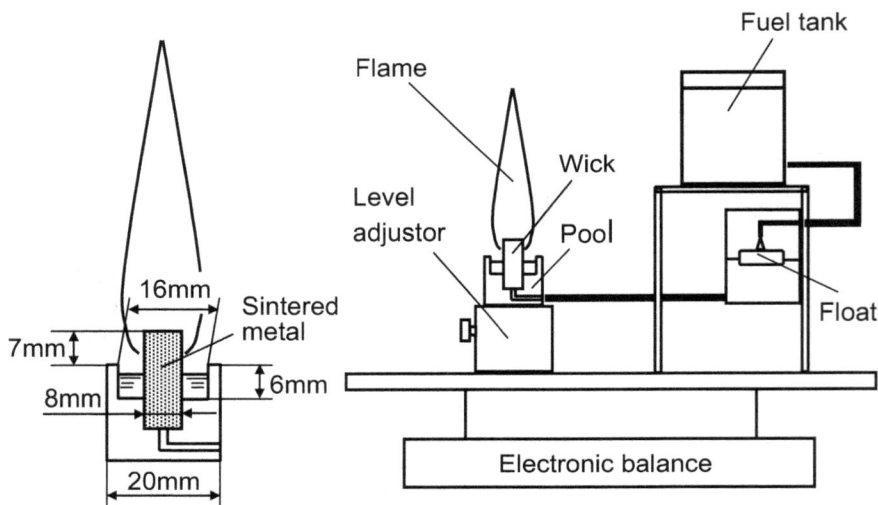

Fig. 1. Schematic of wick combustion burner

2.2 Laser diagnostic system

Laser diagnostic techniques are able to probe combustion products nonintrusively. Laser-induced fluorescence (LIF) and laser-induced incandescence (LII) are attractive techniques for combustion diagnostic and can be used to obtain information about PAHs (Hayashida, 2006) and soot (Shaddix, 1996), respectively. We measured the two-dimensional distribution of the PAHs-LIF in diffusion flames, and laser-induced incandescence (LII) was also used to visualize the soot distribution. Figure 2 shows schematic of the optical arrangement. The laser diagnostic system consisted of an Nd: YAG laser (Spectron Laser Systems, SL856G), a dye laser (Lumonics, HD-300B), and a doubling unit (Lumonics, HT-1000). The laser light was formed into a light sheet (0.5 mm×46 mm) by cylindrical lenses and was introduced into a target flame. Laser-induced emissions were detected by an ICCD camera (Andor

Technology, DH-534-18F-03), which was oriented perpendicular to the laser beam direction. The LIF and LII images were obtained by averaging 20 laser shots. For the PAHs-LIF measurement, the Nd: YAG laser was used to pump the dye laser (Rhodamine 590), producing a beam at 563 nm, and the doubling unit was used to double the dye laser output to produce 281.5 nm radiation. For the LII measurement, second harmonic generation (532 nm) of the Nd: YAG laser was used.

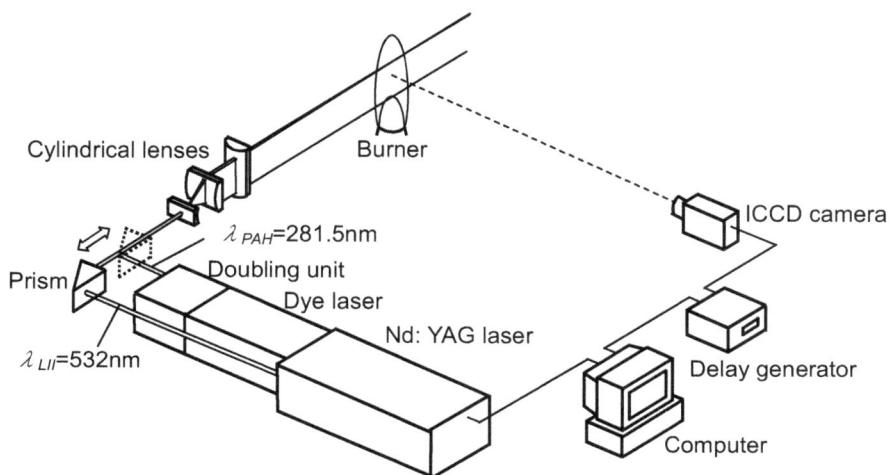

Fig. 2. Schematic of laser diagnostic system

3. Effects of fuel properties on diffusion combustion

3.1 Test fuels

Six types of fuel, with different distillation and compositional properties, were used. The physical properties of the test fuels are shown in Table 1. The calorific value of each fuel was same level, because the difference of net calorific value between test fuels was within 2.5%. Regarding fuel F, its smoke point (>50 mm) implies that the sooting tendency of fuel F was much smaller than that of the other fuels. Distillation characteristics indicate that fuel C was light, whereas fuel D was heavy in the test fuels. Fuel F was comparatively light, and fuel A, B and E had similar distillation characteristics. Influence of sulphur and nitrogen compounds on combustion was negligibly-small, because the contents of sulphur and nitrogen were extremely low.

Table 2 shows the results of the composition analyses of the test fuels. Although content of n-paraffin was not much difference between test fuels (28.6~34.4 vol%), there was considerable difference of i-paraffin content such as the fuel F (57.8 vol%) and fuel C (12.1 vol%). Content of naphthene hydrocarbons was low in fuel F (12.4 vol%) and comparatively low in fuel A (23.5 vol%). Aromatic hydrocarbons were much contained in fuel C (23.3 vol%), and were little contained in fuel E (9.9 vol%). Fuel F did not have any aromatics. Most of the aromatic components of the test fuels were one-ring aromatics; two-ring aromatics were very rare.

	Fuel name	A	B	C	D	E	F
	Density (15°C) [g/cm³]	0.7912	0.8007	0.8011	0.8000	0.7945	0.7547
Distillation characteristics	Initial boiling point [°C]	149.0	156.5	148.5	149.5	154.5	157.0
	5 vol% [°C]	163.5	169.5	160.5	165.5	168.5	165.5
	10 vol% [°C]	166.0	173.5	161.0	169.5	172.5	166.0
	20 vol% [°C]	173.0	179.5	165.5	178.0	179.5	169.5
	30 vol% [°C]	180.0	186.0	168.5	185.5	184.5	173.0
	40 vol% [°C]	187.0	193.0	172.0	194.0	191.0	177.5
	50 vol% [°C]	195.5	201.0	175.0	203.0	198.0	183.0
	60 vol% [°C]	204.5	210.5	179.0	213.0	204.5	190.5
	70 vol% [°C]	215.0	221.0	183.5	225.0	212.5	200.5
	80 vol% [°C]	227.5	233.5	189.5	237.0	222.5	214.5
	90 vol% [°C]	243.0	247.0	198.0	252.0	236.0	231.0
	95 vol% [°C]	253.0	257.5	205.5	262.0	247.0	240.0
	97 vol% [°C]	259.0	262.5	209.5	268.0	253.0	244.0
	End point [°C]	267.0	271.5	221.5	273.5	260.0	246.5
	Percent recovery [vol%]	98.5	98.5	98.5	98.5	98.5	98.5
	Percent residue [vol%]	1.0	1.0	1.0	1.0	1.0	1.0
	Percent loss [vol%]	0.5	0.5	0.5	0.5	0.5	0.5
	Kinematic viscosity (30°C) [mm²/s]	1.361	1.475	1.096	1.501	1.455	1.322
Elemental analysis	Sulphur [mass ppm]	6	7	21	32	3	<1
	Nitrogen [mass ppm]	<1	<1	<1	<1	<1	<1
	Carbon [mass%]	86.0	86.1	86.3	86.0	85.7	84.5
	Hydrogen [mass%]	14.0	13.9	13.4	13.9	14.2	15.2
	Net calorific value [J/g]	43380	43280	43000	43310	43440	44100
	Smoke point [mm]	23.0	23	21.5	22.0	28.0	>50
	Freezing point [°C]	-45.0	-45.0	-70.5	-42.5	-49.5	< -70.0
	Flash point [°C]	43.5	47.0	40.5	45.5	46.5	46.0

Table 1. Physical properties of test fuels used in the diffusion combustion experiment

Component	Composition [vol%]					
	A	B	C	D	E	F
n-paraffins	34.4	32.0	28.6	30.4	29.5	29.8
i-paraffins	22.8	20.0	12.1	15.5	29.0	57.8
Mono-naphthenes	18.7	21.8	30.7	27.4	24.7	10.1
Di-naphthenes	3.9	5.2	4.1	6.6	5.4	2.3
Poly-naphthenes	0.9	1.2	1.2	1.2	1.4	0.0
Alkylbenzenes	12.7	10.8	19.9	10.6	6.1	0.0
Mono-naphtheno benzenes	5.3	6.5	1.6	6.1	3.3	0.0
Di-naphtheno benzenes	0.4	0.6	0.0	0.6	0.2	0.0
Poly-aromatics	1.0	1.9	1.8	1.5	0.3	0.0

Table 2. Composition of test fuels used in the diffusion combustion experiment

3.2 Flame temperature and flame luminosity

Figure 3 shows photographs of the test flames. Flame lengths of fuel C and F were longer than that of the other fuels because of relatively lower distillation temperature. Soot emissions from the flame tip were confirmed in every flame; amount of soot emission of fuel F was very low.

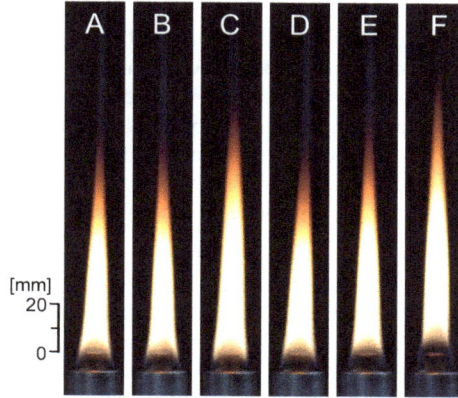

Fig. 3. Photographs of test flames

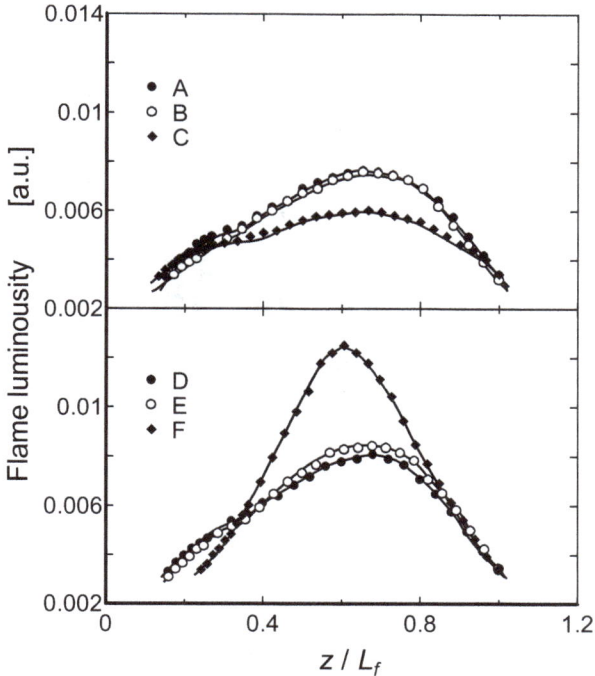

Fig. 4. Axial profiles of flame luminosities

Luminosity of the test flames at the centreline was measured by a CdS cell (wavelength sensitivity 400~800 nm, peak sensitivity 580 nm) through lens and pinhole. Obtained results were shown in Fig. 4. Note that measurement position was indicated as z/L_f. Here, z is distance from the wick and L_f is flame length from the wick; L_f is defined as flame luminosity disappearance position. Since electric resistance of CdS cell decreased with increasing the flame luminosity, the luminosity was expressed by inverse of CdS resistance. Flame luminosity of fuel F was particularly high, whereas luminosity of fuel C was lower than other fuels. The peak luminosity of fuel A, B, D and E was high in order of fuel E, D, A and B, and this order was corresponding to descending order of aromatic contents. According to the theory of black-body radiation, intensity of radiating body (i.e. soot particles) increases with temperature; thus the obtained luminosity would be reflected by flame temperature. Since soot is produced by the incomplete combustion of hydrocarbon fuel, it seems that the high luminosity of fuel F, which was lowest soot emission, was due to its high flame temperature resulting from the least incomplete combustion.

Figure 5 shows the temperature distributions of the test flames obtained by a two-color thermometer (Mitsui optronics, Thermera-seen). Here, two-color thermometer based on the two-color method (Zhao, 1998) measures the radiated energy of soot particles between two narrow wavelength bands, and calculates the ratio of the two energies, which is a function of the temperature. As for the black region in the figure, soot did not exist, or temperature and radiated energy of soot might be below the detection limit. Obtained result reveals that the temperature of flame F was particularly high.

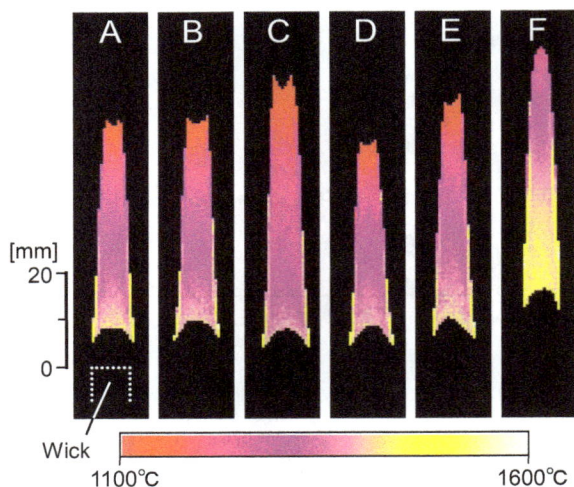

Fig. 5. Temperature distributions of the test flames

Figure 6 shows temperature profiles on centreline of the flame indicated in Fig. 5. The flame temperature of fuel F which did not contain aromatic compounds was the highest. Flame temperature was tendency to decrease in order of increasing aromatic components contained in the fuel, because soot generation within the flame might be increased with increasing the content of aromatics in the fuel.

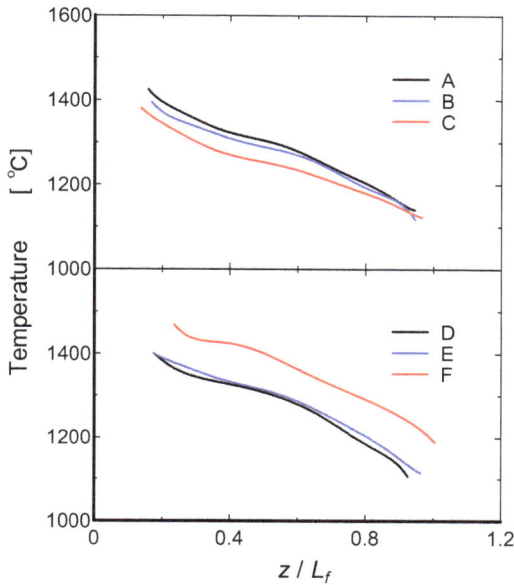

Fig. 6. Axial temperature profiles of the test flames

3.3 Concentration distributions of PAHs and soot

PAHs are considered to be precursors of soot particles formed in a flame, because it bridges the mass gap between fuel molecules and soot particles (Hepp, 1995). Formation and growth of PAHs arise from chemical reactions beginning with pyrolysis of fuel, and then inception of soot particles occurs by coagulation of grown PAHs. To estimate the effect of fuel properties on the soot emission and the soot generation characteristics, concentration distributions of soot and PAHs was investigated.

Figure 7 shows planar images of LII and PAHs-LIF obtained from the test flames. Since the concentrations of fuel F was significantly lower than the other fuels, another color scale was supplied in the figure. Except for fuel F, high concentration of LII at outer edge of the flame reveals that the soot was presence cylindrically in the flame. Regarding the fuel F, although the presence of soot was cylindrically in the lower part of the flame, upper part was not such distribution. LII distributions indicate that soot particles quickly formed in the case of fuel C and D. Since these fuels contain relatively much two-ring and poly-aromatics, soot particles might be promptly formed from those aromatics.

In each flame, PAHs-LIF was detected just after the wick, and the intensity decreases with increasing the distance from the wick. Regarding the fuel F, PAHs-LIF appeared comparatively strong between $z=10\sim20$ mm. A very low PAHs-LIF intensity of fuel F implies a very low PAHs concentration. It was confirmed that PAHs-LIF intensity became stronger with the aromatic contents increases.

Figure 8 shows the intensity profiles of PAHs-LIF and LII on the flame axis. Note that the PAHs-LIF intensity of fuel F was indicated as multiply the original data by 5. The peak of the PAHs-LIF was located at just after the wick in any flames. Since fuel F did not contain any aromatics, the PAHs would be formed by pyrolysis of paraffin components and subsequent reactions within the wick. However, PAHs-LIF intensity of fuel F at just after the wick was much lower than that of the other fuels, thus the PAHs-LIF intensity detected at just after the wick of the other fuels might be derived mainly from the originally contained aromatics in the fuel. The PAHs-LIF intensity rapidly decreases with distance from the wick; this implies that the growth of PAHs occurred by condensation polymerization of PAHs and thereby the number of PAHs molecules decreases.

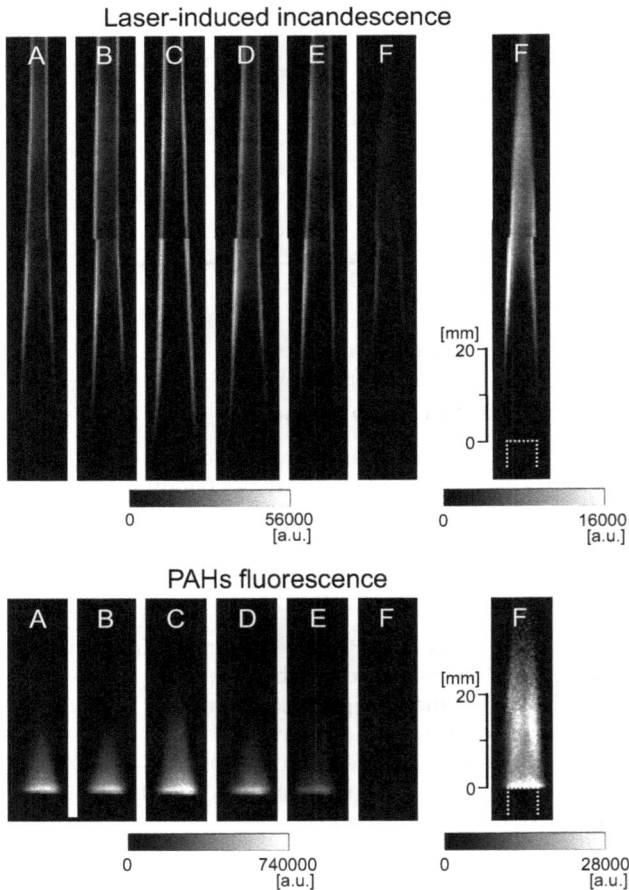

Fig. 7. Planar images of PAHs-LIF and LII

In the case of fuel F, increase of PAHs-LIF from z/L_f =0.15 suggested that the PAHs were newly formed within the flame. The same occurred within the other flames, but impact on the profile was small due to the relatively large LIF intensity of the originally contained PAHs. The disappearance location of the PAHs-LIF coincides with the location where the LII intensity rapidly increases. This figure clearly shows that the soot was formed via PAHs.

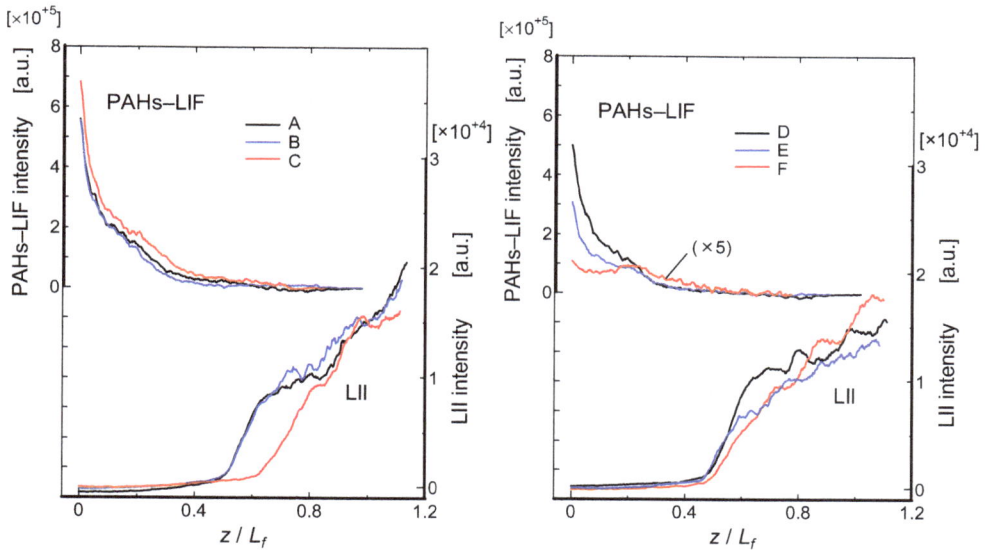

Fig. 8. Axial intensity profiles of PAHs-LIF and LII

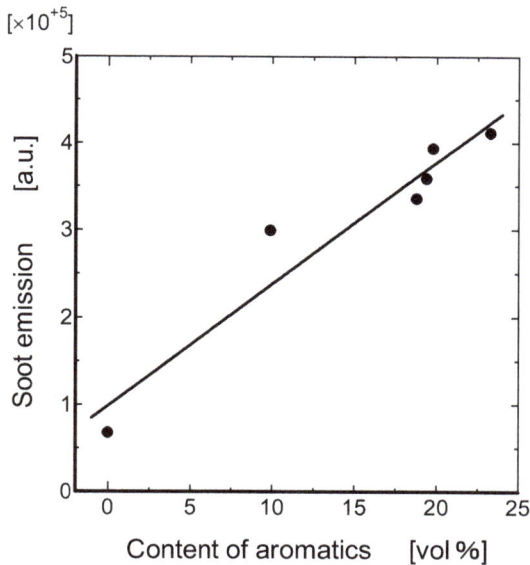

Fig. 9. Relationship between soot emission and content of aromatics

Figure 9 shows the relationship between soot emission from the test flames and content of aromatics in the test fuels. Soot emission was derived from total LII intensity at z=65mm, where combustion reaction seemed to be finished. Soot emission linearly increased with increasing content of aromatics.

4. Effects of fuel properties on deposit accumulation

4.1 Test fuels

Five types of kerosene fraction, with different distillation and compositional properties, were used. The physical properties of the test fuels are shown in Table 3. Regarding fuel K, the end point (246.5 °C) and density (0.7547 g/cm³) were lower than those of the other fuels.

Test item		Test method	G	H	I	J	K
Density (15°C)[g/cm³]		JIS K2249	0.7910	0.7895	0.7884	0.7980	0.7547
Distillation characteristics	Initial boiling point [°C]		153.5	147.0	146.5	145.5	156.0
	5 vol% [°C]		166.0	161.0	161.5	158.0	164.5
	10 vol% [°C]		167.5	165.5	165.0	160.5	165.5
	20 vol% [°C]		174.5	172.5	171.5	167.0	168.5
	30 vol% [°C]		181.0	179.5	178.0	174.0	172.5
	40 vol% [°C]		187.5	187.0	184.5	182.0	177.5
	50 vol% [°C]		194.5	194.5	192.0	191.5	182.5
	60 vol% [°C]		202.5	204.0	201.0	201.5	190.5
	70 vol% [°C]	JIS K2254	212.0	214.0	210.5	211.5	201.0
	80 vol% [°C]		223.0	227.0	225.5	223.0	215.0
	90 vol% [°C]		236.0	244.5	239.0	238.5	232.5
	95 vol% [°C]		246.0	259.5	254.0	249.0	242.0
	97 vol% [°C]		252.0	270.5	264.0	256.0	246.5
	End point [°C]		256.0	276.5	270.0	259.5	246.5
	Percent recovery [vol%]		98.5	98.5	98.5	98.5	98.5
	Percent residue [vol%]		1.0	1.0	1.0	1.0	1.0
	Percent loss [vol%]		0.5	0.5	0.5	0.5	0.5
Kinematic viscosity (30°C) [mm²/s]		JIS K2283	1.347	1.365	1.335	1.373	1.334
Net calorific value [J/g]		JIS K2279	43380	43430	43440	43390	44100
Smoke point [mm]		JIS K2537	23.0	24.0	24.0	26.0	>50
Freezing point [°C]		JIS K2276	-48.5	-45.5	-49.0	-68.0	-66.0
Flash point [°C]		JIS K2265	45.0	43.5	44.0	39.5	46.0
Elemental analysis	Sulphur [mass ppm]	JIS K2541	8	4	2	<1	<1
	Nitrogen [mass ppm]	JIS K2609	<1	<1	<1	<1	<1
	Carbon [mass%]	JPI Method	85.8	85.9	85.6	85.8	84.6
	Hydrogen [mass%]	JPI Method	14.0	14.1	14.1	14.2	15.4
CH ratio		—	0.51437	0.51132	0.50953	0.50713	0.46107

Table 3. Physical properties of test fuels used in the deposit accumulation experiment

Table 4 shows the results of the composition analyses of the test fuels. The results confirmed that fuel K consisted only of saturated hydrocarbons (i.e., paraffins and naphthenes). In contrast, other fuels contained between 7.6~19.5 vol% aromatic hydrocarbons. Most of the aromatic components of the fuels were one-ring aromatics; two-ring aromatics were very rare. In addition, the following low-stability components are also shown in Table 4: naphtheno benzenes, olefin hydrocarbons (bromine number), dienes (diene value), and organic peroxides (peroxide number). Naphtheno benzenes, compounds that consist of naphthene and aromatic rings, were contained in higher concentrations in fuel G (5.4 vol%); these molecules were not contained in fuel K. The bromine number was largest in fuel J (28.8 mgBr$_2$/100g) and lowest in fuel K (4.4 mgBr$_2$/100g). The diene values of fuels J and K indicated 0.05 gI$_2$/100g and 0.01 gI$_2$/100g, respectively; dienes were not detected in other fuels. The peroxide numbers demonstrate that none of the test fuels contained organic peroxides.

Component	G	H	I	J	K
Aromatic HC [vol%]	19.5	17.2	16.9	7.6	0.0
Unsaturated HC [vol%]	0.1	0.0	0.0	0.0	0.0
Saturated HC [vol%]	80.4	82.8	83.1	92.4	100.0
1-aromatics [vol%]	19.1	17.0	16.7	7.5	0.0
2-aromatics [vol%]	0.4	0.2	0.2	0.1	0.0
3+-aromatics [vol%]	0.0	0.0	0.0	0.0	0.0
1-naphtheno benzenes [vol%]	5.0	4.8	4.2	2.5	0.0
2-naphtheno benzenes [vol%]	0.4	0.5	0.3	0.0	0.0
3+-naphtheno benzenes [vol%]	0.0	0.0	0.0	0.0	0.0
Bromine number [mgBr$_2$/100g]	9.4	9.0	10.1	28.8	4.4
Diene value [gI$_2$/100g]	0.00	0.00	0.00	0.05	0.01
Peroxide number [mg/kg]	0.0	0.0	0.0	0.0	0.0

Table 4. Composition of test fuels used in the deposit accumulation experiment

4.2 Relationship between fuel properties and deposit accumulation

Tar-like deposits, formed by the thermal decomposition and polycondensation of the fuel, adhered to and accumulated on the upper part of the wick during combustion. To investigate the effects of fuel properties on tar-like deposit accumulation, tar-like deposits were accumulated on the wick by prolonged combustion. The deposit mass was obtained by subtracting the mass of unused wick from the mass of the wick that had accumulated deposits. In this process, if unburnt fuel remained in the wick with deposits, an accurate deposit mass could not be obtained. To eliminate remaining fuel, the wick was pulled up from the pool with a flame, and combustion was continued until all fuel was used. Obtained deposit was visually confirmed as tar-like deposit, but soot particle might have slightly adhered on the wick.

Figure 10 shows temporal change of deposit mass. In this experiment, one wick was sequentially used in the measurement of one test fuel. As seen in the figure, the deposit growth rate decreased after a certain period of time except fuel K. This result implies that the accumulation of deposits proceeded in two stages. Deposits were first rapidly accumulated in voids of the wick, which gradually saturate over time; we called this period the "internal accumulation mode". Once the voids became saturated with deposits, deposits subsequently accumulated on the outer surface of the wick; we called this phase the "surface growth mode". Although the deposit growth rate in the internal accumulation mode varied somewhat due to individual differences of the wick, the growth rate of the surface growth mode had reproducibility. Regarding fuel K, the deposit accumulation would not have attained to the surface growth mode owing to its extremely low deposit growth rate.

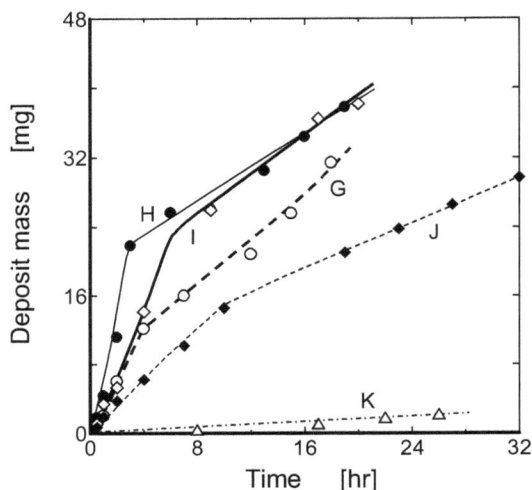

Fig. 10. Temporal change of deposit mass

As the deposit growth rate in the internal accumulation mode was influenced by individual differences of the wick, the relationship between deposit mass and fuel consumption was investigated in the surface growth mode. The results are shown in Fig. 11. Note that the results reported here for fuel K are not for the surface growth mode. Straight lines in the figure were obtained by the least-squares method, and the slopes correspond to the deposit accumulation ratio (the percentage of conversion from fuel to deposit). Deposit accumulation ratios are also indicated in the figure.

Obtained results indicate that the deposit mass of each fuel increased proportionally with increasing fuel consumption. Fuel G displayed the highest deposit accumulation ratio (0.00749 %), followed by fuel H (0.00580 %) and I (0.00574 %). Fuel K (0.00064 %) exhibited an extremely small accumulation ratio. Compared with the fuel compositions shown in Table 4, fuel G, with its high deposit accumulation ratio, was found to contain higher levels of aromatics and naphtheno benzenes; conversely, fuel K, with its low deposit accumulation ratio, contained no aromatics or naphtheno benzenes. Note that there was no correlation between the bromine number, diene value, and accumulation ratio. The extremely low accumulation ratio of fuel K might be caused by several factors: (1) the amount of low-stability components in fuel K was very low, (2) pyrolysis of saturated hydrocarbons was hard to occur within the range of the internal temperature of the wick, and (3) deposits formation from the pyrolysis products of saturated hydrocarbons took relatively long time.

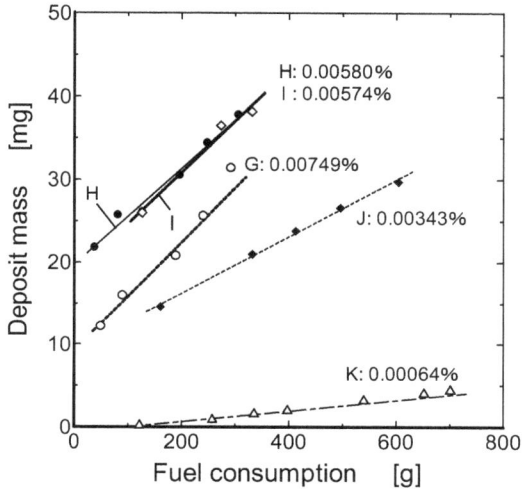

Fig. 11. Relationship between fuel consumption and deposit mass

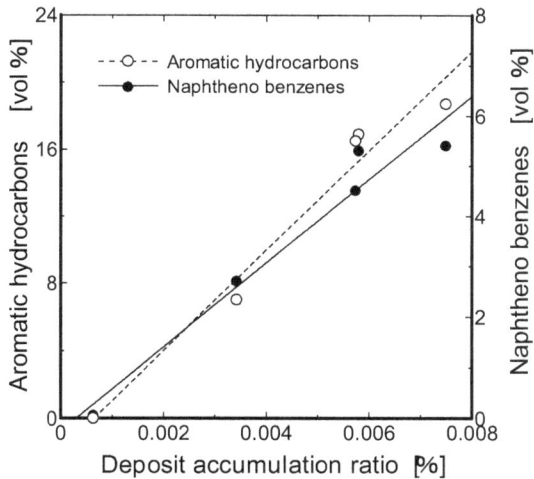

Fig. 12. Relationship between aromatic hydrocarbons, naphtheno benzenes and deposit accumulation ratio

The relationship between aromatic hydrocarbons, naphtheno benzenes and the deposit accumulation ratio is presented in Fig. 12. The deposit accumulation ratio was strongly associated with the content of aromatics and naphtheno benzenes. As naphtheno benzenes are one of the aromatics, more naphtheno benzenes may be contained in the aromatic-rich fuel. It is well known that naphtheno benzenes are more susceptible to thermal decomposition at lower temperatures than other hydrocarbons; thus, pyrolysis and polycondensation of naphtheno benzenes would be performed within the wick. Finally, naphtheno benzenes were transformed into tar-like deposits. Conversely, because the pyrolysis of aromatics was difficult to perform due to their high thermal stability, initial aromatics contained in the fuel transformed into heavy molecules and tar-like deposit without pyrolysis. Note that the sooting tendency became stronger with increasing aromatic content in the fuel; soot adherence to the wick possibly affected the results shown in Fig. 12. Although the deposits did not necessarily originate from aromatics and naphtheno benzenes, the deposit accumulation ratio of fuel K was extremely low and no correlation was found between the bromine number, diene value, and deposit accumulation ratio. These results suggested that most of the deposits originated from the naphtheno benzenes and/or aromatic hydrocarbons.

4.3 Deposit analysis

Deposits that accumulated on the wick were extracted by diethyl ether and chloroform solutions and dried after evaporation of the solutions. The collected deposits were then analyzed with a simultaneous thermogravimetry/differential thermal analysis (TG-DTA) instrument (TA Instruments, SDT2960). Analysis was performed by the following steps:

1. The sample was held at room temperature for 5 minutes and then heated to 250 °C.
    (atmosphere: N_2, flow rate: 100 mL/min)
2. The sample was held at 250 °C for 10 minutes and then heated to 550 °C.
    (atmosphere: N_2, flow rate: 100 mL/min)
3. The sample was held at 550 °C for 20 minutes.
    (atmosphere: air, flow rate: 100 mL/min)

[heating rate; 50°C/min]

Table 5 shows the mass loss of the deposits in each step. Considering the boiling point of hydrocarbons, the mass loss of each step corresponds to (1) the kerosene fraction, (2) the heavy kerosene component and light polycondensation products of low molecular weight, and (3) the heavy polycondensation product. The residue was considered to be a carbonized deposit and soot particles.

Mass loss	G	H	I	J	K
Room temp.~250°C (N_2) [wt%]	15.0	12.5	12.5	15.1	14.2
250~550°C (N_2) [wt%]	22.2	23.2	18.0	21.4	26.4
550°C (air) [wt%]	49.5	51.1	61.4	62.8	60.0
Residue [wt%]	13.4	13.2	8.1	0.8	0.0

Table 5. Results of TG-DTA analysis

The results of TG-DTA analysis suggested that most of the deposit components were heavy polycondensation products formed by thermal decomposition and polycondensation of the fuel within the wick. Residues of fuel G, H and I, which contained higher amounts of aromatics and naphtheno benzenes, were higher than that of the other two fuels; this result is probably due to an increase in the carbonized deposit formed at the wick surface and/or soot adhesion on wick. Additionally, some sulphur compounds might be contained in the residue because these fuels contain sulphur, as indicated in Table 3. In contrast, the deposits of fuel K contained no residue, and the mass loss in the range of 250~550 °C was relatively high. This fact demonstrates the very low deposit growth rate of fuel K, as described previously. The deposits of fuel J, with its very high bromine number and diene value, displayed a relatively large mass loss of the heavy polycondensation product; this observation may be due to the contribution of a tar-like deposit that originated from olefins and dienes (Zanier, 1998).

5. Conclusions

In this chapter, effects of fuel properties on diffusion combustion and deposit accumulation were studied experimentally. Obtained results could contribute to a design of fuel properties for reduction of the pollutant emissions from diffusion combustion of fossil fuels, and for suppression of the deposit accumulation within a combustion device.

5.1 Effects of fuel properties on diffusion combustion

Laminar diffusion flames of the test fuels were formed by using the wick combustion burner. Flame temperature and flame luminosity of each flame were measured. Furthermore, to investigate the effect on soot formation of fuel properties, concentration distributions of PAHs and soot were measured by LIF and LII. The main results are summarized as follows:

1. The peak luminosity of flame was high in order of decreasing aromatic components contained in the fuel.
2. Flame temperature tended to decrease with increasing aromatic components contained in the fuel.
3. In each flame, the PAHs-LIF intensity rapidly decreases with distance from the wick. The disappearance location of the PAHs-LIF coincides with the location where the LII intensity rapidly increases.
4. Soot emission increased with increasing content of aromatic hydrocarbons in the fuel.

5.2 Effects of fuel properties on deposit accumulation

Deposit accumulation processes were investigated by using the wick combustion burner. Deposit accumulation rate was estimated from the mass of deposit which accumulated on wick and the fuel consumption, and components of the deposit were estimated from the TG-DTA analysis. The main results are summarized as follows:

1. Except for fuel K, deposit accumulation per unit time (deposit growth rate) decreased after a certain time after ignition.
2. The conversion percentage from fuel to deposit (deposit accumulation ratio) increased with greater contents of aromatics and naphtheno benzenes in the fuels.
3. The primary components of deposits were heavy polycondensation products (50~60 wt%). Heavy kerosene components, light polycondensation products (about 20 wt%), and kerosene fractions (about 15 wt%) were also identified in the deposits.

6. Acknowledgement

This work was a cooperative research project with the Japan Petroleum Energy Center. We thank Prof. M. Arai and Prof. K. Amagai of Gunma University for helpful suggestions.

7. References

Hayashida, K., Amagai, K., Satoh, K. & Arai, M. (2006). Experimental Analysis of Soot Formation in Sooting Diffusion Flame by Laser-Induced Emissions. *Journal of Engineering for Gas Turbines and Power, Transactions of the ASME*, Vol.128, No.2, pp.241-246, ISSN 0742-4795

Hepp, H., Siegmann, K. & Sattler, K. (1995). New Aspects of Growth Mechanisms for Polycyclic Aromatic Hydrocarbons in Diffusion Flames. *Chemical Physics Letters*, Vol.233, No.1-2, pp.16-22, ISSN 0009-2614

Iwama, K. (2005). The Diversification of Energy Source in the New Era. *Journal of the Japanese Association for Petroleum Technology*, Vol.70, No.2, pp.125-131, ISSN 0370-9868

Kidoguchi, Y., Yang, C. & Miwa, K. (2000). Effects of Fuel Properties on Combustion and Emission Characteristics of a Direct-Injection Diesel Engine. *SAE paper 2000-01-1851*

Kök, M.V. & Pamir, M.R. (1995). Pyrolysis and Combustion Studies of Fossil Fuels by Thermal Analysis Methods. *Journal of Analytical and Applied Pyrolysis*, Vol.35, No.2, pp.145-156, ISSN 0165-2370

Shaddix, C.R. & Smyth, K.C. (1996). Laser-Induced Incandescence Measurements of Soot Production in Steady and Flickering Methane, Propane, and Ethylene Diffusion Flames. *Combustion and Flame*, Vol.107, No.4, pp.418-452, ISSN 0010-2180

Zanier, A. (1998). Thermal-Oxidation Stability of Motor Gasolines by Pressure d.s.c.. *Fuel*, Vol.77, No.8, pp.865-870, ISSN 0016-2361

Zerda, T.W., Yuan, X., Moore, S.M. & Leon y Leon, C.A. (1999). Surface Area, Pore Size Distribution and Microstructure of Combustion Engine Deposits. *Carbon*, Vol.37, No.12, pp.1999-2009, ISSN 0008-6223

Zhao, H. & Ladommatos, N. (1998). Optical Diagnostics for Soot and Temperature Measurement in Diesel Engines. *Progress in Energy and Combustion Science*, Vol.24, No.3, pp.221-255, ISSN 0360-1285

3

Fuel-N Conversion to NO, N_2O and N_2 During Coal Combustion

Stanisław Gil
Silesian University of Technology
Poland

1. Introduction

Pressurised combustion is a very attractive clean coal technology due to increased energy efficiency and abated emission of pollutants resulting from application of combined cycles. There is a general agreement that nitric oxide emissions decrease with enhanced combustion pressure. However, for nitrous oxide, the reported results show a contradictory influence of pressure.

The mode of NO and N_2O formation during coal combustion is far from being understood. The most difficult problem is the pathway of fuel-N conversion for primary nitrogenous species. The issue becomes simpler in case of char combustion, but even here fundamental questions remain. The uncertainty of NO modelling is emphasized by there being two quite different models to explain NO emission during char combustion: char-N is converted to NO with subsequent reduction of NO through a reaction with the char inner pores; alternatively, char-N is converted to HCN with negligible conversion to NO within the pores and subsequent conversion to NO outside the particle.

2. Combustion of nitrogen compounds contained in coal

2.1 Nitrogen in hard coal

Fuel-nitrogen is found in plants, animal proteins and nitrogen-rich bacteria. Reactions of amines with carboxylic groups or aldehyde groups resulted in nitrogenous species present in coal. Coal is a heterogeneous, complex mineral where nitrogen is typically bound to organic matter. During a coalification cycle, nitrogen content in coal substance only slightly changes so the nitrogen fraction (in per cent) in the substance increases as peat loses oxygen (Stańczyk, 1991). Nitrogen content in coal is 0.5% to 2% (Tingey & Morrey, 1973), reaching maximum for elemental carbon $C^{waf} = 85\%$ (Rybak, 1996). Coal typically contains 1% to 2% of nitrogen with bituminous coals usually containing 1.5–1.75% and anthracites mostly containing less than 1%.

The presence of nitrogen in coal has not been fully understood and described yet. There is far better knowledge of the structures of sulphur and oxygen than those of fuel-nitrogen. Due to difficulties encountered in investigations of nitrogen content in solid fuels, indirect methods of analysis are used and structures present in coal extracts or high-temperature coal tars are determined. There are few methods of direct analysis of nitrogen-containing groups in coal.

The knowledge of nitrogenous species in coal would allow for a more effective application of this material in many processing technologies. The significance of the problem is clearly seen in the amounts of nitric oxide emissions during coal combustion (Stańczyk, 1991).

Analyses of coal extracts or analyses of pyrolysis, oxidation or hydrogenation products yield data on nitrogen in coal. In coal extracts and tar (a coal depolymerisation product), basic and neutral nitrogen compounds are found. Their fractions depend on the nitrogen atom environment. During pyrolysis, a conversion of nitrogen compounds occurs: some of basic nitrogen compounds are formed during the process while some are released as ammonia. This depends on the process temperature and coal humidity. In acids, amines and nitrogen contained in a six-membered ring are dissolved. Five-membered rings do not dissolve. Solubility decreases with the increase in a molecular weight and the presence of oxygen functional groups (Attar & Hendrickson, 1982; Stańczyk, 1991).

2.2 Nitrogenous species in coal

In coal substance, there are the following nitrogen functional groups: pyridine and pyridine-derivative nitrogen, pyrrole and pyrrole-derivative nitrogen, nitriles, amines and amides. Nitrogen-containing molecules differ in size and occur as mono- and polycyclic compounds up to nine fused aromatic rings. In a heteroaromatic compound molecule, there is mostly a single nitrogen atom, but there may be two or even three atoms and the repetitive structures are tri- to pentacyclic compounds such as carbazole and acridine (Ostman & Colmsjo, 1988). A majority of nitrogen compounds in coal may be nitriles; however, during pyrolysis, they react with hydrogen to produce basic nitrogen; a product of some nitrile reactions may be ammonia. The presence of amine and cyano groups has not been proved but there is no evidence of their absence either as small amounts of amine groups are probably present due to their reactivity (Attar & Hendrickson, 1982). Studies by Burchill and Welch (1989) demonstrate the issue of nitrogen in coal in a slightly different manner and indicate pyrrole nitrogen to be dominant in coal. The contents of pyrrole and pyridine nitrogen change also during the coalification process so the pyrrole-pyridine nitrogen ratio changes with the degree of coalification. Pyrrole nitrogen reaches maximum in coal with $C^a = 84\%$, while pyridine nitrogen – in coal with $C^a = 90\%$. Hence, it may be concluded that heterocyclic six-membered structures are more stable in later coalification stages. The studies of 182 coals with coalification degrees $C^a = 79–95\%$ showed that mean nitrogen contents reach their maximum in coals with coalification degrees $C^{waf} = 84–85\%$. The increase in N/C ratio within $C^a = 79–81\%$ is explained by decarboxylation which is completed in coals with a slightly smaller coalification degree than $C^a = 80\%$. A decrease in N content in coals with the coalification degree above $C^a = 85\%$ occurs due to a poorer stability of nitrogenous species than that of the main aromatic structures in the coal matrix. A structural formula of one nitrogen binding in the coal matrix is shown in Figure 1 below (van Krevelen, 1981).

Maceral studies revealed non-uniform nitrogen content in macerals. The highest amounts of nitrogen are in vitrinite and then in liptinite, while the lowest amounts are found in inertinite where there is the highest level of pyridine nitrogen (van Krevelen, 1981; Given et al., 1984).

At present, two standardised methods used for quantitative determination of nitrogen in organic compounds are the Kjeldahl method and the Dumas method. In the Kjeldahl method (Krzyżanowska&Kubica, 1978), analysed organic matter is decomposed by sulphuric acid at 393 K to 423 K. Nitrogen in the matter is converted into ammonium sulphate which is next decomposed by NaOH to produce ammonia. The amount of ammonia is determined using a titration method with 0.01n H$_2$SO$_4$ and the acid-base indicator. In the Dumas method (Krzyżanowska & Kubica, 1978), a mixture of analysed substance and a catalyst (cupric oxide) is combusted in a quartz tube purged with carbon dioxide. Nitrogen oxides that form during combustion are decomposed into elemental nitrogen over incandescent copper. A received mixture of carbon dioxide and nitrogen is then introduced into a KOH-filled nitrometer where CO$_2$ is absorbed and the volume of nitrogen is determined. A factor that strongly influences result accuracy is the temperature of combustion. At present, C, H, N elemental analysers are applied with an electronic data processing programme and an autosampler for multi-sample analysis. The most common C, H, N elemental analysers are produced by Perkin - Elmer, Carlo Erba Strumentazione and Hewlett Packard companies. The above techniques of elemental analysis have been standard methods for many years.

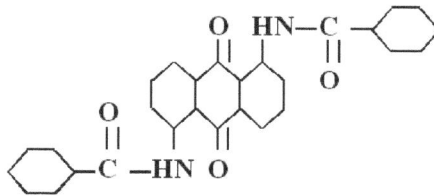

Fig. 1. A structural formula of one nitrogen binding in the coal matrix

2.3 Nitrogen conversion during coal combustion

A coal particle is introduced into a furnace for pulverised coal and heated up to 1770–1970 K in 1 ms due to heat collection from surrounding gases as well as fire and furnace wall radiation (Stańczyk, 1991). Within 10 ms, volatile matter is released from the particle, ignited and combusted within the subsequent 10–100 ms. Remaining char is combusted within 300 ms (Pershing & Wendt, 1979). Thus, there are three combustion stages:

- rapid heating to release volatile matter
- homogenous combustion of the volatile matter
- heterogenous combustion of char

Nitrogen oxide emissions depend on the heating rate and the presence of the above combustion stages. During combustion under fluid conditions, the presence of the specific stages and their duration are different due to larger coal particles and considerably lower furnace temperature 1100–1170 K, but the listed combustion stages still may be taken into consideration. Some papers show that a high heating rate and small particles may result in heterogeneous particle ignition or simultaneous ignition of volatile matter and a solid (Jüntgen, 1987).

Pershing and Wendt (1979) studies reveal that under typical pulverised coal combustion conditions, about a half of coal nitrogen undergoes pyrolysis. Conversion of the nitrogen into NO$_x$ is higher than conversion of char-nitrogen and constitutes about 60–80% of total NO$_x$ emissions. Many studies show that main factors affecting the extent of nitrogen emissions are: reaction stoichiometry and (less significant) nitrogen content in coal (Pereira el al., 1974; Cliff & Young, 1985; Midkiff & Atenkirch, 1988). Heterogeneous and homogenous oxidation of coal nitrogen is included in the model proposed by Midkiff and Atenkrich (1988). In Figure 2, a scheme of coal nitrogen distribution in the process of combustion is presented. The product of primary pyrolysis, nitrogen-containing volatile matter, undergoes secondary pyrolysis to produce HCN, NH_3, CN and N_2. HCN, NH_3 and CN are oxidised to NO_x and N_2O. One part of N_2 is formed directly during pyrolysis, while the other is formed through NO_x reduction by hydrocarbon radicals or in a reaction with CO. During the first 4 ms of combustion, 43% of converted nitrogen was transformed into N_2, while 57% into NO_x and N_2O. Thus, about half of nitrogen would be directly converted into N_2, while the other part into HCN and NH_3 which then would be oxidised into NO_x and N_2O (Peck et al., 1984).

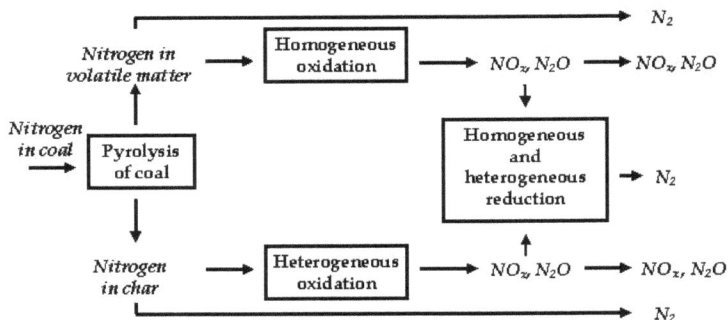

Fig. 2. Distribution of coal nitrogen during combustion

The composition of nitrogen-containing volatile matter (HCN, NH_3, NO_x) is influenced by coal types. Studies related to oxygen deficiency (λ = 0.5–0.8) revealed that in case of anthracite, only NO_x (and not HCN or NH_3) was formed in the amount of 17.5%. For bituminous coals, the amount of HCN is higher than that of NH_3 and increases with the increase in coal volatile matter content to produce 6–11% of NO_x. Low-rank coals (subbituminous coals and lignites) release the highest amounts of NH_3 and HCN, but less NO_x than bituminous coals. The studies of model nitrogen-containing liquid combustion showed a similar conversion of different types of compounds into NO_x (Stańczyk, 1991).

Many studies suggest that emissions of specific coal nitrogen compounds during combustion are strongly related to the reaction stoichiometry. Air deficiency promotes N_2 formation, while its excessive amounts lead to NO_x formation (Bruisma el al., 1988).

Changes in air excess in furnaces mainly affect the extent of NO_x (formed from nitrogen contained in the volatile matter) emission. The main effect of all furnace modifications aimed at multi-stage combustion is limitation of NO_x (formed from nitrogen released from coal during the primary pyrolysis) emission. Char-nitrogen, however, is insensitive to these procedures and NO_x emissions (formed from char-nitrogen) cannot be limited through furnace aerodynamic modifications (Preshing & Wendt, 1977, 1979).

There is no agreement among researchers on the effects of the nitrogen content in coal on its conversion into NO_x. In general, the increase in nitrogen content in coal results in enhanced NO_x emissions. However, coals with the same nitrogen contents and the same degree of coalification may significantly differ with respect to nitrogen oxide emissions (Preshing & Wendt, 1977).

3. Review of studies on formation of nitrogen compounds during coal combustion

Reduction of carbon dioxide and nitrogen oxide emissions in pressurised combustion methods is a result of intensive research (Gajewski, 1996). Release of fuel-nitrogen from coal during pressurised combustion, the rate and degree of its conversion to NO and N_2O are the data necessary for understanding of formation and destruction mechanisms for these pollutants.

The published studies on the mechanism of fuel-nitrogen release show a lot of discrepancies and contradictions. In case of pressurised combustion, there are no comprehensive kinetic models of NO and N_2O formation available. Few papers, mentioned below, require further investigations on modelling of nitrogen conversion in pressurised combustion processes.

Weiszrock et al. (1997) conducted studies on pressurised coal combustion in a laser reactor. At various O_2 levels, the increase in pressure resulted in NO emission reduction, but N_2O emissions increased.

Aho et al. (1995) as well as Aho and Pirkonen (1995) demonstrated reduction of conversion into N_2O, which is contradictory to the findings of Weiszrock et al. (1997), while Lu et al. (1992) did not find any effects of the combustion pressure on nitrogen conversion into N_2O.

Croiset et al. (1996, 1998) proposed a model of NO and N_2O formation during pressurised combustion of char. The model was based on eight surface reactions with [C], [CO], [CN] and [CNO] active centres. Modelling of nitrogen oxide emissions using kinetic constants calculated by the authors demonstrates reduction of NO levels with enhanced pressure. Croiset et al. noted that at increased temperature and pressure, reduction of NO to N_2O is a more important reaction than reduction of NO to N_2. In their calculations, the authors of the model omitted the inner structure and made the reaction kinetic constants dependant on pressure, which arouses some doubts.

For 0.2 MPa, De Soete et al. (1999) complemented the model with the efficiency of particle volume application in the reaction of [CO] and {CNO] active centre formation. They also compared kinetic coefficients of individual reactions published by Croiset et al. (1996, 1998) for "Westerholt" coal (at 0.2 MPa) and by de Soete (1990) for "Prosper" coal (at 0.1 MPa), but they did not make any comments on kinetic coefficients for higher pressures presented by Croiset et al. (1998).

Tomeczek and Gil (2000, 2001) as well as Gil (2000) proposed their model of NO and N_2O formation and reduction in pressurised char combustion based on chemical reactions presented by de Soete (1990) and Croiset et al. (1998). They determined their own kinetic coefficient of NO reduction over char where the activation energy of char was 79 kJ/mol within the pressure range of 0.2 MPa to 1.0 MPa. In case of the other coefficients of heterogeneous reactions, they considered the efficiency of particle volume application and made the coefficients non-pressure dependant, which resulted in a good agreement of the model and the conducted experiments within a large pressure range (0.2–1.5 MPa). Moreover, they estimated a total concentration of active centres (not determined yet) and included it in kinetic equations. However, the model requires further investigations and experimental studies.

In case of NO and N_2O modelling, the initial pathway of fuel-nitrogen conversion into NO during coal combustion is extremely difficult to explain. During char coal combustion, the problem seems less difficult, but the basic questions remain. It should be noted that experimental results of NO emissions during char combustion may be modelled in various ways, which means a substantial uncertainty of understanding of NO formation from char-nitrogen.

Visona and Stanmore (1996, 1999) showed that while modelling NO emissions, comparable results may be obtained through various pathways of fuel-nitrogen conversion:

- in reactions with O_2 in particle pores, fuel-nitrogen is converted to NO
- nitrogen released as HCN diffuses to a particle surface and is converted into NO in the particle boundary layer.

The researchers found that both pathways of fuel-nitrogen release are possible at about 1750 K.

Molina et al. (2000) compared models of coal particle combustion and pathways of fuel-nitrogen conversion into nitrogen oxides presented by Wendt and Schulze (1976), Shimizu et al. (1992), Goel et al. (1994), Visona and Stanmore (1996) and Soete et al. (1999).

The models proposed by Visona and Stanmore (1996) as well as by de Soete et al. (1999) are described above. A feature of the model presented by Wendt and Schulze (1976) is carbon oxidation to CO and char-nitrogen oxidation to NO in pores of heterogeneous particles. During diffusion towards the outer surface of the particle, CO undergoes homogenous oxidation to CO_2, and NO homogenous reduction occurs which was assumed according to the Zeldowicz opposing reaction (1946). The mechanism is based on a reburning phenomenon that occurs in particle pores. An important factor for the mechanism is kinetics of CO oxidation adopted from Howard et al. (1973). The diameter-length ratio of the pores was demonstrated to determine a degree of char-N conversion into NO at about 1000 K. At temperatures far above 1000 K, oxygen is basically consumed at pore entrances so diffusion into pores becomes less significant and nitrogen conversion depends on phenomena occurring in the pore boundary layer. In this case, it was demonstrated that CO concentration in the pore neighbourhood is a key factor in the nitrogen conversion into NO. The authors did not perform any experimental verification of the developed mathematical model. Within the investigated combustion range, char strongly depends on initial coal; thus, in a mathematical model aimed at predicting NO concentrations, such factors as individual char characteristics and its changes during particle burning should be considedred.

Shimizu et al. (1992) presented the simplest model of nitric oxide emissions comprising one combustion reaction and two NO formation and reduction reactions. In coal particles, oxygen directly reacts with fuel nitrogen to produce NO; then it is reduced to N_2 in a homogenous reaction with CO. Chan et al. (1983) and Goel et al. (1994) are the only researchers to include also the reaction of NO with CO catalysed by char. Studies by Chan et al. (1983) revealed that in the presence of CO, the rate of heterogeneous NO reduction is enhanced, particularly up to a CO/NO ratio of about 3; if the ratio is higher, a poor stabilisation occurs.

Goel et al. (1994) extended the model presented by de Soete (1990). They proposed a mechanism of N_2O formation through a reaction with an active centre [NCO], and N_2O reduction in a heterogeneous reduction reaction with elemental carbon C. They described NO decomposition through heterogeneous reduction over char and a homogenous reaction of NO with CO. Moreover, the authors proposed a mechanism where oxygen breaks a boundary active centre [CN] to form [N] which subsequently may react with O_2 to produce NO or with NO to produce N_2O, which was also suggested by Krammer and Sarofim (1994). A disadvantage of this model is that the experiments with one coal type were subsequently used for determination of kinetic constants.

In available literature, there are also some other models of nitrogen oxide formation and reduction described that were not compared in the paper by Molin et al. (2000).

Aria et al. (1986, 1986) presented a model where fuel-nitrogen was assumed to release as NH and NO in a heterogeneous reaction with O_2, and NO reduction was assumed to occur in a homogenous reaction in a boundary layer of a particle. Although the combustion process was assumed to occur on the external layer of a particle and in its pores, porosity changes were not made dependant on a particle burn-out level, which is suggested by Visona and Stanmore (1996, 1999). Also, the preparation method of char which was subsequently chilled arouses some doubts.

Tullin et al. (1993) assumed that NO is produced as a result of a heterogeneous reaction of oxygen with char-nitrogen, while N_2O is formed in a heterogeneous reaction of NO with fuel-nitrogen, and both compounds are reduced during a heterogeneous reaction with elemental carbon C. The authors found that N_2O decreased with the increase in the particle burn-out level, which is contradictory to the observations made by de Soete (1990) and Croiset et al. (1998) who showed that the increase in N_2O emissions was proportional to the coal burn-out level. Tullin et al. (1993) as well as Goel et al. (1994) verified the proposed model for one coal type.

Sarofim et al. (1995) described the porosity effects on NO and N_2O release from fuel-nitrogen. They observed a dependency on the surface area of reaction and diffusion in pores during conversion of fuel-nitrogen into NO and N_2O. Moreover, they showed that the conversion degree of fuel-nitrogen is proportional to the particle burn-out. In their investigations, the authors ignored a large fraction of mesopores involved in the reaction and they entirely neglected macropores whose role in diffusion of gases is significant.

Despite considerable progress, understanding of the mechanisms of nitrogen compound conversions during coal combustion is still unsatisfactory (Benson, 1968; Anthony el al., 1976; Pottigisser, 1980; Bliek el al., 1985). There are no reliable data on the kinetics of nitrogen conversion at increased pressures. In view of the development of pressurised combustion techniques, there is an urgent need of detailed studies on the issue. In the published studies on the mechanism of fuel-nitrogen release and its conversion in a particle, measuring points (mostly recorded under isothermal conditions when the final heating temperature was reached) were applied. Experiments where samples are analysed during heating allow for more precise determination of kinetic constants because for individual reactions, it is possible to separate temperature and time dependencies. A majority of researches agree that formation of nitrogen oxides should be described on the basis of internal diffusion and a chemical reaction in pores. In case of pressurised combustion, there are no comprehensive analyses in the form of kinetic equations (Mallet, 1995) so the issue requires systematic studies on nitrogen conversion in pressurised combustion processes. A literature review shows that while determining kinetic constants of fuel-nitrogen conversion into nitrogen oxides, individual characteristics of investigated char porous structures should be considered.

From the conducted analysis, it may be concluded that a significant part of experimentally confirmed differences in kinetic constants of char-nitrogen conversion into nitrogen oxides result from omission of individual characteristics of investigated char porous structures that strongly affect the mechanism of nitrogen conversion.

4. Mechanisms of nitrogen oxide formation during coal combustion

In the process of coal combustion, nitrogen oxides are produced during homo- and heterogeneous reactions of the air, volatile matter and char. In combustion gases, nitrogen oxides occur as NO, N_2O and NO_2 species. In Figure 3, a simplified scheme of the mechanism of their formation is presented. The basic component of nitrogen oxide emissions in high-temperature processes is NO. Complex mechanisms of nitrogen oxide formation during coal combustion have been classified using the source of nitrogen and divided into two basic groups (Zeldowicz, 1946; Fenimore & Jones, 1957; Fenimore, 1971; Johnsson et al., 1992; Bowman, 1973; Sarofim & Pohl, 1973; de Soete, 1975; Malte & Pratt, 1977; Levy et al., 1978; Fenimore & Fraenkel, 1981; Heyd, 1982; Miller et al., 1984; Glarborg et al., 1986; Fong & Peters, 1986; Niksa, 1988; Cheng et al., 1989; Miller & Bowman, 1989; Muzio et al., 1990; Glarborg et al., 1992; Williams et al., 1992; Tomeczek & Gradoń, 1997; Williams et al., 1997):

- fuel processes: oxidation of nitrogen compounds that are chemically bound with the fuel organic matter
- thermal processes: reactions of atmospheric nitrogen with atomic oxygen that is produced at high temperatures.

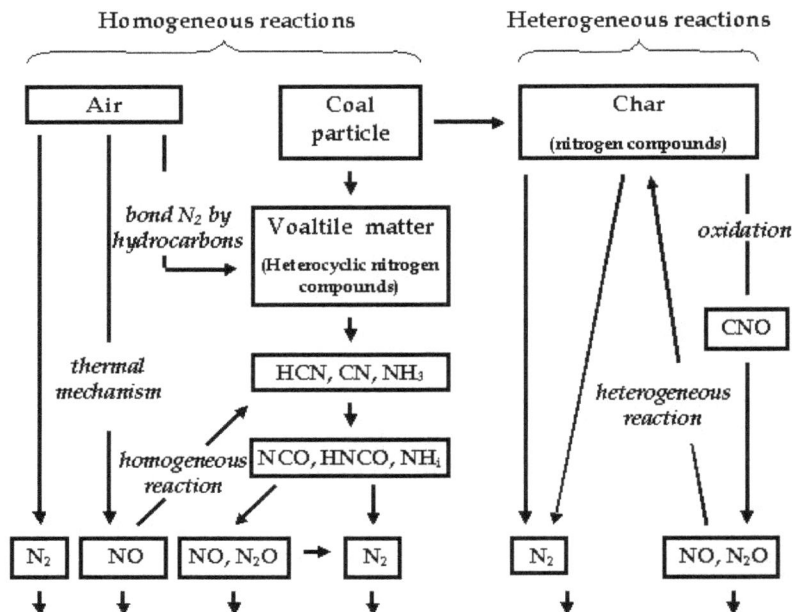

Fig. 3. A simplified scheme of nitrogen oxide formation during coal combustion

According to the above processes, the amounts of forming oxides depend on combustion conditions. High oxygen concentrations during the initial phase of coal combustion promote formation of fuel-nitrogen oxides (Attar and Hendrickson, 1982).

4.1 Homogeneous mechanism

Historically, the first homogeneous mechanism of NO formation is a thermal mechanism. The thermal mechanism of nitrogen oxide formation from molecular nitrogen is a sequence of chemical reactions that occur independently of the combustion process.

Zeldowicz mechanism

During his studies on gaseous flames at high temperatures, Zeldowicz (1946) observed that formation of thermal NOs cannot result from direct collisions of N_2 and O_2 molecules according to a global reaction:

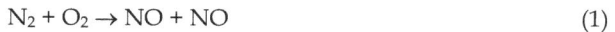

$$N_2 + O_2 \rightarrow NO + NO \tag{1}$$

He proposed a double-reaction mechanism:

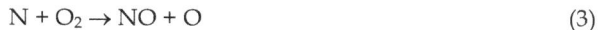

$$N_2 + O \rightarrow NO + N \tag{2}$$

$$N + O_2 \rightarrow NO + O \tag{3}$$

which is triggered by a reaction of molecular oxygen dissociation:

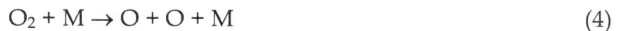

$$O_2 + M \rightarrow O + O + M \tag{4}$$

As the mechanism led to lower calculated results than the experimental ones, it was supplemented by a reaction (Fenimore & Jones, 1957; Fenimore, 1971):

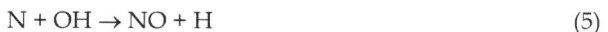

$$N + OH \rightarrow NO + H \tag{5}$$

The rate of NO formation resulting from the Zeldowicz mechanism (1946) mainly depends on the reaction kinetics (5) as well as concentrations of oxygen atoms and nitrogen molecules in the particle neighbourhood. Kinetic constants of the above reactions are presented in scientific papers (Arai et al., 1986; Miller & Bowman, 1989). A detailed review of the kinetic constants of thermal NOs formation are presented in the Ph.D. dissertation by Gradoń (2003), while the issues of the thermal mechanism are included in the papers by, among others: Fenimore, 1971; Johnsson et al., 1992; Bowman, 1973; Sarofim & Pohl, 1973; de Soete, 1975; Malte & Pratt, 1977; Fenimore & Fraenkel, 1981; Glarborg et al., 1986; Miller & Bowman, 1989; Muzio et al., 1990; Glarborg et al., 1992; Williams et al., 1992; Tomeczek & Gradoń, 1997.

Extended thermal mechanism

After a series of experiments conducted in a flow reactor at 1653 K to 1798 K, Tomeczek and Gradoń (2003) proposed a concept of extended thermal mechanism. Measured NO concentrations were far higher than those calculated according to the Zeldowicz mechanism (1946). The authors proposed the extended thermal mechanism based on five reactions: two reactions [(2) and (3)] were adopted from the Zeldowicz mechanism (1946) and three reactions [(6), (7) and (8)] were adopted from the N_2O mechanism developed by Malte and Pratt (1977):

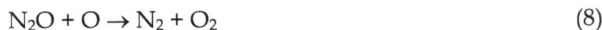

$$N_2 + O \rightarrow NO + N \tag{2}$$

$$N + O_2 \rightarrow NO + O \tag{3}$$

$$N_2 + O + M \rightarrow N_2O + M \tag{6}$$

$$N_2O + O \rightarrow NO + NO \tag{7}$$

$$N_2O + O \rightarrow N_2 + O_2 \tag{8}$$

The mechanism is also triggered by dissociation of an oxygen molecule (Reaction 4). Altering the rates of the (7) and (2) reactions, Tomeczek and Gradoń demonstrated that at temperatures below 1770 K, NO is primarily formed through the N_2O mechanism (Reactions 6 and 7) where (7) is a key reaction. For temperatures above 1770 K, NO is mainly formed through the Zeldowicz mechanism (Reactions 2 and 3) which is controlled by the (2). The kinetic constants of the extended thermal mechanism are included in the papers by Tomeczek & Gradoń, 1997 and Gradoń, 2003.

Mechanism of NO_2 formation

During combustion, nitrogen dioxide is formed through conversion of primarily generated NO.

Convection of oxygen atoms and NO molecules from the high-temperature flame zone to the cooler post-flame zone results in NO_2 formation. The mechanism is based on the following chemical reactions (Cernansky & Sawyer, 1975; Miller & Bowman, 1989):

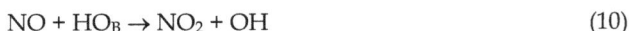

$$O + NO + M \rightarrow NO_2 + M \tag{9}$$

$$NO + HO_B \rightarrow NO_2 + OH \tag{10}$$

The presence of HO_B in these regions may result from diffusion of this compound from high- to low-temperature regions. At the atmospheric pressure, the reaction (9) becomes significant only below 500 K (Heyd, 1982; Fong et al., 1986; Niksa, 1988). In high-temperature regions, NO_2 undergoes rapid re-reduction to NO as a result of a reaction with O and H radicals. At temperatures above 900 K and high concentrations of O and H radicals, NO_2 reduction reactions are dominant:

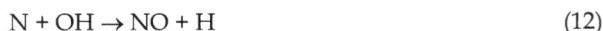

$$NO_2 + O \rightarrow NO + O_2 \tag{11}$$

$$N + OH \rightarrow NO + H \tag{12}$$

In case of solid fuel combustion, the presence of NO_2 in combustion gases that leave a reactor depends on the presence of low-temperature regions in the post-flame zone. The NO_2 fraction of the total NO_x emission in high-temperature reactors is usually small and falls below a few per cent (Cernansky & Sawyer, 1975). The observed higher NO_2 concentrations seem to result from reactions in the measuring probe where rapid cooling of combustion gases during sampling in the flame zone occurs (Miller & Bowman, 1989).

4.2 Heterogeneous mechanism

As during coal combustion a majority of nitrogen oxides is formed from nitrogen chemically bound with the fuel, heterogeneous reactions of nitrogen release and its conversion to oxides in the flame are of the greatest importance.

Mechanism of fuel-nitrogen oxide formation

Nitrogen that is chemically bound with the fuel releases from coal during devolatilisation or during char combustion. In case of devolatilisation, nitrogen mainly moves to the gaseous phase as hydrogen cyanide HCN. Due to rapid oxidation, the compound is primarily transformed into NCO and NH_i radicals. Miller et al. (1984) presented a cycle of fuel-nitrogen transformations into NO or N_2 in the following simplified scheme:

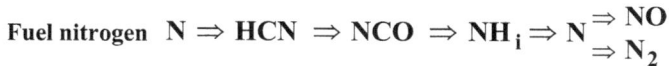

$$\textbf{Fuel nitrogen } \textbf{N} \Rightarrow \textbf{HCN} \Rightarrow \textbf{NCO} \Rightarrow \textbf{NH}_i \Rightarrow \textbf{N} \begin{matrix} \Rightarrow \textbf{NO} \\ \Rightarrow \textbf{N}_2 \end{matrix}$$

In general, mechanisms of fuel-nitrogen oxide formation are assumed to poorly depend on temperature, contrary to thermal NO formation (Miller & Bowman, 1989). Fuel-NO_x plays a key role during coal combustion within the temperature range of 1500–2000 K (Rybak, 1996).

The volatile matter is a complex mixture of combustible and non-combustible gases such as: carbon monoxide and dioxide, water vapour, saturated and unsaturated hydrocarbons, sulphur compounds, carbon black and nitrogen compounds (N_2, HCN, NH_3). The volatile matter composition changes with temperature and devolatilisation duration. Once fuel-nitrogen is released as HCN from coal, the compound is rapidly transformed into amino groups NH_i (i = 0, 1, 2, 3) which subsequently form NO or N_2 through a series of intermediate reactions (Rybak, 1996). Miller et al. (1984) concluded that intermediate chemical reactions which controlled the oxidation rate of HCN were reactions with atomic oxygen. This mechanism is also important when there is oxidant deficiency during combustion. According to many authors (Lavoie et al., 1970; Fraihaut et al., 1982; Miller et al., 1984; Miller & Bowman, 1989), among the mentioned reactions, the most important ones seem to be homogeneous reactions of NCO and NH_i formation:

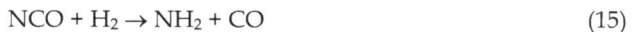

$$HCN + O \rightarrow NCO + H \tag{13}$$

$$NCO + H \rightarrow NH + CO \tag{14}$$

$$NCO + H_2 \rightarrow NH_2 + CO \tag{15}$$

The amino groups NH_i (NO precursors) result also from the following reactions:

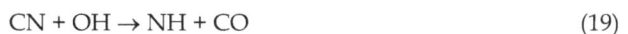

$$HCN + OH \rightarrow HNCO + H \tag{16}$$

$$HNCO + H \rightarrow NH_2 + CO \tag{17}$$

$$HCN + OH \rightarrow CN + H_2O \tag{18}$$

$$CN + OH \rightarrow NH + CO \tag{19}$$

In this mechanism, nitrogen oxide is produced in a reaction of NH_4 and NCO amino groups with O and OH radicals.

Summing up, NO concentration in combustion gases depends on the combustion process organisation as the temperature and oxygen concentration in the combustion zone determine the transition pathways for the reactions of NH_i amino groups with O, H and OH.

After volatilisation, a considerable amount of fuel-nitrogen remains in char. The amounts of nitrogen compounds remaining in the char depend on the devolatilisation temperature and level as well as, indirectly, on the ratio of oxygen excess. A part of the total amount of coal nitrogen remaining in the char increases with decreased combustion temperature and decreased ratio of the oxygen excess. Post-devolatilisation nitrogen species that are bound in the char are probably five-membered pyrrole groups which, due to temperature, are transformed into more stable heterocyclic five-membered structures (Rybak, 1996). At present, there is little knowledge of conversions of nitrogen compounds that remain in char. The fraction of nitrogen oxides resulting from char-nitrogen compounds may constitute about 20–30% of the total amount of generated NOs (Pershing & Wendt, 1979). Time necessary for oxidation of char-nitrogen compounds during char burning is far longer than the combustion duration of the volatile matter. Study results show that during char combustion, char-nitrogen compounds may undergo significant devolatilisation in a parallel manner to their oxidation. Calculated kinetic constants of the above reactions are presented in the papers by Miller et al., 1984; Arai et al., 1986; Glarborg et al., 1986; Thorne et al., 1986; Miller&Bowman, 1989.

There are also clues that a significant part of NOs that are formed during char combustion undergoes reduction to N_2 due to a contact with the char surface. If CO is absent, reduction of NO that is formed from the char may be controlled by NO chemisorption on the char surface according to the following reaction (Chan et al., 1983):

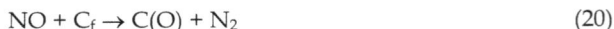

$$NO + C_f \rightarrow C(O) + N_2 \tag{20}$$

The effects of the above heterogeneous reaction become less significant when temperature rises and H_2O and O_2 levels increase. A part of nitrogen oxide may also undergo reduction as a result of a heterogeneous reaction with surfaces of carbon black particles (Pohl & Sarofim, 1977; Cheng et al., 1989; Kordylewski et al., 1996). The kinetic constant of NO reduction to N_2 in the presence of CO is presented in the papers by Arai et al., 1986; Williams et al., 1997.

Mathematical modelling of kinetics of char-nitrogen conversion to NO and N_2O species during coal combustion was performed by many researchers (Wendt & Schulce, 1976; Arai et al., 1986; de Soete, 1990, Tullin et al., 1993; Goel et al., 1994; Williams et al., 1997). They developed their models using a few to more than ten chemical reactions and determined kinetic constants of NO and N_2O formation. The NO formation kinetics in a heterogeneous reaction during char combustion is extremely complex so the mathematical models significantly simplified the issue (Wendt & Schulce, 1976; Phol & Sarofim, 1977; Arai et al., 1986). De Soete (1990) was the first researcher to develop a heterogeneous model of NO and N_2O formation and decomposition on the basis of surface reactions with active centres [CN] and [CNO].

De Soete's model

De Soete's (1990) model is based on ten reactions of NO and N_2O formation and reduction:

$$O_2 + 2[C] \rightarrow 2[CO] \tag{21}$$

$$[CO] \rightarrow CO + C_f \tag{22}$$

$$2[CO] \rightarrow CO_2 + [C] + C_f \tag{23}$$

$$O_2 + [C] + [CN] \rightarrow [CO] + [CNO] \tag{24}$$

$$[CNO] \rightarrow NO + [C] \tag{25}$$

$$NO + [C] \rightarrow 0.5N_2 + [CO] \tag{26}$$

$$[CN] + [CNO] \rightarrow N_2O + 2[C] \tag{27}$$

$$N_2O + [C] \rightarrow N_2 + [CO] \tag{28}$$

$$NO + 2[C] \rightarrow [CO] + [C \ldots N] \tag{29}$$

$$2[C \ldots N] \rightarrow N_2 + 2[C] \tag{30}$$

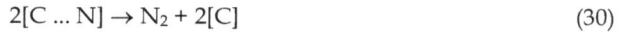

A disadvantage of this model is omission of diffusion in pores; however, the author noted that above 800 K, NO and N_2O formation is controlled by oxygen diffusion in char pores. In their model, Goel et al. (1994) subsequently extended the stage of NO and N_2O formation and destruction, adding more reactions as well as introducing "intermediates" and the active centre [NCO]. In their work, they assumed a constant effective diffusion coefficient or they determined it in the procedure of matching the model to the experiments. It means that there is little probability of precise results of modelling of nitrogen oxide emissions during coal combustion without previous experiments.

A model developed by Goel et al.

The authors (Goel et al., 1994) described their model with fifteen reactions and divided it into stages "a" to "f". The "a" stage is CO and CO_2 formation:

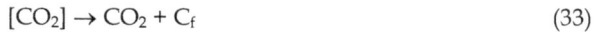

$$0.5 \, O_2 + [C] \rightarrow [CO] \tag{31}$$

$$[CO] \rightarrow CO + C_f \tag{22}$$

$$0.5 \, O_2 + [CO] \rightarrow [CO_2] \tag{32}$$

$$[CO_2] \rightarrow CO_2 + C_f \tag{33}$$

The "b" stage is NO formation:

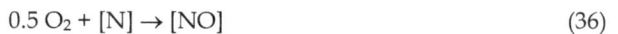

$$0.5 \, O_2 + [CN] \rightarrow [CNO] \text{ or } [NCO] \tag{34}$$

$$[NCO] \rightarrow CO + [N] \tag{35}$$

$$[CNO] \rightarrow NO + [C] \tag{25}$$

$$0.5 \, O_2 + [N] \rightarrow [NO] \tag{36}$$

$$[NO] \rightarrow NO + C_f \tag{37}$$

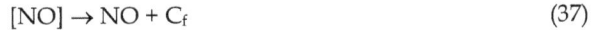

The "c" and "d" stages are related to NO reduction:

$$NO + [C] \rightarrow 0.5\ N_2 + [CO] \tag{26}$$

$$NO + C_f \rightarrow [NO] \tag{38}$$

$$CO + C_f \rightarrow [CO] \tag{39}$$

$$[NO] + [CO] \rightarrow 0.5\ N_2 + CO_2 + 2\ C_f \tag{40}$$

The "e" stage is N_2O formation:

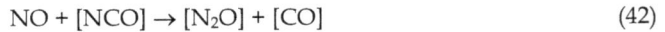

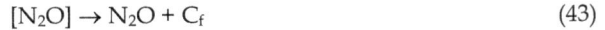

$$NO + [N] \rightarrow [N_2O] \tag{41}$$

$$NO + [NCO] \rightarrow [N_2O] + [CO] \tag{42}$$

$$[N_2O] \rightarrow N_2O + C_f \tag{43}$$

The "f" stage is N_2O reduction:

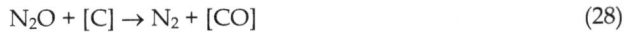

$$N_2O + [C] \rightarrow N_2 + [CO] \tag{28}$$

In the model developed by de Soete et al., the combustion mechanism was based on three reaction: two of them were related to CO generating and the other one referred to CO_2 formation. Goel et al. (1994) described CO and CO_2 formation with four reactions: the first two were related to CO formation and the next two referred to CO_2 formation. De Soete developed the mechanism of NO formation with the use of two reactions of CNO formation and decomposition. In the study by Goel et al., NO release was described by five reactions where nitrogen oxide was formed either from CNO or through the active centre [NO]. De Soete described NO reduction be means of one reaction of NO with an active carbon centre [C], while Goel et al. described it with the use of four reactions where the first one was the same as de Soete's and the next three led to formation of N_2, CO_2 and a free active site. In de Soete's model, N_2O generating was based on one reaction of the active centre [CNO] destruction, while Goel et al. described the process by means of three reactions with NO transition to the active centre [N_2O] and, subsequently, to a gaseous species N_2O. De Soete and Goel et al. described N_2O reduction in the same way using one reaction of N_2O with the active carbon centre [C]. In de Soete's model (1990), NO and N_2O formation is controlled by oxygen adsorption: at temperatures below 800 K, the adsorption is in the kinetic region, while at higher temperatures, it is controlled by oxygen diffusion in pores. The author determined kinetic constants for two chars within 800–1300 K which, in case of oxygen adsorption over the char matrix, depend on the depth of oxygen penetration into the particle. The author did not specify a porous structure of the investigated chars. Goel et al. (1994) extended de Soete's model (1990), allowing for N_2O formation in the absence of oxygen through a reaction with an "intermediate" [NCO]. N_2O destruction only occurred in a heterogeneous reaction with elemental C in the char matrix. In the mechanism of NO destruction, they included both heterogeneous and homogeneous reaction of NO with CO catalysed by char. The proposed mechanism promotes N_2O formation in sites with high NO concentrations. A disadvantage of this model is a stable diffusion constant which neglects variations of porous structure.

5. Kinetics of heterogeneous nitrogen oxide formation during pressurised char combustion

In the mechanism, concentrations of free C_f and occupied active centres are considered among: [C], [CN], [CO] and [CNO]. Initially, it is assumed that char contains free active centres, C_f, and active centres occupied by carbon atoms and by carbon-nitrogen bonds [C] and [CN], respectively, concentrated on the surface. The organically bound nitrogen can react heterogeneously either to produce NO, N₂O or N₂. During the presented research, the mechanism proposed by Croiset et al. (1998) was chosen for testing mainly because the rate constants were evaluated by Croiset et al. from enhanced pressure experiments:

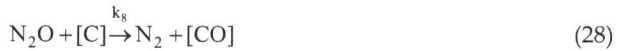

$$N_2 + 2[C] \xrightarrow{k_1} 2[CO] \tag{21}$$

$$[CO] \xrightarrow{k_2} [CO] + C_f \tag{22}$$

$$2[CO] \xrightarrow{k_3} CO_2 + [C] + C_f \tag{23}$$

$$O_2 + [C] + [CN] \xrightarrow{k_4} [CO] + [CNO] \tag{24}$$

$$[CNO] \xrightarrow{k_5} NO + [C] \tag{25}$$

$$NO + [CNO] \xrightarrow{k_6} N_2O + [CO] \tag{44}$$

$$NO + [C] \xrightarrow{k_7} 0.5N_2 + [CO] \tag{26}$$

$$N_2O + [C] \xrightarrow{k_8} N_2 + [CO] \tag{28}$$

The total concentration of the occupied active centres is equal to

$$S = [C] + [CO] + [CN] + [CNO] \tag{45}$$

If the fractions of the occupied active centres are defined as (Croiset et al. 1998):

$$\theta_C = [C] / S \tag{46}$$

$$\theta_{CO} = [CO] / S \tag{47}$$

$$\theta_{CN} = [CN] / S \tag{48}$$

$$\theta_{CNO} = [CNO] / S \tag{49}$$

then we can express the rate of gaseous species formed during the combustion by means of equations:

$$\dot{R}_{CO} = k_2 \theta_{CO} S mc \, \eta_1, \tag{50}$$

$$\dot{R}_{CO_2} = k_3 \theta_{CO}^2 S mc \, \eta_1, \tag{51}$$

$$\dot{R}_{NO} = k_5 \theta_{CNO} S mc \, \eta_4 - k_6 \theta_{CNO} S mc \, \eta_4 p_{NO} - k_7 (\phi - \theta_{CO}) S mc \, \eta_7 p_{NO}, \tag{52}$$

$$\dot{R}_{N_2O} = 2k_6 \theta_{CNO} S mc \, \eta_4 p_{NO} - 2k_8 (\phi - \theta_{CO}) S mc \, \eta_8 p_{N_2O}, \tag{53}$$

$$\dot{R}_{N_2} = (\phi - \theta_{CO})(k_7 \eta_7 p_{NO} + 2 k_8 \eta_8 p_{N_2O}) S mc. \tag{54}$$

For modelling, a crucial problem is the concentration of occupied active centres which cannot be eliminated from the kinetic equations. In the past, that problem was overcome by determination of products of the surface concentration of active centres and the rate constant. It may be successful when we are only interested in the char particle reaction rate. However, in order to model individual species, separation of active site concentrations from the rate constants is necessary. Considering all the difficulties, it is proposed to assume the concentration of the total active centres as a number of carbon atoms in a monolayer of active centres on the particle internal surface. For a particle of an internal surface area A, diameter d and porosity ε, the total concentration of the occupied active centres can be described using the equation by Gil (2002):

$$S = KAd^3(1-\varepsilon), \tag{55}$$

where $K = 1.95 \cdot 10^{28}$ a.c. (active centres) m^{-5} or $K = 32.32$ kmol$_{a.c.}$ m^{-5}.

It was assumed that the total fraction of the active centres [C] and [CO] is a dimensionless, equal value (Croiset et al., 1998):

$$\vartheta_C + \vartheta_{CO} = \phi \tag{56}$$

The total fraction of the active centres [CN] and [CNO] is:

$$\vartheta_{CN} + \vartheta_{CNO} = 1 - \phi \tag{57}$$

where ϕ is defined as the probability that a C atom will not bind to a nitrogen atom during combustion (Croiset et al., 1998):

$$\phi = 1 - \frac{N_K^a}{C_K^a} \frac{M_C}{M_N} \tag{58}$$

In the kinetic equations of the above mechanism, there are two key parameters: rate constants k_i of reactions (21–26, 28, 44) and the mode of reaction expressed by the effectiveness factor η_1, yielding a part of the particle volume available for the i-th reaction. The rate constants of reaction (26) were found to be independent of pressure within the range of 0.2–1.5 MPa (Tomeczek & Gil, 2001). Based on this finding, it was assumed that for all the other reactions, the rate constants were independent of pressure and they were evaluated using Croiset et al. (1998) data as a mean value within the experimental pressure range of 0.2–1.0 MPa. A very difficult problem of the effectiveness η factor for the particular reactions was simplified by the assumption: $\eta_1 = \eta_4$, $\eta_6 = \eta_4$ and $\eta_7 = \eta_8 = 1$. Because calculation of the values η_1 and η_4 for the dual site reactions (21) and (24) is not easy, the combustion reactions were replaced by an overall reaction $O_2 + [C] \rightarrow CO_2$ with the rate constants of reaction (21). This enabled calculation of the effectiveness factor from the equation:

$$\eta_1 = \eta_4 = 3[(\tanh(\text{Th}))^{-1} - (\text{Th})^{-1}] / (\text{Th}), \tag{59}$$

where the Thiele module is:

$$\text{Th} = 0.5d\sqrt{kp / D_e}, \tag{60}$$

and $k = k_1\rho(1-\varepsilon)/\gamma M_C$, $\gamma = p/RT$, ρ - char density, M_C – carbon molar mass, p – pressure, R – gas constant, T – temperature. The effective diffusion coefficient was expressed as a function of the oxygen-helium molecular diffusion coefficient D and the Knudsen diffusion coefficient D_K through the particle pores of tortuosity τ (Laurendeau, 1979):

$$D_e = \varepsilon / (D^{-1} + D_K^{-1})\tau^2 . \tag{61}$$

In order to calculate the rate of CO, CO_2, NO, N_2O and N_2 formation by means of equations (50–54), it is necessary to find the fractions of the occupied active centres [CO] and [CNO] from the following kinetic equations:

$$\frac{d\theta_{CO}}{dt} = 2k_1(\varphi - \theta_{CO})^2 p_{O_2} - k_2\theta_{CO} - 2k_3\theta_{CO}^2 + k_4(\varphi - \theta_{CO})(1 - \varphi - \theta_{CNO}) \cdot$$
$$\cdot \eta_4\eta_1^{-1}p_{O_2} + k_6\theta_{CNO}\eta_4\eta_1^{-1}p_{NO} + k_7(\varphi - \theta_{CO})\eta_7\eta_1^{-1}p_{NO} + \tag{62}$$
$$+ k_8(\varphi - \theta_{CO})\eta_8\eta_1^{-1}p_{N_2O},$$

$$\frac{d\theta_{CNO}}{dt} = k_4(\varphi - \theta_{CO})(1 - \varphi - \theta_{CNO})p_{O_2} - k_5\theta_{CNO} - k_6\theta_{CNO}p_{NO} . \tag{63}$$

The 90s experiments were conducted at pressures of 0.2 – 1.5 MPa in a pressurised reactor. A single layer of char particles 0.1– 0.8 g of initial diameter 0.9 mm was placed in a canthal tray and heated from the bottom and the top at a rate of 100 Ks up to the final temperature 1073–1373 K. The final heating temperature and the samples residence time at this temperature were automatically controlled. The maximum deviation from the adjusted temperature within the isothermal period was smaller than ±40K. The temperature of the sample was measured by a NiCr-NiAl thermocouple and continuously recorded. The char was produced at 1373 K in atmospheric pressure helium during 20-minute devolatilisation of subbituminous "Siersza", "Janina" and "Piast" coals, whose characteristics (water content in the analytical state – W^a, volatile matter content in the water ash free state – V^{waf}, ash content in the water free state – A^{wf}, carbon content in the water ash free state – C^{waf}, hydrogen content in the water ash free state – H^{waf}, sulphur content in the analytical state – S^a, nitrogen content in the water ash free state – N^{waf}, solid particle porosity – ε_0) are presented in Table 1. A mixture of oxygen and helium of initial volume content $O_2 = 21\%$ was used. After reactor cooling, the concentrations of NO and N_2O were analysed as well as a mass of the remaining char and its nitrogen content. Gas sampling was calibrated on an empty reactor.

Enriched coal	W^a	V^{waf}	A^{wf}	C^{waf}	H^{waf}	S^a	N^{waf}	ε_0
	%	%	%	%	%	%	%	%
Siersza	2.6	37.7	6.9	77.0	5.1	2.5	1.51	15.8
Janina	3.3	33.9	8.1	77.6	3.3	1.1	1.34	14.6
Piast	5.7	30.8	7.1	84.2	3.9	1.2	1.42	15.1

Table 1. Characteristics of coals

The rate constants of reactions (21–26, 28, 44) are presented in Table 2. The constants of reactions (21) and (24) were evaluated using Croiset et al. (1998) data, assuming average values of k_i within the experimental pressure range 0.2–1.0 MPa.

Rate constant	Author	Pressure	$k = k_0 \cdot \exp(-E \cdot R^{-1} \cdot T^{-1})$
k_1, $s^{-1} \cdot Pa^{-1}$	Croiset et al., 1998	0.2 - 1.0 MPa	$6.4 \cdot 10^{-6} \cdot \exp(-42 \cdot R^{-1} \cdot T^{-1})$
k_2, s^{-1}	Gil, 2003	0.2 - 1.5 MPa	$8.4 \cdot 10^{2} \cdot \exp(-90 \cdot R^{-1} \cdot T^{-1})$
k_3, s^{-1}	Gil, 2003	0.2 - 1.5 MPa	$8.2 \cdot 10^{5} \cdot \exp(-51 \cdot R^{-1} \cdot T^{-1})$
k_4, $s^{-1} \cdot Pa^{-1}$	Croiset et al., 1998	0.2 - 1.0 MPa	$8.1 \cdot 10^{-4} \cdot \exp(-58 \cdot R^{-1} \cdot T^{-1})$
k_5, s^{-1}	Gil, 2003	0.2 - 1.5 MPa	$2.2 \cdot 10^{2} \cdot \exp(-84 \cdot R^{-1} \cdot T^{-1})$
k_6, $s^{-1} \cdot Pa^{-1}$	Tomeczek & Gil, 2001	0.2 - 1.5 MPa	$1.1 \cdot 10^{2} \cdot \exp(-110 \cdot R^{-1} \cdot T^{-1})$
k_7, $s^{-1} \cdot Pa^{-1}$	Tomeczek & Gil, 2001	0.2 - 1.0 MPa	$2 \cdot 10^{-4} \cdot \exp(-79 \cdot R^{-1} \cdot T^{-1})$
k_8, $s^{-1} \cdot Pa^{-1}$	Gil, 2003	0.2 - 1.0 MPa	$4 \cdot 10^{-5} \cdot \exp(-75 \cdot R^{-1} \cdot T^{-1})$

Table 2. Rate constants of reactions

The initial properties (surface area – A_0, carbon content in analytical state – C^a_0, nitrogen content in analytical state – N^a_0, solid particle porosity – ε_0, real density – ρ_0) of "Siersza", "Janina" and "Piast" chars used for testing of the mechanism and the kinetic constants are presented in Table 3. The constant value of char pores tortuosity was $\tau = 1.83$ (Tomeczek & Mlonka, 1998). During particles burning, the internal surface area A as well as the porosity ε altered because most combustion reactions take place within the pores (Stanmore, 1991): $A = A_0 (1 + 2.5 B) (1 - B)$ and $\varepsilon = \varepsilon_0 + B (1 - \varepsilon_0)$, where B is the particle burn-out.

Char	A_0	C^a_0	N^a_0	ε_0	ρ_0
	m^2kg^{-1}	%	%	%	kg^1m^{-3}
Siersza	$27.1 \cdot 10^3$	79.3	0.91	40.0	1210
Janina	$16.7 \cdot 10^3$	80.1	0.79	59.3	1390
Piast	$12.5 \cdot 10^3$	86.6	0.78	45.7	1320

Table 3. Characteristics of chars

In Figures 4 and 5, comparisons of the model curve and experimental points for enriched "Siersza", "Janina" and "Piast" coals are shown. The rates of nitric and nitrous oxide

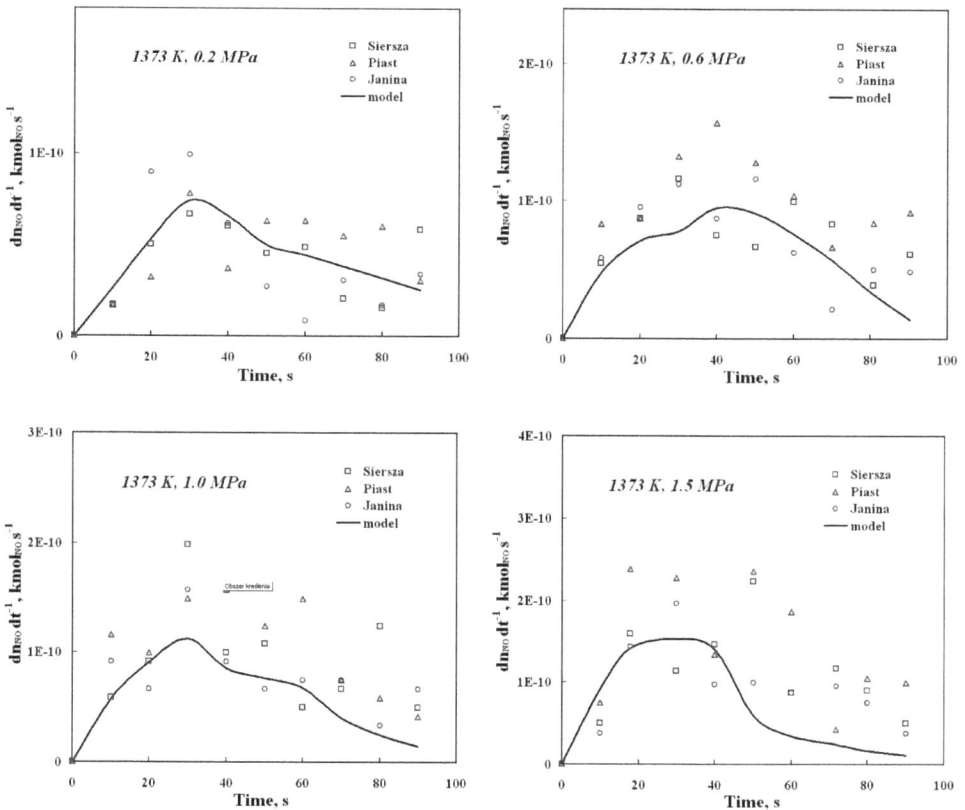

Fig. 4. Measured and modelled (kinetic constants Table 2) rates of nitrous oxide emission as a function of time during char combustion

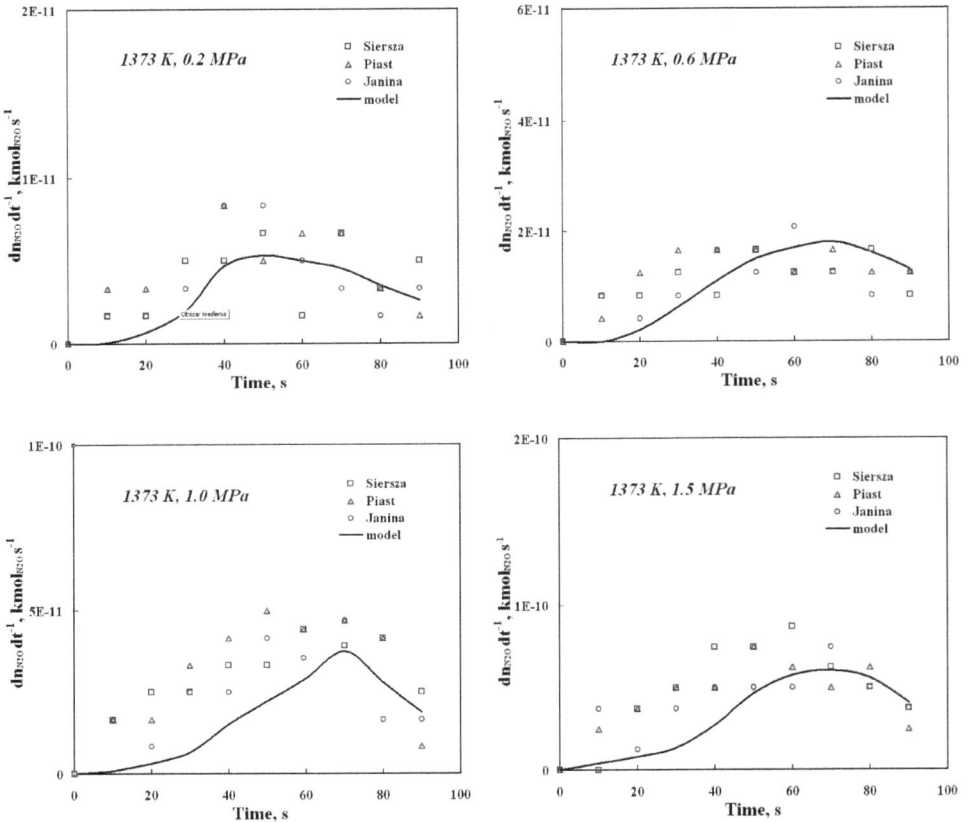

Fig. 5. Measured and modelled (kinetic constants Table 2) rates of nitrous oxide emission as a function of time during the char combustion

emissions were presented as a function of time during char combustion for the pressure range of 0.2 MPa - 1.5 MPa and 1373 K. The agreement of the kinetic model and the experimental points is better for "Siersza" than for "Janina" and "Piast" coals because the kinetic data in Table 2 were evaluated only on the basis of the first. The average correlation coefficients for the studied chars within the pressure range of 0.2–1.5 MPa at 1373 K were approximately 0.80 for NO emission and 0.75 for N_2O emission. The best compatibility of the modelled NO emission rate and recorded experimental values was achieved for the constant $k_{07} = 2 \cdot 10^{-2}$ $s^{-1}Pa^{-1}$, and then drastically decreased correlation of N_2. The model is not significantly sensitive to the activation energy of reaction (R7), E_7, since a change in the range of 30 kJ mol^{-1} to 150 kJ mol^{-1} did not give a clear difference in the results. By changing the preexponential factor of constant k_{08} in the range of $4 \cdot 10^{-5}$ s^{-1} Pa^{-1} to $2.1 \cdot 10^3$ s^{-1} Pa^{-1}, a clear increase in N_2O emission, and significantly reduced emissions of N_2 could be obtained. The model was very sensitive to the change of activation energy E_8 because its small growths caused a very large increase of N_2O emissions.

Pressurised combustion abates char-N conversion into NO and enhances conversion into N_2O. Within the analysed ranges of pressure and temperature, reduction of NO by char is controlled by both chemical kinetics and diffusion in pores, with the activation energy equal to 79 kJ/mol. The rate constants of the heterogeneous mechanism of char-N conversion are independent of the combustion pressure if the chars are produced at the same pressure. About 60% of char-N is converted into N_2 through reduction of NO and N_2O over char; however, modelling of this pathway needs further investigations.

6. References

Aho, M.J., Paakkinen, K.M., Pirkonen, P.M., Kilpinen, P. & Hupa M. (1995). The effects of pressure, oxygen partial pressure, and temperature on the formation of N_2O, NO and NO_2 from pulverized coal. *Combustion and Flame* Vol.102: 387-400.

Aho, M.J. & Pirkonen, P.M. (1995). Effects of pressure, gas temperature and CO_2 and O_2 partial pressure on the conversion of coal-nitrogen to N_2O, NO and NO_2. *Fuel* Vol. 74: 1677-1681.

Anthony, D.B., Howard, J.B., Hottel, H.C. & Meissner, H.P. (1975). Rapid devolatilization of pulverized coal.. *15th Symposium (International) on Combustion*. The Combustion Institute, Pittsburgh, pp. 1303-1317.

Arai, N. & Hasatani, M. (1986). Model of simultaneous formation of volatile-NO and char-NO during the packed-bed combustion of coal char particles under a NH3/O₂/Ar gas stream. *21st Symposium (International) on Combustion*. The Combustion Institute, Pittsburgh, Presented in the Poster Session.

Arai, N., Hasatani, M., Ninomiya, Y., Churchill, S.W. & Lior, N. (1986). A comprehensive kinetic model for the formation of char-NO during the combustion of a single particle of coal char. *21st Symposium (International) on Combustion*. The Combustion Institute, Pittsburgh, pp. 1207-1216.

Attar, A. & Hendrickson, G.G. (1982). *Coal Structure*. Academic Press, New York.

Benson, S.W. (1968). *Thermochemical Kinetics*. John Wiley & Sons, New York.

Bliek, A., van Poelje, W.M., van Swaaij, W.P.M. & van Beckum, F.P.H. (1985). Effects of intraparticle heat and mass transfer during devolatilization of single coal particle. *AICHE Journal*, Vol. 31: 1666-1681.

Bowman, C.T. (1973). Kinetics of nitric oxide formation in combustion processes. *14th Symposium (International) on Combustion*. The Combustion Institute, Pittsburgh, pp. 729-738.

Bruisma, O.S.L., Geertsma, R.S., Oudhuis, A.B.J.,Kapteijn, F. & Moulijn J.A. (1988). Measurement of C, H, N - release from coals during pyrolysis. *Fuel* Vol. 67: 1190-1196.

Buchill, P. & Welch, L.S. (1989). Variation of nitrogen content and functionality with rank for some UK bituminous coals. *Fuel* Vol. 68: 100-104.

Chan, L.K., Sarofim, A.F. & Beer J.M. (1983). Kinetics of the NO-Carbon reaction at fluidized bed combustor conditions. *Combustion and Flame*, Vol. 52: 37-45.

Chen, Z., Lin, M., Ignowski, J., Kelly, B., Linjewile T.M. & Agarwal P.K. (1995). Mathematical modeling of fluidized bed combustion. 4: N_2O and NO emissions from combustion of char. *Fuel* Vol. 80: 1259-1272.

Cheng, M.T., Kirsch M.J. & Lester T.W. (1989). Reaction of nitric oxide with bound carbon at flame at temperatures. *Combustion and Flame* Vol. 77: 213-217.

Cliff, D.J. & Youg, B.C. (1985). NO_x generation from the combustion of Australian brown and subbituminous coals. *Fuel* Vol. 64: 1521-1524.

Croiset, E., Heurtaebise, C., Rouan, J.P. & Richarad, J.R. (1998). Influence of pressure on the heterogeneous formation and destruction of nitrogen oxides during char combustion. *Combustion and Flame* Vol. 112: 33-44.

Croiset, E., Mallet, Ch., Rouan, J.P. & Richard, J.R. (1996). The influence of pressure on char combustion kinetics. *26th Symposium (International) on Combustion*. The Combustion Institute, Pittsburgh, pp. 3096-3102.

Fenimore, C.P. (1971). Formation of nitric oxide in premixed hydrocarbon flames. *13th Symposium (International) on Combustion*. The Combustion Institute, Pittsburgh, pp. 373-380.

Fenimore, C.P. & Fraenkel, H.A. (1981). Formation and interconversion of fixed-nitrogen species in laminar diffusion flames. *18th Symposium (International) on Combustion*. The Combustion Institute, Pittsburgh, pp. 143-149.

Fenimore, C.P. & Jones, G.W. (1957). Nitric oxide decomposition at 2200-2400 K. *The Journal of Physical Chemistry* Vol. 61: 654-657.

Fong, W.S., Peters, W.A. & Howard, J.B. (1996). Kinetics of generation and destruction of pyridine extractables in a rapidly pyrolysing bituminous coal. *Fuel* Vol. 65: 251-254.

Freihaut, J.D., Zabielski, M.F. & Seery, D.J. (1982). A parametric investigation of for release in coal devolatilization. *19th Symposium (International) on Combustion*. The Combustion Institute, Pittsburgh, pp. 1159-1167.

Gil, S. (2000). Research over the influence of pressure on kinetics of coal nitrogen conversion during pyrolysis and combustion. *PhD Dissertation*. Silesian University of Technology, Katowice.

Gil, S. (2002). Influence of combustion pressure on fuel-N conversion to NO, N_2O and N_2. *Karbo* No. 9: 272-275.

Gil, S. (2003). Heterogeneous destruction of nitrogen oxides by char. *18th International Symposium on Combustion Processes*. Ustroń, pp. 34-35.

Gil, S., (2003). Influence of pressure on the rate of nitrous oxide reduction by char. *12th International Conference on Coal Science*. Cairns, pp. 274 (full manuscript 13B2 on CD).

Gil, S. (2003). Kinetics of heterogeneous nitrogen oxides formation during pressurized char combustion. Grant KBN PAN Nr 4 T10B 029 22, 1075/T10/2002/22.

Given, P.H. , Spackman, W., Davis, A., Zoeller, J., Jenkins, R.G. & Khan, R. (1984). Chemistry of some maceral concentrates from British coals: provenance of samples and basic compositional data. *Fuel* Vol. 63: 1665-1659.

Glarborg, P., Lilleheie, N.I., Byggstøyl, S., Magnussen, B.F., Kilpinen, P. & Hupa, M. (1992). A reduced mechanism for nitrogen chemistry in methane combustion. *24th Symposium (International) on Combustion*. The Combustion Institute, Pittsburgh, pp. 889-898.

Glarborg, P., Miller, J.A. & Kee, R.J. (1986). Kinetic modeling and sensitivity analysis of nitric oxide formation in well-stirred reactors. *Combustion and Flame* Vol. 65: 177-202.

Goel, S., Morihara, A., Tullin, C.J. & Sarofim, A.F. (1994). Effect of NO and O_2 Concentration on N_2O Formation during Coal Combustion in a Fluidized-Bed Combustor: Modeling Results. *25th Symposium (International) on Combustion*. The Combustion Institute, Pittsburgh, pp. 1051-1059.

Gradoń, B. (2003). A role of nitrous oxide in modelling of NO emissions derived from combustion of gaseous fuels in high-temperature furnaces. *Zeszyt Naukowy nr 67 (Hutnictwo)*, Silesian University of Technology, Gliwice.

Heyd, L.E. (1982). Weight loss behavior of coal during rapid pyrolysis and hydropyrolysis. *MSc Thesis*. Princeton University, Princeton.

Howard, J.B., Williams, G.C. & Fine, D.H. (1973). Kinetics of Carbon Monoxide Oxidation in Post Flame Gases. *14th Symposium (International) on Combustion*. The Combustion Institute, Pittsburgh, pp. 975 - 986.

Johnson, J.L. (1975). Relationship between the gasification reactivities of coal char and the physical and chemical properties of coal and coal char. *American Chemical Society, Division of Fuel Chemistry* Vol. 20: 85-101.

Johnson, J.L. (1987). *Kinetics of coal gasification*. John Wiley & Sons, New York.

Johnsson, J.E., Glarborg, P. & Dam–Johansen, K. (1992). Thermal dissociation of nitrous oxide at medium temperatures. *24th Symposium (International) on Combustion*. The Combustion Institute, Pittsburgh, pp. 917-923.

Jüntgen, H. (1987). Coal characterization in relation to coal combustion. *Erdöl und Khole* Vol. 40: 153.

Krammer, G.F. & Sarofim, A.F. (1994). Reaction of Char Nitrogen During Fluidized Bed Coal Combustion – Influence of Nitric Oxide and Oxygen on Nitrous Oxide. *Combustion and Flame* Vol. 97: 118-124.

Van Krevelen, D.W. (1981). *Coal*. Elsevier, Amsterdam.

Krzyżanowska, T. & Kubica, K. (1978). Determination of nitrogen in coal and its liquefaction products. *Koks Smoła Gaz* No. 4: 119-123.

Laurendeau, N.H. (1979). Heterogeneous Kinetics of Coal Char Gasification and Combustion. *Progress in Energy and Combustion Science* Vol. 4: 221-270.

Levy, J.M., Lomgwell, J.P. & Sarofim, A.F. (1978). MIT Energy Lab., Report to Energy an Environmental Res. Corp. MIT, Cambridge.

Lu, Y., Jahkola, A., Hippinen, I. & Jalovaara, J. (1992). The emissions and control of NO_x and N_2O in pressurized fluidized bed combustion. *Fuel* Vol. 71: 693-699.

Mallet, C., Rouan, J.P. & Richard, J.R. (1995). Influence of Pressure on Blends on the Combustion Rate and Pollutant Emissions of Blends of Char and Charcoal. *8th International Conference on Coal Science*. Oviedo, pp. 779 - 782.

Malte, P.C. & Pratt, D.T. (1977). Hydroxyl radical and atomic oxygen concentrations in high-intensity turbulent combustion. *16th Symposium (International) on Combustion.* The Combustion Institute, Pittsburgh, pp. 145-155.

Midkiff, C.K. & Atenkirch, R.A. (1988). Including heterogeneous combustion in first-order and distributed activation energies models of coal nitrogen release. *Fuel* Vol. 67: 459-463.

Miller, J.A. & Bowman, C.T. (1989). Mechanism and modeling of nitrogen chemistry in combustion. *Progress in Energy and Combustion Science* Vol. 15: 287-338.

Miller, J.A., Branch, M.C., Mc Lean, W.J., Chandler, D.W., Smooke, M.D. & Kee, R.J. (1984). The conversion of HCN to NO and N_2 in H_2-O_2-HCN-Ar flames at low pressure. *20th Symposium (International) on Combustion.* The Combustion Institute, Pittsburgh, pp. 673-684.

Molina, A., Eddings, E.G., Pershing, D.W. & Sarofim, A.F. (2000). Char Nitrogen Conversion: Implications to Emissions from Coal-Fired Utility Boilers. *Progress in Energy and Combustion* Vol. 26: 507-531.

Muzio, L.J., Montgomery, T.A., Samuelsen, G.S., Kramlich, J.C., Lyon, R.K. & Kokkinos, A. (1990). Formation and measurement of N_2O in combustion systems. *23rd Symposium (International) on Combustion.* The Combustion Institute, Pittsburgh, pp. 245-250.

Niksa, S. (1988). Modeling the devolatilization behavior of high volatile bituminous coal. *22nd Symposium (Int.) on Combustion.* The Combustion Institute, Pittsburgh, pp. 105-114.

Östman, C.E. Colmsjö, A.L. (1988). Isolation and classification of polycyclic aromatic nitrogen heterocyclic compounds. *Fuel* Vol. 67: 396-400.

Peck, R.E., Altenkirch, R.A. & Midkiff, K.C. (1984). Fuel - nitrogen transformations in one - dimensional coal – dust flames. *Combustion and Flame* Vol. 55: 331-340.

Pereira, F.J., Beer, J.M., Gibbs, B. & Hedley, A.B. (1974). NO_x emission from fluidized-bed coal combustions. *15th Symposium (International) on Combustion.* The Combustion Institute, Pittsburgh, pp. 1149-1156.

Pershing, D.W. & Wendt, J.O.L. (1977). Pulverized coal combustion: the influence of flame temperature and combustion on thermal and fuel NO_x. *16th Symposium (International) on Combustion.* The Combustion Institute, Pittsburgh, pp. 389-399.

Pershing, D.W. & Wendt, J.O.L. (1979). Relative contributions of volatile nitrogen and char nitrogen to NO_x Emissions from pulverized coal flames. *Independent Engineering Chemical Process Design and Development* Vol. 18: 60-67.

Pohl, H. J. & Sarofim, A. R. (1977). Devolatilization and oxidation of coal nitrogen. *16th Symposium (International) on Combustion.* The Combustion Institute, Pittsburgh, pp. 491-501.

Pottgiesser, C. (1980). Pyrolyse von Steinkohlen in einem Druckbereich von 1 bis 100 bar. *PhD Dissertation.* RWTH Aachen University, Aachen.

Rybak, W. (1996). Coal structure and pollutant emissions during combustion. *PKBP Conference on Research and Technology.* Ustroń, pp. 13-26.

Sarofim, A.F., Kandas, A.W. & Goel, S. (1995). Char nitrogen: effect of coal type pore structure on NO/N$_2$O evolution. *25th Symposium (International) on Combustion*. The Combustion Institute, Pittsburgh, pp. 1125-1130.

Sarofim, A.F. & Pohl, J.H. (1973). Kinetics of nitric oxide formation in premixed laminar flames. *14th Symposium (International) on Combustion*. The Combustion Institute, Pittsburgh, pp. 739-754.

Shimizu, T., Sazawa, Y., Adshiri, T. & Furusawa, T. (1992). Conversion of Char-Bound Nitrogen to Nitric Oxide During Combustion. *Fuel* Vol. 71: 361-365.

De Soete, G.G. (1990). Heterogeneous N$_2$O and NO formation from bound nitrogen atoms during coal char combustion. *23rd Symposium (International) on Combustion*. The Combustion Institute, Pittsburgh, pp. 1257-1264.

De Soete, G.G., Croiset, E. & Richard, J.R. (1999). Heterogeneous formation of nitrous oxide from char-bound nitrogen. *Combustion and Flame* Vol. 117: 140-154.

Stańczyk, K. (1991). Chemistry of nitrogen in coal and formation of nitrogen oxides during coal and char combustion – a review. *Koks Smoła Gaz* No. 11: 260-264.

Stanmore, B.R. (1991). Modeling of combustion behavior of petroleum coke. *Combustion and Flame* Vol. 83: 221-227.

Szarawara, J. & Skrzypek, J. (1980). *The basics of chemical reactor engineering*. PWN, Warszawa.

Tingey, G.L. & Morrey, J.A. (1973). Coal structure and reactivity. A Battelle Energy Program Report. Battelle Pacific Northwest Laboratories, Richland, Washington.

Tomeczek, J. & Gil, S. (2000). Nitrogen oxide reduction rate during char pressurised combustion. *VIIIth Research Seminar*. Silesian University of Technology, Katowice, pp. 157-160.

Tomeczek, J. & Gil, S. (2001). Modelling of nitrogen oxide emissions during char pressurised combustion. *IXth Research Seminar*. Silesian University of Technology, Katowice, pp. 217-220.

Tomeczek, J. & Gil, S. (2001). Influence of pressure on the rate of nitric oxide reduction by char. *Combustion and Flame* Vol. 126: 1602-1606.

Tomeczek, J. & Gradoń, B. (1997). The role of nitrous oxide in the mechanism of thermal nitric oxide formation within flame temperature range. *Combustion Science and Technology* Vol. 125: 159-180.

Tomeczek, J. & Mlonka, J. (1998). The parameters of a rondom pore network with spherical vesicles for coal structure modelling. *Fuel* Vol. 77: 1841-1844.

Tullin, C.J., Goel, S., Morihara, A., Sarofim, A.F. & Beer, J.M. (1993). NO and N$_2$O formation for coal combustion in a fluidized bed: Effect of carbon conversion and bed temperature. *Energy & Fuels* Vol. 7: 796-802.

Visona, S.P. & Stanmore, B.R. (1996). Modeling NO$_x$ Release from a Single Coal Particle, II. Formation of NO from Char-Nitrogen. *Combustion and Flame* Vol. 106: 207 -218.

Visona, S.P. & Stanmore, B.R. (1999). Modeling nitric oxide formation in a drop tube furnace burning pulverized coal. *Combustion and Flame* Vol. 118: 61-75.

Weiszrock, J., Schwartz, D., Gadiou, R. & Prado, G. (1997). NO$_x$ emission during pressurized coal combustion. *4th European Conference Industrial Furnaces and Boilers*. Espinho-Porto.

Wendt, J.O.L. & Schulze, O.E. (1976). On the Fate of Fuel Nitrogen During Coal Char Combustion. *AICHE Journal* Vol. 22 (No. 1): 102-110.

Williams, A., Pourkashanian, M., Jones, J.M. & Rowlands, L. (1997). A review of NO_x formation and reduction mechanisms in combustion systems with particular references to coal. *Combustion and Emission Control III*. The Institute of Energy, Elsevier, London, pp. 1-26.

Williams, A., Woolley, R. & Lawes, M. (1992). The formation of NO_x in surface burners. *Combustion and Flame* Vol. 89: 157-166.

Zeldowicz, Y.B. (1946). The oxidation of nitrogen in combustion and explosions. *Acta Phisicochemica USSR* Vol. 21: 577-628.

4

Fossil Fuel Power Plant Simulators for Operator Training

José Tavira-Mondragón, Guillermo Romero-Jiménez
and Luis Jiménez-Fraustro
Electric Research Institute
Mexico

1. Introduction

Before the 1970´s the use of simulators to train the operation personnel of the power plants was not widely diffused. In these times, the operators acquire their skills by working head to head with some experienced operators in the actual plant, so they learned all the knowledge of their mentor, this means, all the virtues and defects of the experienced people. As expected, the trainees also receive the classic classroom lessons with the aim to complement their training. The training finished when the manager of the plant decided the trainee was ready to operate and control the plant. In the majority of the cases, the main problem of this kind of training was that the operator just learned the typical actions related with the start-up of the equipment and operation of the plant in nominal conditions. Therefore, operators had not been trained in abnormal situations, where they needed to act rapidly to keep the power plant in safety conditions. Naturally any operative mistake could lead to a unit trip, equipment damage or risk to staff with all the economic looses related with this type of problems. During the 1970´s, in the United States, the nuclear power industry made the commitment of including simulators as a part of the training programs of their nuclear power plant operators, gradually the use of simulators in the nuclear power industry gained worldwide acceptance. In 1979 a major accident occurred at Three Mile Island (TMI) Unit 2 in Middletown, Pennsylvania resulted in a critical assessment of the preparedness of operations staff to respond to the accident. It is commonly believed that the incident at TMI would not have occurred if the operators had been properly trained. This accident prompted a complete re-evaluation of the nuclear industry's operator training programmes (Perkins, 1985). Events like this reinforced the growth of the rising industry of simulators and that extended its application to the fossil fuel power plants too. Specifically in this segment, the Electric Power Research Institute [EPRI] (1993) carried over a cost-benefit analysis of simulators used at fossil fuel power plants, where the identified benefits were: availability savings, thermal performance savings, component life savings, and environmental compliance savings. Additionally EPRI reported that approximately 20% of forced plant outages were direct result of operator or maintenance error. Therefore, reducing operator controllable outages through training on a simulator can significantly reduce operating costs. Additional quotes about operators errors (Serious Games LLC, 2006), establishes that "One manufacturing analyst estimated, human error leading to abnormal situations costs the UK process industry $1.4 billion a year" and "In the last 25 years, the largest 100 accidents in the hydrocarbon-chemical processing industry cost $7.52 billion in losses; operator error accounted for 21% of these events at an average of $75 million per loss".

The modern distributed control systems of the power plants provide to the operator with the elements to get a power generation stable, safe and reliable, but as a consequence, there is a reduction of the operator's confidence to carry out unusual manoeuvres, e.g. a start-up in manual mode or the requested actions after a feed water pump trip. Training simulators help operators to practice this type of manoeuvres. The main advantage of a simulator as a training tool is that the operator does not need to touch the actual unit to learn to operate it in a broad range of possible scenarios. These scenarios include normal operations like unit start-up from cold iron to full load and shutdown. Also can be defined scenarios for malfunctions in which the trainee practice the suitable operative actions when in the simulated unit there are events like: trips of pumps and turbine, tube ruptures, and "faulty" instrumentation. In other words, the operators use the simulator to practice their normal operation procedures and to practice infrequent evolutions and faulted conditions. Therefore, one of the most important parts of the training programmes of power plant operators is carried out trough simulators, a big number of these simulators are of the type called full-scope, these simulators incorporate detailed modelling of those systems of the referenced plant with which the operator interfaces in the actual control room environment. Usually, replica control room operating consoles are included (International Atomic Energy Agency [IAEA], 2004). In these simulators, the responses of the simulated unit are identical in time and indication to the responses received in the actual plant control room under similar conditions. A significant portion of the expense encountered with this type of simulators is the high fidelity simulation software that must be developed to drive it. The completeness of training using a full-scope simulator is obviously much greater than that available on other simulator types since the operator is performing in an environment that is identical to that of the control room. Experienced operators can be effectively retrained on these simulators because the variety of conditions, malfunctions, and situations offered do not cause the operator to become bored with the training or to learn it by rote (Instrument Society of America [ISA], 1993). Therefore, full-scope simulators are recognized worldwide as the only realistic method to provide real-time and hand-on training of operators. Also the simulators can be utilized to validate the normal operating procedures, to conduct engineering studies and to train plant technical supporting personnel.

However, the expense of developing this kind of simulators, the necessity of training a bigger number of the operation staff and the search of better training has driven the development of different training tools, for instance, there are part-task simulators, where the users are only trained in a particular system of the power plant (e.g. feed-water system, steam turbine, etc.). There are also compact simulators, where the users can practice the majority of the main operation actions required in a power plant, but the operation interfaces are of a generic type and not necessarily are similar to the ones the operators utilize in their actual power plant. In many cases, the part-task and compact simulators are portables, so they are transported to the power plants, in this way, the operators can practice onsite, these simulators can be utilized with the assistance of an instructor, in a free-hands context, or with the guidance of an expert system. In spite of the shortcomings of these simulators , there are some clearly identified benefits of using a variety of training simulators, which are: the ability to train on malfunctions, transients and accidents; the reduction of risk to plant equipment and personnel; the ability to train personnel on actual plant events; a broader range of personnel receiving effective training; and individualized instruction or self-training being performed effectively on simulation devices designed with these capabilities in mind. The use of simulators has proven trough the years to be one of the most effective and confident ways by training power plant operators. Using simulators, operators can learn how to operate the plant more efficiently, lowering the heart rate and reducing the power required by plant auxiliary equipment (Hoffman, 1995).

The main features and types of training simulators, the importance of a well structured training programme to maximize the benefits of a simulator and the two most important paradigms for the mathematical modelling are discussed in the following sections.

2. Training simulators

According to The Free Dictionary (thefreedictionary, 2008), a simulator is defined as "any device or system that simulates specific conditions or the characteristics of a real process or machine for the purposes of research or operator training". In the context of training of power plant operators, a simulator is a system composed of a Human Machine Interface (HMI) which replicates the operation consoles of the actual plant and a computer that executes mathematical models, which "replicate" unit performance. These simulators are based on the mathematical modelling of dynamic systems and their expected responses have a real-time functioning. Usually the training sessions are guided by an instructor which establishes the initial condition, starts the simulation, and supervises the actions of the trainees. This concept is shown in a simplified way in Figure 1.

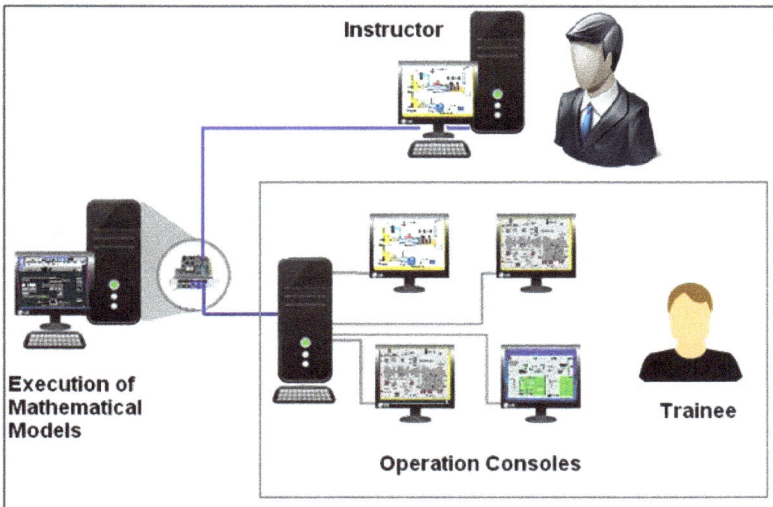

Fig. 1. Schematic representation of a training simulator.

2.1 Simulators with different scopes

According to the training objectives and the available hardware, there are different types of simulators; a brief review of them is done in the next sections.

2.1.1 Full-scope

A full-scope simulator can be defined as an exact duplicate of a power plant control room, containing duplicates of all actual controls, instruments, panels, and indicators. The unit responses simulated on this apparatus are identical in time and indication to the responses received in the actual plant control room under similar conditions. A significant portion of the expense encountered with this type of simulator is the high fidelity simulation software that must be developed to drive it (ISA, 1993). Due to the HMI of the trainee must be a "copy" of the control room of the power plant, it was ordinary that a simulator had the same control boards of the actual unit, which naturally involved a big expense for its construction. Figure 2 shows two control board simulators, for a 300 MW fossil fuel power plant (left side) and for a 350 MW fossil fuel power plant (right side).

On the other hand, in recent years the power increase of computers, their reliability and variety of graphical interfaces (Yamamori et al., 2000), added to the continued search to cut costs caused a new technological trend. In this trend, the power plants have replaced their former control boards with a local area network of Personal Computers (PC) with graphical user interfaces. In this way, new or modernized power plants have a HMI, where all the supervising and operation actions are carried out through interactive processes diagrams.

Fig. 2. Control board simulators.

Naturally, the operators of these plants need a suitable training because they face a complete change in their operation paradigm, and because of this, the training simulators also require a HMI as the one in the actual plant, an example of these simulators is in Figure 3.

The technological revolution affected the hardware and software components of the simulator, for instance, Zabre and Román (2008) describe the evolution on hardware, operating systems and software for the power plant simulators developed by the Electric Research Institute of Mexico in the last 30 years. Table 1 shows the main features of the hardware-software platforms for different simulators. In fact, this revision is focused over the Mexican market but it is very representative of the world scale evolution of the hardware-software platforms for simulators. The typical architecture of this type of simulators is given in section 4.

Fig. 3. Simulator with interactive process diagrams.

Item	1980s	Latest 1990s	Nowadays
Main Computer	Gould/Encore	Workstation	Local Area Network of Personal Computers
HMI	Control Boards	Control Boards and Interactive Process Diagrams	Interactive Process Diagrams
Operative System	MPX-32	Unix TRU64	Windows XP/Windows 7
Software	Fortran compiler	Fortran and C compilers Oracle DB GE DataViews	Fortran and C compilers MS Access/SQL Server MS Visual Studio Adobe Flash
Examples of simulators	300 MW fossil fuel (no DCS) power plant 350 MW fossil fuel (with DCS) power plant	Upgrade of 300 MW fossil fuel Update of 350 MW fossil fuel	350 MW coal fired power plant 450 MW combined cycle power plant

Table 1. Evolution of hardware and software platforms.

Typical operations to carry out by the trainees in a simulator are:

- Unit start-up from cold iron or hot standby up to nominal power.
- Boiler pressurization.
- Turning, acceleration and synchronization of the turbine-generator group.
- Unit shutdown.
- Any operative action feasible in the actual plant.

The simulator also can reproduce abnormal or emergency situations due to a deficient operation of the trainee or due to a malfunction inserted by the instructor. In the last case, during the specification of the simulator must be defined a malfunctions group. These malfunctions are of two types: binary and analogue. The first group contains malfunctions like: pump trips, fail position of valves, etc. The analogue malfunctions have a degree of severity (usually normalized from 0 to 100%) and their severity can be selected by the instructor. Examples are tubes rupture of steam lines and fouling factor of heat exchangers.

2.1.2 Part-task simulators

A part-task simulator is focused only on specific plant systems. These systems are represented with features of a full-scope simulator, in this way, detailed mathematical modelling of the referenced plant systems is included and just a part of the actual control room is duplicated with all key instrumentation, controls and alarm signals. The systems not included in the HMI are simulated with a reduced scope or no simulated and considered as always "on service", just to satisfy the interactions of the main systems. For instance, Figure 4 shows a part-task simulator to train operators in turning and acceleration of the steam turbine (Burgos, 1993). This simulator includes the portion of the control board corresponding to the steam valves (throttling, governing, stop and intercept) and the related instrumentation and control.

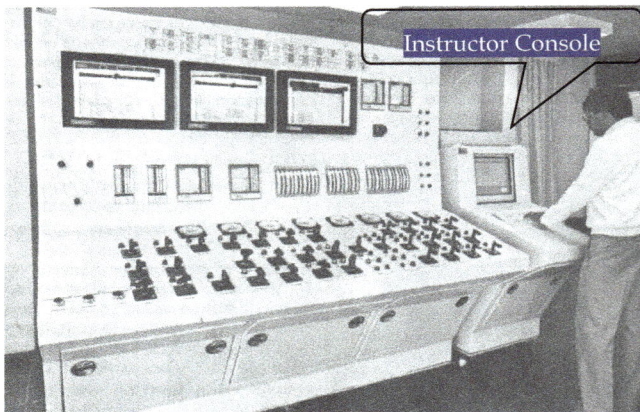

Fig. 4. Part task simulator

The systems connected to the steam turbine (including the thermal model for rotor and casings), lubrication oil, control oil and required controls are simulated in a full-scope context, but other systems like main steam, feed water and main condenser are simulated with a reduced scope. Therefore these simulators are beneficial to improve the knowledge and provide training in particular areas of the power plant. Tavira-Mondragón et al. (2006) describes a part-task simulator for five subsystems of a fossil fuel power plant with a HMI based on interactive process diagrams. The systems considered are: electric network, auxiliary services, boiler pressurization, turning turbine and power increasing from minimum up to nominal power. This simulator is portable to be transported to the power plants.

A different approach for a simulator is the proposed by Pevneva et al. (2007), these authors present a unified training simulator for the personnel of the boiler-turbine and chemical departments with the purpose of perfecting interaction skills between these areas.

2.1.3 Compact simulators

Compact simulators are frequently generics, this means they reproduce the behaviour of a specific power plant, but the rated power and the HMI for the trainee no necessarily are the same of the actual plant. However, they include mathematical modelling of wide scope which allows simulating plant conditions from cold iron up to nominal power. In this way they are mainly utilized to train novice operators and field personnel. Fray and Divakaruni (1995) claim this kind of simulators can be parameterized with unit-specific design and operating data for power units of specific generation and inclusive it is possible to include the emulation of an exact replica of the plant control system. This type of modifications necessarily increment the initial cost of the simulator and such modifications must be done by very specialized personnel.

Currently, with the power of modern computers, a complete power plant simulator can be installed in a laptop and easily transported, the main problem is the use of the simulator by the trainee because its interface is reduced, in the best-case scenario to a equipment with three displays (Figure 5), and with the number of actions required to operate the simulated system, a suitable training can be a complex problem without a effective method to do it.

Fig. 5. Compact simulator

2.1.4 Classroom simulators

These simulators usually include detailed mathematical modelling and they can be divided in two groups: graphical and multi-user. Graphical simulators are based on a representation of the HMI in graphical form (display units or virtual images). These simulators provide a low-cost alternative to other simulators requiring the use of control room hardware (IAEA, 1998); therefore they have been used preferably in the nuclear power industry due to the great number of control boards of these plants. On the other hand, multi-user simulators are installed in a local area network and they are used as a complement of the training courses for operators of fossil power plants (Tavira-Mondragón et al., 2005; Romero-Jiménez et al., 2008). In these simulators, the instructor has a console where he simultaneously directs the simulation sessions for each one of the trainees. With his console, the instructor establishes the same or different training exercises for each one of the students and supervises them in an individual way from the same interface. Each one of the trainees has his own console to operate the simulated power plant in an independent way of the other students.

2.1.5 Virtual simulators

With the increase in processing capacity, computational speed, advances in computer technology and the sophistication in modelling, it is becoming more feasible to develop systems based on Virtual Reality (VR) or Augmented Reality (AR). These systems implement hardware interfaces which deliver a more stimulating experience to the trainee, via the experimentation of "realistic" sensations. In an available dynamic simulator can be implemented a three-dimensional visor in order to train maintenance people on local operations and to complement actual training programmes with three-dimensional view of equipment like: turbines, boilers, electric generator, etc. With these systems, trainees practice the operation of the process, and at the same time, they get a better understanding of the physical and chemical phenomena occurred and obtain detailed knowledge about the equipment. Martínez-Ramírez et al. (2011) present a prototype of a VR training system for a thermal power plant.

2.1.6 Web simulators

According to the simulators architectures previously discussed, the simulators are installed in a training centre or on site if the simulator is easily portable. In the case of the simulators where more than a PC is involved, the computers are connected in a local area network. Therefore, the communication between the interactive processes diagrams and the mathematical models is based on a proprietary protocol and implemented over TCP/IP with the aim to link two points within a local network. Such diagrams are part of a conventional Windows application, which can only be executed on a computer with the environment provided by the operating system and interconnected to the same network, so it cannot be transmitted through the Internet and take advantage of modern information technologies like cloud computing.

Cloud computing is the delivery of computing as a service rather than a product, whereby shared resources, software and information are provided to computers and other devices over a network (typically the Internet). Cloud computing provides computation, software, data access, and storage services that do not require end-user knowledge of the physical location and configuration of the system that delivers the services. The concept of cloud computing fills a perpetual need of information technologies: a way to increase capacity or add capabilities on the fly without investing in new infrastructure, training new personnel, or licensing new software. Cloud computing encompasses any subscription-based or pay-per-use service that, in real-time over the Internet, extends information technologies existing capabilities. Cloud computing providers deliver applications via the internet, which are accessed from a web browser, while the business software and data are stored on servers at a remote location (Wikipedia, 2011a).

In the current transition process to new computational paradigms, where Internet plays a important role, the use of modern techniques for the development of software applications is a key strategy to guide these technologies, and training simulators require to adapt their platforms to support suitable graphical interfaces for the final user and implement the Internet communication mechanisms called web services. As a result of this process, it is obtained a web user interface which allows interacting with a simulator from a remote location via a HMI with similar features to the ones available in the simulators of the training centres. In such way, the simulator is available to any computer with an internet connection and a web browser with the plugins required for the application. Figure 6 shows the services provided by cloud computing and its comparison with the services required for a full-scope simulator.

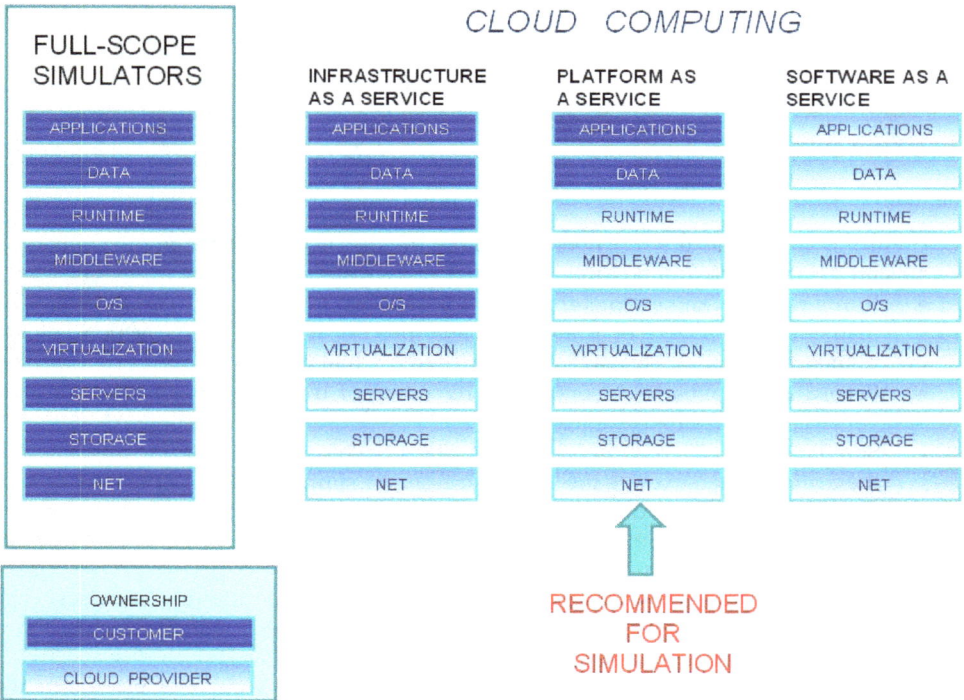

Fig. 6. Cloud computing services.

2.2 Instructor role and training programmes

As it was mentioned at the beginning of the chapter, the instructor of a training session is in charge of directing and supervising to the trainees. The role of the instructor depends of the simulator type and the trainees´ knowledge, in this way, the instructor needs to balance the complexity of training practices with the needs of the trainees and provide them of the additional information and explanations to get a complete understanding of the phenomena involved. The instructor console is the instructor interface to conduct the training session; this interface is usually a graphic display with pull-down menus and icons for an easy access to the functions required. Figure 7 shows a typical instructor console interface.

Fig. 7. Main interface of the instructor console

The main functions of the instructor console are as follow:

- Control Menu. It has the functions that allow the instructor to manage the simulation session. The available options are:
  - Run/Freeze. The instructor starts or freezes a dynamic simulation. The mathematical models of the control and process respond to the actions of the trainee in a comparable way as it occurs in the actual plant.
  - Record and Playback. A continuous recording of trainee actions for later replay which can be repeated automatically
  - Simulation speed. In its default mode, the simulator is executed in real-time, but the instructor can execute the simulator faster or slower than real-time. This option is especially important when the instructor wants the trainees analyze a fast transient, allowing him to simulate it slower. On the other hand, a slow thermal process like turbine iron-heating can be simulated faster.
  - Automatic Exercises. The instructor can create automatic training exercises, each one of them can include: initial conditions, malfunctions, remote actions, and a time sequence. The exercises are stored for their subsequent use.
- Initial Conditions. They allow establishing the state of the simulated process at the beginning of the session and creating new ones. The options are:
  - Selecting an initial condition (snapshot) to start the simulation session, for instance: *Cold start, Ready to roll turbine, Full load.*
  - Recording a new initial condition or erasing an old initial condition;
  - Specifying the time interval of automatic snapshots.
- Instruction Functions Menu. They contain functions which alter the simulated process.

- The option of Malfunctions is used to introduce/remove an equipment malfunction at any time during the simulation session. Examples of malfunctions are: pump trips, heat exchanger tubes rupture, and control valve obstructions. All the malfunctions are grouped in systems and subsystems for easy location. For the binary malfunctions, the instructor has the option of defining its time delay and its duration. For analog malfunctions, besides the former time parameters, the instructor can also define their intensity. Additionally it is possible to select any instrument of the HMI and make it faulty, so it shows an unreliable indication to the trainee.
- The instructor has the option of Remote Functions to simulate the operative actions not related with automated equipment. These operative actions are associated with the local actions performed in the actual plant by an auxiliary operator. Examples are: to open/close valves and to turn on/off pumps or fans. Similarly to the malfunctions, they are grouped in systems and subsystems.
- The option of External Parameters allows the instructor to modify the external conditions to the process. These conditions are: atmospheric pressure, room temperature, voltage and frequency of the external electric system and fuel composition.
- Tracking and Miscellaneous Menus. They contain additional functions to help the instructor to get information of the simulator behaviour and to manage the access to the simulator.

A panoramic view of the instruction room with three instructors guiding the training sessions in three different simulators is shown in Figure 8.

Fig. 8. Instruction room

With the aim of providing to the operation personnel of the knowledge and abilities to operate in a safe and reliable way a power plant, the training programmes usually include: classroom lessons, practice in the simulator and "on the job experiences". The design and execution of a successful training programme involves a variety of people coordinating their efforts to achieve the desired results.

The ADDIE model is a suggested methodology to develop the training programmes for the power plant operators (EPRI, 2005). The acronym ADDIE signifies the phases of the training development process: Analysis, Design, Development, Implementation and Evaluation. A brief description of these phases is as follow:

- The Analysis phase is the base of the instructional design model. In this phase the personnel in charge of the programme identifies the learning problem, establishes the training objectives, and the trainee necessities. This phase may include specific research techniques such as trainee analysis and task analysis. During the trainee analysis, the training specialist examines the current knowledge and skills of trainees and determines what they already know and their abilities. The training specialist uses this information to create a course that focuses on trainees needs. On the other hand, in the task analysis a training specialist is able to create a competency map for trainees. The results of this phase often include educational goals and a list of tasks.

- The Design phase uses the results of the analysis phase to plan a strategy for the development of the training. During this phase, the training specialist outlines a systematic process to achieve the educational goals previously identified, he specifies too, the organization of the course, topics to be covered (and the delivery format), activities, exercises and evaluation of the trainees. In the phase are defined the required practice in simulator and its type, for instance, to achieve objectives related with familiarization of the HMI, a classroom simulator is a good option, and to reach objectives of developing abilities for a safer operation when the trainee faces a malfunction, a full-scope simulator is the best choice. So the specification of the required features of the simulator becomes a key point of this stage.

- The Development phase is the creation of the content and learning materials based on the design stage. Therefore, the end result are the media and its content, this includes written lessons, software (e.g. computer-based instruction) and hardware-software (e.g., simulators), all of this with the aim of getting the training objectives. In the case of a fossil fuel power plant simulator is a common practice to acquire it instead of developing it, due to the complexity and the time required to build a simulator.

- The Implementation phase is the training impartation, whether based in the classroom, simulator or in the job. From an ideal point of view, the instruction would be efficient and effective, but it can be one of the hardest parts of the system because in this phase many of the failures and virtues of the previous stages are manifested, besides the trainers faculties to transmit the required knowledge to trainees is a very important element of the learning process. However, this phase must promote understanding of the material by trainees, and the reaching of the programme objectives.

- The Evaluation phase measures how well the training objectives were achieved. This evaluation is carried out in each one of the previous stages (Formative) to improve the training process before the implementation of the final version of the programme. There is also an evaluation performed after the final version is implemented (Summative), which gives an overall assessment of the training process, this serves to determine if the programme requires notorious changes or just a fine-tune, in this evaluation the feedback of trainees is very important.

The interaction of the different phases of the ADDIE is depicted in Figure 9. The main benefit in using a structured phased approach is that to the end of the process, the training objectives will be more likely achieved.

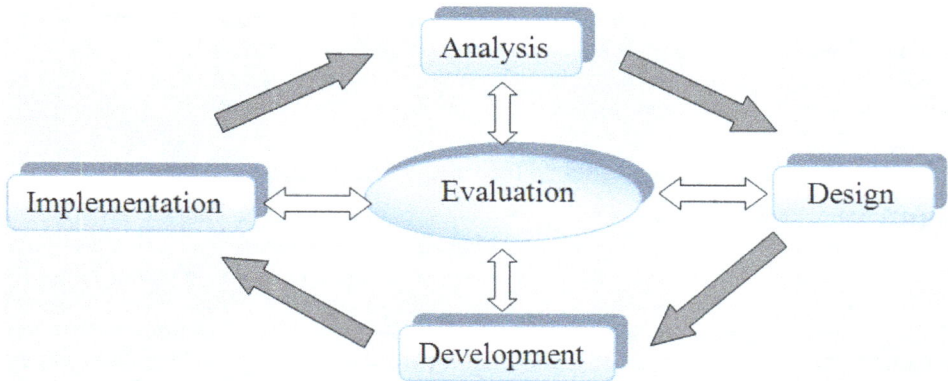

Fig. 9. ADDIES phases

2.3 Simulation and the plant lifecycle

The use of simulation throughout the plant life cycle, from the design until training and operation stages, has the potential of providing significant benefits. The cumulative effect of cost savings and improved operating rate can be substantial and typically return the initial simulation investment within the first year of operation. It will also continue to contribute to profitability based on better operation practices. Ahmad et al. (2010) present a case study where the use of an operator training simulator and the incorporation of advanced process control reduce the cost of a project in 49 Millions USD. The application of dynamic simulation throughout the plant lifecycle can provide the following benefits:

- Identify process and control constraints at the conceptual design phase.
- Begin start-up of a new or modernized unit sooner.
- Train operators on safe operation procedures.
- Transfer best practices to new operators with hands-on practice.
- Avoid or minimize incidents and recover faster from abnormal situations.
- Satisfy government regulations.
- Provide simulation applications to improve operation and control
- Safety training and crisis handling.
- Reduce equipment trips.
- Improve process know-how.
- Increase process controllability and reliability.
- Expand component life and reduce the risk of equipment damage
- Start-up faster and reduce heat rate.
- Operate closer to environmental limits.
- Better operation during start-ups, power increasing and shutdowns.
- Better response to plant transients, abnormal situations and emergencies.
- Verify the operation in design conditions.
- Stimulate the teamwork and communication.
- Reduced costs through evaluation and process optimization
- Improve plant safety

Besides the referred documents of the ISA, IAEA and EPRI, additional information about simulators and training programmes can be founded in EPRI (1998) and IAEA (1996, 2003). The documents published by IAEA deal about the training of operators of nuclear power plants but many of the key concepts are applicable to the fossil industry.

3. Application of expert systems to training simulators

Expert Systems (ES) are computer programs that incorporate a large amount of knowledge in a very specific field and are used to give advice or solve problems. The use of ES became a viable solution to real problems since the 1980's, since then, the use of ES has proliferating to many technological sectors, as it is demonstrated in the review of Liao (2005). Olmstadt (2000) presents a definition of ES which synthesize many of the different definitions available in the literature, this is: ES use human expertise (the result of deliberate practice on standard tasks over many years) to answer questions, pose questions, solve problems, and assist humans in solving problems. They do so by using inferences similar to those a human expert would make, to produce a justified, sound response in a brief period of time. When questioned, they should be able to produce the rules and processes that show how they arrived at the solution. The main parts of the ES are:

- Knowledge base. It contains the knowledge of the facts and experiences of experts in a particular domain, i.e., it contains general knowledge about the expert domain.
- Inference engine. It is responsible of modelling the process of human reasoning. This engine works with the information contained in the knowledge base.
- User interface. This represents the method through the ES interacts with the user. This may require designing the interface using menus, dialog boxes, forms, graphics, etc.

According to the capabilities of ES, their use in training power plant simulators has been explored as intent of minimizing the instructor role. Seifi and Seifi (2002) developed their own intelligent tutoring system for a fossil fuel power plant simulator, while Arjona et al. (2003) utilize CLIPS as foundation for their tutoring system of a part-task simulator for a steam turbine. The C Language Integrated Production System (CLIPS) is probably the most widely used expert system tool because it is fast, efficient and free (Wikipedia, 2011b). CLIPS is an inference engine initially developed to facilitate the representation of knowledge to model human expertise, it provides a cohesive tool for handling a wide variety of knowledge with support for three different programming paradigms: rule-based, object-oriented and procedural. Rule-based programming allows knowledge to be represented as heuristics, or "rules of thumb", which specify a set of actions to be performed for a given situation. Object-oriented programming allows complex systems to be modelled as modular components, which can be easily reused to model other systems or to create new components. In the procedural approach, CLIPS can be called from a procedural language, perform its function, and then return control back to the calling program (CLIPS, 2011).

3.1 Knowledge acquisition and its representation

A critical task in the development of a knowledge-based system is the knowledge acquisition, which is the process of collecting information from any source (expert knowledge, book, manuals, etc) needed to build the system. In the case of a simulator a good reference to carry out this process are the available training procedures and the operation manuals. Usually these documents describe the objectives of the training session, the instructor actions (initial condition, malfunctions, etc) and the required actions to operate the unit (to turn on pumps, to open valves, etc), all this information helps to build a very complete knowledge base. These documents include normal operation (start-up, normalization and shutdown operations, for each one of the power plant systems) and in many cases abnormal operations. Additional tests can be performed in the simulator with the aim of getting supplementary information, mainly in the cases of malfunctions, where there are not enough documented records.

The acquired knowledge must be formalized and ordered with the aim of being useful to the ES; this process is named "Knowledge representation". One of the most common methods to represent knowledge is the production rules. In this method, the knowledge is divided into small fractions of knowledge or rules. A rule is a conditional structure that logically relates the information contained in the part of antecedent with other information contained in the part of the consequent. A very important feature is that the knowledge base is independent of the inference mechanism used to solve problems. Thus, when the stored knowledge become obsolete, or when new knowledge is available, it is relatively easy to add new rules, delete old ones or correct existing errors. Therefore, there is no need of reprogramming all the expert system. The rules are stored in hierarchical sequence logic, but this is not strictly necessary. It may be in any sequence and the inference engine will use them in the right order to solve a problem. This approach is also called IF-THEN rules and some of its main benefits are their modularity and that each rule defines a relatively small and independent piece of knowledge. However, the process of coding the rules can be a cumbersome chore for personnel little familiar with this kind of responsibilities. Tavira-Mondragón et al. (2010b) describes a graphic tool which serves to build training exercises for a combined cycle power plant simulator, with no guidance of a human instructor. This editor contains a group of blocks where each block represents a rule (or a group of rules), and each block is customized by their characteristic parameters. Figure 10 shows in a schematic way, the graphic representation of the malfunction insertion during a training

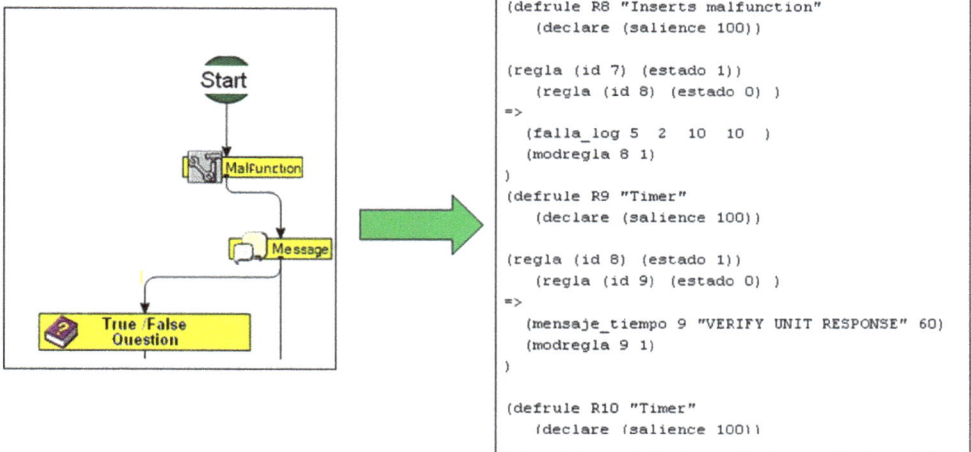

Fig. 10. Knowledge representation in CLIPS.

session and the corresponding rules generated by the editor, which are used during the execution of CLIPS as tutoring system. In such way, the ES is responsible of tracking the status of the simulation to determine the group of rules that should be fired. Due to its inference engine, and according to the configuration of the simulation exercise, the ES is able to modify the simulation process, because it can insert malfunctions, modify values of selected process variables, and change the status of the simulation without the intervention of a human instructor.

The use of ES is especially suitable for training standalone systems, because these systems incorporate: a simulator of a power plant, an intelligent tutor to guide the training session, and besides it can include the trainee evaluation and study material in some multimedia format as theoretical support of the training objectives. Naturally, the HMI for the trainee must be designed bearing in mind that the user, in addition of its operation interfaces, will need "a window" to observe the tutor messages.

4. Hardware-software architecture of a simulation system

According to the different simulator types described in the second section, the required hardware is characteristic for each one of them; therefore the next description of hardware is based on a full-scope replica simulator.

4.1 Hardware architecture

The hardware requirements are exemplified in Figure 11, where there are four PC interconnected through a fast Ethernet local area network. Each PC must have the processor and memory required to execute the simulator smoothly, and to support high processing demand functions like execution faster than real-time. The monitors of the operator consoles must be of a similar size of the ones in the actual power plant. In the case of the Figure 11, the configuration depicts monitors of 20" and 50". During the training session, the trainee

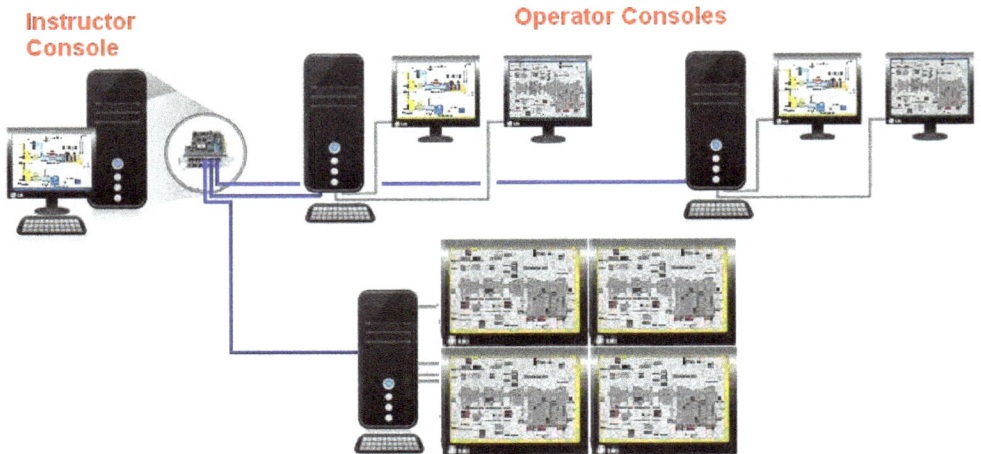

Fig. 11. Hardware architecture.

can use any one of his consoles to supervise and control any process of the power plant. The instructor console is provided with two monitors, hence, besides of using the instructor functions described previously, he can display any screen of the operator consoles with the purpose of watching any operative action carried out by the trainee. In many architectures of simulators is common to find an additional PC, the maintenance station (not shown in Figure 11), which serves as a backup if the instructor console fails or as a test station, this means that any software modification is tested and validated in this station before any change is carried out in the simulator.

4.2 Software architecture

The simulation software is designed with the purpose that the response of simulator is comparable with the results observed in the reference plant under similar conditions. As expected, besides the mathematical models, it is required the execution software or simulation environment. Tavira-Mondragón et al. (2010a) describes the software architecture for a simulation environment. The software architecture of the simulation environment has four main parts: the real-time executive, the operator module, the instructor console module, and mathematical models. Each one of these modules can be hosted in the same or in different PC, and they are connected through the TCP/IP protocol under Windows operating system. A brief description of each module is shown in the following paragraphs (the mathematical models are discussed in the next section).

- Real-Time Executive. The real-time executive module coordinates all simulation functions, so it includes the mathematical model launcher, the managers for: interactive process diagrams, global area of mathematical models and instructor console. Additionally it includes data base drivers and the main sequencer, which sequences all the simulator functions in real-time.
- Operator Module. The operator module is in charge of the operator HMI and manages the information flow with the executive system. The HMI consists of interactive process diagrams, which are animations with static and dynamic parts. The static part is constituted by a drawing of a particular flow diagram whereas the dynamic part is configured with graphic components stored in a library which are related to each one of the plant's equipment, e.g., pumps, valves, motors, etc. These components have their own properties and they are established during the simulation.
- Instructor Console Module. This module carries out all the tasks related to the graphical interface of the instructor and a module to dynamically update the instructor console with the simulation information.

4.3 Main features of the human machine interface for trainees

The better option for the operator console is to emulate via software the consoles of the actual plant, this represent the less cost option compared with the acquisition of such consoles, in this way a graphic imitation of the actual HMI provides a suitable operation interface. This HMI is a graphical application based on a multi-window environment with interactive process diagrams, these diagrams are organized in hierarchical levels following the organization of the power plant systems, i.e. boiler, turbine, etc. There are two main types of diagrams: information diagrams and operation diagrams. The first type shows values of selected variables. The values are presented as bar or trend graphs. The trainee uses the operation diagrams to control and monitor the whole process, with them he operates pumps, fans valves, and also he can modify set points of automatic controls and carry out any feasible operation in a similar way as he would do in the actual power plant. When the trainee needs to perform an action, he selects the suitable pictogram with the cursor, and then a pop-up window appears with the corresponding operation buttons. At any time the trainee can open all the pictograms he wants, and can do this in any operation console. The operation diagrams also have value windows; they show a pop-up window with the value of one variable (e.g., boiler drum level, turbine speed, etc.) and its operation range. The trainee easily visualizes the off-service equipment because it is shown in white and the equipment on-service has a specific colour depending on its working fluid. To this end, green equipment handles water, blue equipment handles air, red equipment handles steam, and so on. Figure 12 shows the operation diagram of combustion gas, where it is open a pop-up window to start a motor.

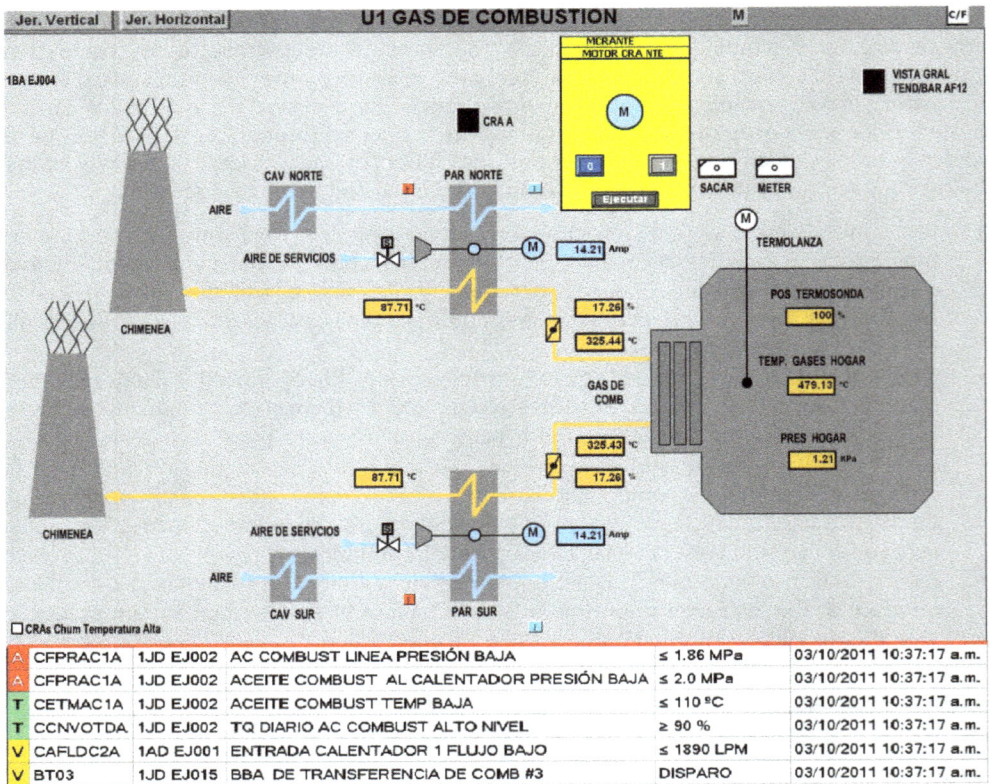

A	CFPRAC1A	1JD EJ002	AC COMBUST LINEA PRESIÓN BAJA	≤ 1.86 MPa	03/10/2011 10:37:17 a.m.
A	CFPRAC1A	1JD EJ002	ACEITE COMBUST AL CALENTADOR PRESIÓN BAJA	≤ 2.0 MPa	03/10/2011 10:37:17 a.m.
T	CETMAC1A	1JD EJ002	ACEITE COMBUST TEMP BAJA	≤ 110 ºC	03/10/2011 10:37:17 a.m.
T	CCNVOTDA	1JD EJ002	TQ DIARIO AC COMBUST ALTO NIVEL	≥ 90 %	03/10/2011 10:37:17 a.m.
V	CAFLDC2A	1AD EJ001	ENTRADA CALENTADOR 1 FLUJO BAJO	≤ 1890 LPM	03/10/2011 10:37:17 a.m.
V	BT03	1JD EJ015	BBA DE TRANSFERENCIA DE COMB #3	DISPARO	03/10/2011 10:37:17 a.m.

Fig. 12. Interactive process diagram

One important improvement of this kind of HMI is its capacity to show to trainee more information (temperatures, pressures, flow rates, etc.) compared to former control board simulators so it is expected that this kind of features help the operator to analyze in a better way a particular phenomena. In the bottom of the diagram displayed in Figure 12 there is a chronologic list of the alarms fired during the simulation, in this way the trainee is always notified of the occurred events. This list must be according to the alarms of the actual power plant.

The main challenge for the simulator users (operators) is the cultural change, because now operators have to utilize a modern tool like a PC instead of a control boards, therefore the operators must forget their former operation habits and adopt novel operation techniques for a fluent and safe navigation in a new HMI.

4.4 Performance criteria and acceptance procedures

Usually the fidelity of a simulator is mainly based on the behaviour of a group of variables called critical parameters. These parameters are related with conservation principles of mass and energy of the power plant and they will be selected only if they can be accurately measured. Any other variable not selected as critical parameter and which is observable in the operator HMI is called no critical. Typical critical parameters are:

- Flow, pressure and temperature of main steam.
- Flow, pressure and temperature of reheat steam.
- Feedwater flow.
- Main condenser pressure.
- Fuel flow.
- Combustion air flow.
- Generated electric power.

According to the previous classification, the criteria to assess the performance of the simulator can be summarized in the following three points:

- In steady state, the maximum variation of the critical parameters is ± 2% and for no critical parameters is ±10%. The value of these parameters must be consistent regarding the information of the reference plant. All of this is only valid for generation states greater than 25% of rated load. Another operation states, for instance, "cold iron" can be verified to assure that all simulator parameters, e.g. temperatures, correspond with the room temperature of the simulation.
- During transients conditions, due to malfunctions or abnormal operations, the simulator must have the same trend as the one reported in the actual plant, under the same operating conditions. Regarding the permitted duration of these transients, it is suggested a maximum time variation of ± 20% between simulation and actual data. In the absence of information, the trends and duration of the transients must be according to the expected behaviour of the existing physical phenomena.
- In any state, the simulator will not violate any physical or conservation law and its real-time operation will be assured.

The acceptance procedures define the required tests to carry out before a simulator can be ready to use it as a part of the training programmes for operators, the execution of these procedures is also a way of verifying if the simulator meets with its specification and scope. These procedures include exhaustive tests of all the hardware and software involved, the required tests can be summarized as:

- Carrying out a complete installation of the software (e.g. operating system, graphic packages, real-time executive, instructor console, mathematical models, etc). In the case of the real-time executive, instructor console and mathematical models, a good practice is carrying out a complete compilation and rebuilding all solution projects with the aim of guarantying a full compatibility between the source and executable codes.
- Verifying the communication among the stations of the local area network.
- Validating each one of the functions of the instructor console and the operator HMI, according to their corresponding specifications.
- Carrying out availability tests with no aborts in any simulator task. This includes a continuous simulation for time periods of at least eight hours with a minimum availability of 95 %.
- Carrying out operative tests from cold iron to full-load generation, shutdown operations and malfunctions. The operative tests must be well documented with their specific objectives and the expected results for each one of the operative manoeuvres.
- The application of the acceptance procedures and the documentation of the found discrepancies are key elements in the final tuning of the simulator, before it can be released for its commercial use.

The general requirements of fossil fuel power plant simulators are well defined by the Instrument Society of America (ISA) and the Electric Power Research Institute (EPRI). These entities provide extensive guides related to the design, development, fabrication, performance, testing, training, documentation and installation of power plant simulators.

5. Mathematical modelling

In training simulators, the mathematical models must be able to reproduce, in a dynamic way, the behaviour of the power plant in any feasible operation, this includes: steady states from cold iron up to full-load generation, and transients states, as a part of operation itself or because of malfunctions. The better way of accomplishing this is using physical modelling techniques, where the conservation of mass, momentum and energy are always fulfilled.

5.1 The procedural approach

The focus of procedural programming is to break down a programming task into a collection of variables, data structures and subroutines. EPRI (1983) published in 1983 an approach named Modular Modelling System (MMS), which provided an economical and accurate computer code for the dynamic simulation of fossil and nuclear power plants. Some of the most important uses of the MMS were: evaluation of plant design, checkout of control systems, operational procedures development, diagnosis of plant performance and training simulator qualification. MMS is based on the methodology of resistive and capacitive components or combinations of them depending of which variables are transmitted between adjacent modules (causality). According to this theory, resistive components are related with the simulation of the elements which involve a pressure variation in the process (valves, pumps, etc) and the storage of mass and energy are neglected. Usually, the behaviour of the resistive elements is represented by algebraic non-linear equations. For instance, in the case of a valve for incompressible fluid, the equation to calculate the flow is obtained from the steady state momentum equation and it is:

$$w = K \, Ap\sqrt{\rho(P_i - P_o)} \tag{1}$$

where: w is the flow, K is the valve conductance, Ap is the valve position, ρ is the density, P_i and P_o are the inlet and outlet pressures.

In the case of elements like pumps and fans, an approach based in the operation curves of the actual equipment is preferred because it gives a complete representation of the flow-pressure behaviour to any operation speed. For instance, in the case of a centrifugal pump, from the nominal data of the head-volumetric flow rate curve (H vs. q), the application of a least squares fitting gives the following expression:

$$\Delta H = a + b \, q + c \, q^2 \tag{2}$$

where: ΔH is the head developed by the pump, q is the volumetric flow rate and a, b, c are the coefficients obtained from the least squares fitting. The application of the pump affinity laws and the relationship for the developed head transforms the former equation in another one in terms of the flow, discharge pressure and pump speed, which is more suitable for the simulation.

$$P_o - P_i = A \, \rho \Omega^2 + B \, \Omega \, w + C \, \frac{w^2}{\rho} \tag{3}$$

where: w is the flow, P_i and P_o are the inlet and outlet pressures, ρ is the density, Ω is the angular speed and A, B, C are the transformed coefficients depending on pump nominal data.

On the other hand, capacitive elements are those which have a storage effect in the process (tanks, metal walls, etc). In this case the equations of the element are based on the lumped parameters approach; this approach simplifies the description of the behaviour of spatially distributed physical systems into a topology consisting of discrete entities that approximate the behaviour of the distributed system under certain assumptions, e.g. perfect mixing, which assumes that there are no spatial gradients in a given physical envelope, so the outlet stream has the same conditions of the fluid inside a control volume. In this way, to model a tank of constant volume with a single-phase fluid, the mass conservation equation yields:

$$\frac{d\rho}{dt} = \frac{w_i - w_o}{V} ; \quad \rho\big|_{t=0} = \rho_0 \tag{4}$$

where ρ is the density, t is the time, w_i and w_o are the inlet and outlet flows and V is the volume. The energy conservation equation with no work is expressed as:

$$\frac{dU}{dt} = w_i h_i - w_o h + Q ; \quad U\big|_{t=0} = U_0 \tag{5}$$

where U is the total internal energy, h_i and h are the inlet and outlet enthalpies and Q is the heat flow rate. For an incompressible fluid, the internal energy is equal to its enthalpy, and with the assumption of perfect mixing, equation (5) is transformed in:

$$\frac{dh}{dt} = \frac{w_i (h_i - h) + Q}{\rho V} ; \quad h\big|_{t=0} = h_0 \tag{6}$$

the heat flow rate can be evaluated as:

$$Q = j A (T_w - T) \tag{7}$$

where A is the area of the heat-transfer surface, T_w is the wall temperature, T is the fluid temperature and j is the heat transfer coefficient. In the previous equations, there are the following types of variables:

- System states. They are the independent variables of the model and they will establish the operation state of the simulator at any time. In the tank model, these variables are ρ and h, which are related with the solution of their ordinary differential equations.
- Initial guess for iterative methods. They are the variables related with the solution of algebraic equations (mainly non-linear) which requires an initial guess to converge to their solution. In the case of a single valve (Equation 1) this cannot be required, but in a complete system of a simulator, e.g. the feed water system for a combined cycle power plant, which have more than 30 valves, 4 pumps, and several pipe fittings, the solution turns more complicated, and usually the problem will be based on the solution of a simultaneous system of nonlinear equations.
- Fluid properties calculations. In the simulation of power plants, the calculation of thermodynamic properties of the water (liquid and steam) is essential for a accurate representation of the phenomena occurred in the power plant. This calculation includes the evaluation of densities, enthalpies, entropies, viscosities, etc. The calculation of properties for lubricating oil, fuel, combustion gas and air are also required.
- Design data of equipment. The physical size and nominal operation data of the actual equipment are very important because they determine the dynamic response of the simulator. For instance, this type of data includes: nominal flow rates, size and type of valves; operation curves of pumps and geometry of tanks.
- Empirical functions. These calculations are related with the use of empirical functions available in the literature like heat transfer coefficients and friction factors.

Many times the conservation equations do not give, in a straight way, the information required for the simulated control elements or for the operator HMI, therefore, it is necessary to introduce additional expressions to transform the Equations (4) and (6) in terms of measurable variables like liquid height (L) and temperature (T). This can be easily made with the definitions of density and heat capacity at constant pressure (Cp) and considering a tank with constant cross-sectional (a), the result is:

$$\frac{dL}{dt} = \frac{w_i - w_o}{\rho\, a}; \quad L\big|_{t=0} = L_0 \tag{8}$$

$$\frac{dT}{dt} = \frac{w_i\left(h_i - Cp\,T\right) + Q}{\rho\, V\, Cp}; \quad T\big|_{t=0} = T_0 \tag{9}$$

for the deduction of the former equations it is assumed that density and heat capacity are constants in the validity range of the model.

In their original version, the use of the MMS requires a simulation language like EASY-5 and ACSL, which are in charge of gathering, sorting and solving all the equations related with the system to simulate (Murthy, 1986). This kind of languages use common procedural languages, such as FORTRAN or C, in this way, the whole system model is a collection of procedure calls and the assembling and connecting of the various components of a large system is performed through a sequence of elementary commands merely specifying the desired topological connections between modules. The language automatically translates these orders into equivalent FORTRAN or C statements and aligns a consistent set of variables names to all quantities transmitted from one module to another. The language organizes the order in which the equations are solved in order to satisfy causality. In other words, all model representations are translated into C or FORTRAN language source code for compilation and execution, with the aim of ensuring a fast performance. The current version of ACSL available for PC keeps its basis on FORTRAN and C languages (acslX, 2010).

Due to the industry of training simulators grew during the 80´s and a big expense was done to create it, each one of the simulators builders have their own mathematical models libraries, and it is common to find that the core of installed simulators still have their mathematical models running in modern versions of FORTRAN compilers.

5.2 The object oriented approach

The Object-Oriented Programming (OOP) is based on to divide a programming task into objects where each one of these objects encapsulates its own data and methods (procedures associated with a class). The most important distinction regarding Procedural Programming (PP) is that this one uses procedures to operate on data structures, whereas the OOP uses procedures and data structures together. In object-oriented modelling, the objects are packages of data and functionalities and the methods can be sent to these objects rather than data only, as in PP. The main disadvantage of OOP compared with the PP is that the last one has a faster execution, which in real-time applications is an essential issue, in past decades, due to restrictions of memory and processing capacity, this arose as a serious problem, but nowadays, with the available computing power, the OOP is a feasible alternative to develop training simulators.

Leva and Maffezzoni (2003) establish some paradigms of the OOP in mathematical modelling; one of the most important is the definition of physical ports as the standard interface to connect a certain component model. In this way, an object is the mathematical model of a power plant component (e.g. valve, pump, etc) and the integration of these objects reflects the physical plant layout. The interactions among the components are satisfied with the flow information of the connectors, which are also related with the physical connections. Figure 13 describes these concepts, each one of the icons represents a physical component and according to this, it has defined their suitable communication ports (small coloured squares). The lines between two icons are equivalents to the actual physical connections. Table 2 shows some connector types.

Although it is not destined to serve as a platform to develop training simulators, Modelica (Modelica, 2011) is representative of this kind of technologies. Modelica is a non-proprietary, object-oriented, equation based language to conveniently model complex physical systems containing, e.g., mechanical, electrical, electronic, hydraulic, thermal, control, electric power or process-oriented subcomponents. Therefore, Modelica is a modelling language rather than a true programming language. The Modelica classes are not compiled in the usual sense, but are translated into objects which are then executed by a simulation engine.

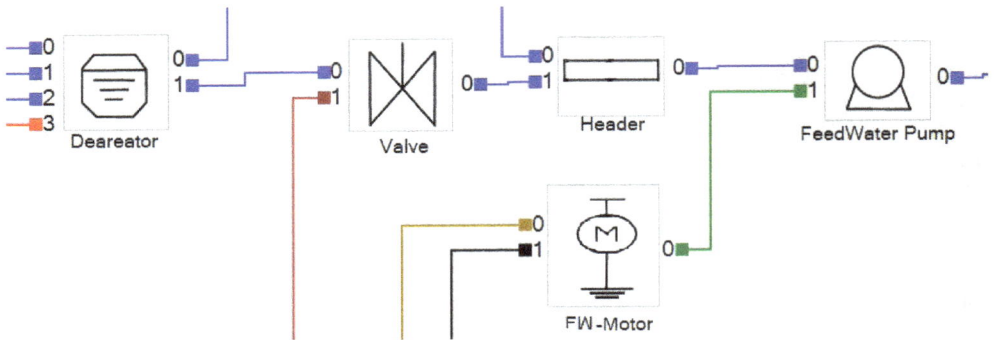

Fig. 13. Object oriented approach

Connector	Information
Hydraulic	• Pressure • Total Flow rate • Liquid fraction • Density • Enthalpy • Composition
Actuator	• Valve position
Mechanical Motion	• Mechanical torque • Angular speed
Thermal	• Temperature • Heat rate
Electric	• Voltage • Current

Table 2. Connector types.

Tavira-Mondragón et al. (2010c) describe an OOP development with the aim of modelling and developing simulators of power plants. Based on the previously discussed concepts of components, ports and connectors; these authors present a specially designed editor to fulfil the requirements of these kind of applications. This editor was designed using the design pattern Model-View-Controller (MVC). In this context, the Model component is the instantiation of each generic component including their connectors, and the result of putting together the required components is the construction of the mathematical model to simulate. The View is the graphical representation of the Model, i.e. each one of the elements of the Model is represented by a graphic icon. The Controller component is responsible of maintaining consistency between the View and Model; therefore, any change to the Vista (user interaction) will be reflected in the Model and vice versa. The result of the editing process is a simulation diagram. During the edition of a complete diagram, the generic elements are linked by mean of a set of connectors in various domains. These connectors specify the information flow among different elements of the diagram. The simulation of a diagram consists in calling in an orderly way the diagrams. During the simulation sequence, and for each one of the diagrams, it is built a data structure, which represents a directed graph, this graph has the topology of the simulated diagram and serves to construct the mathematical solution of the system (Solver). For instance, in a flow-pressure network, the nodes are the points of the network where the process lines are joined or split, if there is not considered accumulation of materials in these nodes, the solution of the continuity equation in steady state for all the nodes yields a system of simultaneous nonlinear algebraic equations.

Depending of their complexity, no matter the selected simulation approach, the mathematical models are constituted by: linear algebraic equations, nonlinear algebraic equations, differential equations or a combination of them. Linear equations are resolved with LU decomposition methods, for instance, the solution of the admittance matrix of the electric network. Methods of Newton-Raphson with relaxation techniques are utilized to resolve the algebraic nonlinear equations, for instance, flow-pressure networks where the Jacobian matrix can be calculated with the analytical first-order partial derivatives with respect to the pressures of the nodes. The differential equations can be resolved with implicit or explicit integration methods, according to the simulator solver. Examples of these equations are (8) and (9) or the speed equation of the electric generator. Tolsma and Barton (2002) highlight several numerical algorithms used to solve common problems that arise during process modelling, including solution of systems of nonlinear equations and numerical integration.

Regarding to general-purpose graphic modelling tools, MATLAB-SIMULINK (Mathworks, 2011) is a remarkable product due to its integrated simulation environment, mathematics libraries and toolboxes (e.g. control, data acquisition and signal processing). Some developments using MATTLAB-SIMULINK are reported by Lu (1999), who presents the simulation of 677 MW coal and gas-fired power plant, with the aim of obtaining a good insight into boiler dynamics and steady state performance. Another development is reported by Alam-Jan et al. (2002). These authors describe a training simulator for operators of a coal-fired power plant, and they claim that their nonlinear models implemented are moderately complex but reproduces the dynamic behaviour of the actual plant over a wide operation range.

Finally can be pointed that, no matter the selected approach or the computational tools used for the integration of a training simulator, the mathematical models must provide a realistic and consistent response in a wide operation range, this include any transient and steady state of the simulated power plant..

Additional issues about mathematical modelling and the application of dynamic simulators are discussed by Cameron et al. (2002).

5.3 Modelling of control systems

The Distributed Control System (DCS) of a power plant is a group of Programmable Logic Controllers (PLC) where all the control algorithms are executed in an automatic way. A DCS is a very complex system involving many thousands of signals and hundreds of diagrams. The control algorithms are organized in components with specific function or task, for instance: PID controllers, high/low detectors, timers, memories set/reset, etc. This organization is represented by means of a network of these components, which communicate information through connections, (Figure 14). These networks are organized in a hierarchical way, in the bottom levels are the basic elements like AND, OR, NOT gates, in the middle level are the diagrams, and finally in the top level are the modules. In this way, the DCS is constituted by a collection of modules.

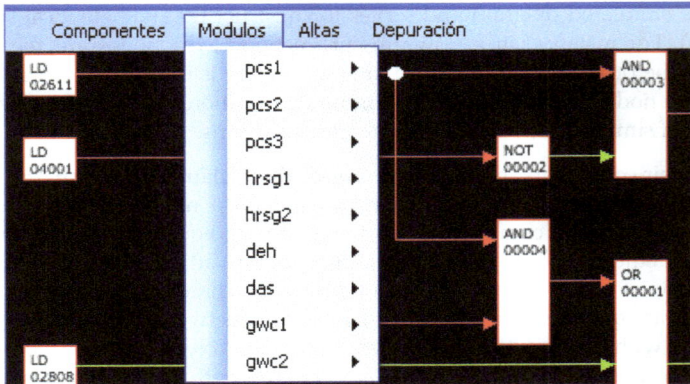

Fig. 14. Control diagram of a DCS

One of the most viable approaches to simulate the DCS is the translation of the control algorithms of the actual power plant; this guarantees a full reproduction of all control loops, alarms and signals to the HMI. In the context of a simulator and according to the methodology described by Romero-Jiménez et al. (2008), the translation procedure involves mainly the next tasks:

- Building the libraries of analogue and logic components (e.g. PID controllers, logical gates, timers, etc.)
- Organizing the component execution sequence.
- Creating structures in order to store component states.
- Designing interfaces with a description outside software implementation, so the code requiring an interface can use any component/object (Polymorphism).
- Design and implementation of a control component database.
- Implementation of an on-line visor to verify and visualize signals, states, inputs, outputs and parameters of components during simulation. This visor allows disabling diagrams, modules or components, in this way, it is possible isolate components and verifying their behaviour.

In the case of small control systems or when the control loops are not included in the DCS for its translation, these control algorithms can be developed by means of a graphical tool like VisSim (VisSim, 2011), using as reference the SAMA diagrams of the actual power plant. VisSim provides almost all the basic modules required to model control systems and generates C code, so it can be easily coupled to the simulator solver. In such way, the SAMA diagrams can be drawn totally in the VisSim environment to reproduce the required control. As expected, it is necessary coding in a manual way the modules with a specific function do not available in the VisSim libraries.

6. Conclusion

Training programmes for power plant operators using dynamic simulator have been used extensively in many parts of the world during the last 30 years, and their direct benefits in unit availability, thermal performance, environmental compliance and safe operation have been proven and documented. It is important to mention that, simulators are a very important part of these programmes, but their value as training tool is maximized when they are integrated in well-designed and structured training courses. The ADDIE model is a suitable methodology to get this goal.

The increase of power computing and the development of friendly graphical user interfaces had two main effects over the simulators; on the one hand, the power plants have replaced their former control boards with personal computers with graphical user interfaces. Naturally, the operators of these plants need a suitable training because they face a complete change in their operation paradigm, and because of this, the training simulators also require a HMI as the ones in the actual plants. On the other hand, a complete simulator can be installed in a single PC, with no demerit of the scope of the mathematical modelling or its real-time functioning. Furthermore, web services and cloud computing extend the training options, because specific training objectives can be fulfilled just with a PC with an internet connection. This kind of applications make possible to reach a big number of trainees with no necessity of: transporting personnel to a training centre, transporting a simulator to different places, or acquiring a simulator. Another important aspect is the inclusion of expert systems in a training simulator. This option is suitable for standalone applications which require reducing or even eliminating the necessity of a human instructor. A convenient knowledge representation of the expert gives to the simulation system all the elements to conduct a training session in an autonomous way.

In the Object-Oriented Programming, an object is the mathematical model of a power plant component and the integration of these objects reflects the physical plant layout. The interactions among the components are satisfied with connectors, which are also related with the actual physical connections; this type of approaches simplifies the construction of simulators and provides a direct relation between the physical and simulated systems.

7. References

acslX (2010). acslX Software for Modeling and Simulation of Dynamic Systems and Processes, 21.09.2011, Available from http://www.acslx.com/products/

Ahmad, A.L.; Low, E.M. & Abd Shukor, S.R. (2010). *Safety Improvement and Operational Enhancement via Dynamic Process Simulator: A Review*, Chemical Product and Process Modeling: Vol. 5, No. 1, Article 25, pp 1-25.

Arjona, M.; Hernández, C. & Gleason, E. (2003). *An Intelligent Tutoring System for Turbine Startup Training of Electrical Power Plant Operators*. Expert Systems with Applications, Vol. 24, No.1, pp. 95-101.

Alam-Jan, S.; Šulc, B. & Neuman, P. (2002). Object Oriented Modeling of a Training Simulator, *Proceedings International Carpathian Control Conference ICCC' 2002*, pp 757-762, May 27-30, 2003. Malenovice, Czech.

Burgos, E. (1993). *Simulador de Rodado de Turbine para el Adiestramiento de Operadores*, Boletín IIE, Vol. 17, No. 4 pp. 167-172.

Cameron, D.; Clausen, C. & Morton, W. (2002). Chapter 5.3 Dynamic Simulators for Operator Training, In: *Software Architectures and Tools for Computer Aided Process Engineering-Computer-Aided Chemical Engineering, Vol. 11)*, Braunschweig, I. & Gani, R., pp. 393-432, Elsevier, ISB N: 0-444-50827-9, The Netherlands.

CLIPS, A Tool for Building Expert Systems, 20.09.2011, Available from http://clipsrules.sourceforge.net

EPRI (1983). *Modular Modeling System (MMS): A Code for the Dynamic Simulation of Fossil and Nuclear Power Plants. Report CS:NP 3016*, Electric Power Research Institute, U.S.A.

EPRI (1993). *Justification of Simulators for Fossil Fuel Power Plants, Technical Report TR-102690*, Electric Power Research Institute, U.S.A.

EPRI (1998). *Simulator Procurement Guidelines for Fossil Power Plants: Simulator Specifications-AD-103790*, Electric Power Research Institute, U.S.A.

EPRI (2005). *Guidelines for the Development of an Initial Systematic Training Program-1009849*, Electric Power Research Institute, U.S.A.

Fray, R. & Divakaruni M. (1995). Compact Simulators Can Improve Fossil Plant Operation, *Power Engineering*, Vol. 99 No. 1, pp. 30-32, ISSN 0032-5961.

Hoffman, S. (1995). A New Era for Fossil Power Plant Simulators, *EPRI Journal*, Vol. 20, No. 5, pp. 20-27.

IAEA (1996). *Nuclear Power Plant Personnel Training and its Evaluation a Guidebook Technical-Reports Series No. 380*, International Atomic Energy Agency, Austria.

IAEA (1998). *Selection, Specification, Design and Use of Various Nuclear Power Plant Training Simulators-IAEA-TECDOC-995*, International Atomic Energy Agency, ISSN-1011-4289, Austria.

IAEA(2003). *Means Of Evaluating And Improving The Effectiveness Of Training Of Nuclear Power Plant Personnel-IAEA-TECDOC-1358*, International Atomic Energy Agency, ISBN-92-0-108204-7, Austria.

IAEA (2004). *Use of Control Room Simulators for Training of Nuclear Power Plant Personnel-IAEA-TECDOC-1411*, International Atomic Energy Agency, ISBN-92-0-110604-1, Austria.

ISA (1993). *Fossil-Fuel Power Plant Simulators–Functional Requirements-ISA-S77.20-1993*, Instrument Society of America, ISBN-1-55617-494-2, U.S.A.

Leva, A. & and Maffezzoni, C. (2003). Modelling of Power Plants, In: *Thermal Power Plant Simulation and Control*, Flynn, D., pp 16-60, Knovel, Available from http://www.knovel.com/web/portal/browse/display?_EXT_KNOVEL_DISPLAY_bookid=1399

Liao, S. (2005). *Expert System Methodologies and Applications – A Decade Review from 1995 to 2004*, Expert Systems with Applications, Vol. 28, No.1, pp 93–103.

Lu, S. (1999). *Dynamic Modeling and Simulation of Power Plant Systems*, Proceedings of the Institution of Mechanical Engineers, Part A: Journal of Power and Energy, Vol. 213, No. 1, pp. 7-22

Martínez-Ramírez, R.; Romero-Jiménez, G. & Martínez-Cuevas, S. (2011). What's New About Enabling Technologies in Power Plant Simulators and Training Systems: Visor3D-SD Prototype (Accepted for publication), *11th IERE General Meeting and The IERE – IIE Latin American Forum*, Oct 31-Nov 3, 2011. Cancún, Q.R. México.

Mathworks (2011). 04.09.2001, Available from http://www.mathworks.com/index.html

Modelica (2011). 01.09.2011, Available from https://modelica.org/

Murthy S. (1986). *The Application of Simulation in Large Energy System Analysis*, Modeling, Identification and Control, Vol. 6, No. 4, pp 231-247.

Olmstadt, W. (2000). Cataloging Expert Systems: Optimisms and Frustrated Reality, *Journal of Southern Academic and Special Librarianship*, pp. 1-11, ISSN 1525-321X.

Perkins, T. (1985). *Simulation Technology in Operator Training*, IAEA bulletin, Autumn-1985, pp18-23

Pevneva, N.; Piskov, V. & Zenkov, A. (2007). An Integrated Computer-Based Training Simulator for the Operative Personnel of the 800-MW Power-Generating Unit at the Perm District Power Station, *Thermal Engineering*, Vol. 54, No. 7, (July 2007), pp. 542-547.

Romero-Jiménez, G.; Jiménez-Fraustro, L.; Salinas-Camacho, M. & Avalos-Valenzuela, H. (2008). 110 MW Geothermal Power Plant Multiple Simulator, Using Wireless Technology, *Proceedings of World Academy Of Science, Engineering And Technology*, pp 154-159, ISSN 1307-6884, Jul 30, 2008. Paris, France.

Romero-Jiménez, G; Jiménez-Sánchez, V. & Roldán-Villasana, E. (2008). Graphical Environment for Modeling Control Systems in Full Scope Training Simulators, *Proceedings of World Academy of Science, Engineering and Technology*, pp. 792-797, ISSN 1307-688, July 30, 2008. Paris, France

Seifi, H. and Seifi, A. (2002). *An Intelligent Tutoring System for a Power Plant Simulator*, Electric Power Systems Research, Vol. 62, No. 3, pp. 161-171.

Serious Games LLC,Plant Simulator (2006). 15.08.2011, Available from http://plantsimulator.com/chooseplantsim.html

The Free Dictionary (2008). 11.08.2011, Available from http://www.thefreedictionary.com/simulator

Tavira-Mondragón, J.; Parra-Gómez, I. & Martínez-Ramírez, R. (2005). 350 MW Fossil Power Plant Multiple Simulator for Operators Training, *Proceeding of the Eight IASTED International Conference Computers and Advanced Techno logy in Education*, pp 492-497, Aug. 29-31, 2005, Oranjestad, Aruba.

Tavira-Mondragón, J.; Parra-Gómez, I.; Melgar-García, J.; Cruz-Cruz, R. & Téllez-Pacheco, J. (2006). A 300 MW Fossil Power Plant Part-Task Simulator, *Proceedings of Summer Simulation Multiconference*, pp. 291-297, Calgary Alberta Can., Jul 30-Aug 3, 2006.

Tavira-Mondragón, J.; Jiménez-Fraustro, L. & Romero-Jimenez, G. (2010a). *A Simulator for Training Operators of Fossil-Fuel Power Plants with an HMI Based on a Multi-Window System*. International Journal of Computer Aided Engineering and Technology, Vol. 2, No. 1, pp. 30-40.

Tavira-Mondragón, J.; Martínez-Ramírez, R.; Jiménez-Fraustro, F.; Orozco-Martínez, R. & Rafael Cruz-Cruz (2010b). Power Plants Simulators with an Expert System to Train and Evaluate Operators, *Proceeding of the World Congress on Engineering and Computer Science 2010, WCECS 2010*, ISBN 978-988-18210-0-3, Oct. 20-22, 2010, San Francisco, Cal., USA.

Tavira-Mondragón, J.; Jiménez-Fraustro, F. & Jiménez-Fraustro, L. (2010c). Graphical Environment to Simulate Power Plants, *Proceeding of the UKSim Fourth European Modeling Symposium on Computer Modeling and Simulation*, pp 289-294, Nov. 17-19 2010, Pisa, Italy.

Tolsma, J. E. & Barton, P. I. (2002). Chapter 3.2 Numerical Solvers, In: *Software Architectures and Tools for Computer Aided Process Engineering-Computer-Aided Chemical Engineering, Vol. 11)*, Braunschweig, I. & Gani, R., pp. 127-164, Elsevier, ISB N: 0-444-50827-9, The Netherlands.

VisSim (2011). 17.09.2011. Available from http://www.vissim.com

Wikipedia, The Free Encyclopedia (2011a). 16.09.2011, Available from

http://en.wikipedia.org/wiki/Cloud_computing
Wikipedia, The Free Encyclopedia (2011b). 20.09.2011, Available from
    http://en.wikipedia.org/wiki/CLIPS
Yamamori, T.; Ichikawa, T.; Kawaguchi, S. & Honma, H. (2000). *Recent Technologies in Nuclear Power Plant Supervisory and Control Systems*, Hitachi Review, Vol. 49, No. 2, p61-65.
Zabre E. & Román R. (2008). Evolution, Tendencies and Impact of Standardization of Input/Output Platforms in Full Scale Simulators for Training Power Plant Operators, *Proceedings of World Academy of Science, Engineering and Technology*, pp. 904-912, ISSN 1307-6884, Jul 2008. Paris, France.

5

Co-Combustion of Coal and Alternative Fuels

Pavel Kolat and Zdeněk Kadlec
VŠB-Technical University Ostrava
Czech Republic

1. Introduction

Energy utilization of alternative fuels including biofuels is one of the main tasks for development of recoverable sources in the world. The research consists of combustion tests in the large CFB boilers and measuring data inside the combustor. Since 1995, 29 large CFB boilers of different designs and power outputs have been in operation in the Czech Republic - as shown in the Table 4. Construction of the boilers, technical documentation, licensing and engineering has been based on foreign experience. Every large power project is always preceded by trial measurements and tests on smaller pilot, trial or if need be model equipment. Due to the great difference in scale, some unexpected measuring equipment behavior or problems must be taken into consideration for co-combustion of coal and alternative fuels.

The present research aiming at characterising the co-combustion under-atmospheric fluidized bed conditions by different physical and chemical characteristics has the following objectives:

- Ash formation upon fluidized bed co-combustion.
- Fate of toxic trace metals upon fluidized bed co-combustion.
- Recommendations for suitability of co-combustion in the atmospheric circulated fluidized bed boilers CFB and minimizing the harmful solid and gaseous emissions.

2. Model research

The model research has been carried out at the Technical University of Dresden. It includes the combustion tests - Table 2. - on experimental pilot equipment - Fig. 1. , 2. with atmospheric circulating fluidized bed for coal and bio-fuels produced from the sewage sludge from WWTP (waste water treatment plant) and biomass, and thermo-analytical study of bio-fuels - Table 1. The modelling had the following aims:

- To determine non-uniformity of combustion in the fluidized bed combustor as influencing the composition of flue gases and specification in terms of minor constituents (NO_x, chlorine compounds, alkalis, etc.).
- Analogically the influence of the size or for that matter the influence of fuel granulometric distribution on the process.
- Chemical composition, crystallographic structures, and mechanical properties of combustion solid products (bottom ash, fly ash, deposits).

- Analytical establishment of sulphur forms in fuel and combustion solid products, as well as element analysis for fuel and biomass.
- To perform leaching tests for combustion solid products.
- Detailed study of mineralogical and chemical composition of bottom ash, fly ash, and the solid emission phase in cyclone, heat exchanger and filter. Fig. 3., 4. Table 3.
- Balance for volatile elements, Cl, S, Hg, Se, semi-volatile elements, V, Ni, Co, As, and some non-volatile elements, Cr a Sn. Based on these balances to calculate the content of these elements in emissions and compare with the results of balance measurements.

Fig. 1. General view on pilot plant

Laboratory studies were focused on a detailed identification of input raw materials (coal, biofuel, limestone) so that the measurements could be reproducible: 1. Raw material input analysis and dependence of combustion solid residues on raw material input. 2. Combustion inaccuracy assessment in actual unit condition (T, gaseous and solid components, velocities, modelling). 3. Balance of combustion elements choice, studying mechanisms of deposit formation and composition. 4. Verifying a redistribution model for a choice of elements between the fuel and solid by-products.

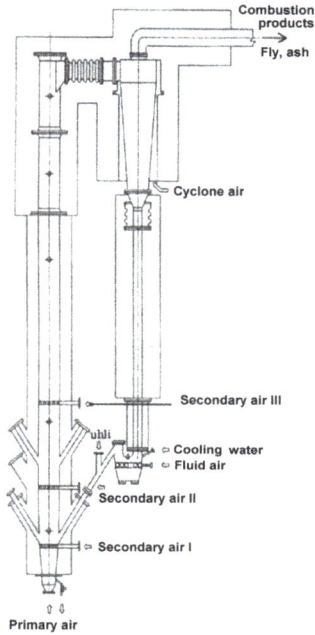

Fig. 2. CFB boiler 300 kW

	Coal	Bio fuel	Water	Ash	C	H	N	S	Volatile combustible	Heating value
	%	%	%	%	%	%	%	%	%	kJ/kg
Brown coal	100	0	14.0	5.2	55.0	3.86	1.03	0.70	45.8	20,599
Mixture	75	25	13.9	8.0	50.4	3.83	1.04	0.65	47.4	18,749
Mixture	50	50	12.6	11.9	45.2	3.79	0.92	0.54	50.6	16,728
Mixture	25	75	12.9	14.0	42.3	3.73	1.08	0.47	51.9	15,688
Biomass	0	100	14.6	13.7	36.6	4.31	1.34	0.24	57.8	13,291

Table 1. The analysis of the fuel mixture

Fig. 3. Scanning electron microscopy (SEM) of combustion residues during experiment - 85 % coal/15 % biomass. Sample 1 - fly ash from heat exchanger. Sample 3 - mixture of biofuel

Fig. 4. Scanning electron microscopy-morphology. Sample 85/15-filter, enlargement 2000x.
Sample 85/15-filter, enlargement 2500x

Test	No. 1	No. 2	No. 3	No. 4	No. 5	No. 6
Coal/biomass M_{pal} amount of fuel	100:0% mass 126 kg brown coal	75%:25% $M_{pal} =$ 132kg	50%:50% $M_{pal} =$ 159,3kg	25%:75% $M_{pal} =$ 180kg	85%:15% $M_{pal} =$ 135kg	0%:100% $M_{pal} =$ 203kg
Fuel delivery	42 kg.h^{-1}	44 kg.h^{-1}	53,1 kg.h^{-1}	44 kg.h^{-1}	42 kg.h^{-1}	67.7 kg.h^{-1}
Thermal output	240.3 kW	229.2 kW	246.7 kW	261 kW	222.2 kW	250 kW
Temperature in fluid bed	870 °C	850 °C	850 °C	804 °C	886 °C	800 °C
Content of the flue gases	$O_2 = 3.2\%$	$O_2 = 5.0\%$ $CO =$ 201ppm $SO_2 = 260$ ppm $NO_x = 197$ ppm	$O_2 = 1.6\%$ $CO =$ 1 887 ppm $SO_2 = 714$ ppm $NO_x = 191$ ppm	$O_2 = 1.9\%$ $CO =$ 1 842 ppm $SO_2 = 950$ ppm $NO_x = 215$ ppm	$O_2 = 2.8\%$ $CO =$ 577 ppm $SO_2 = 967$ ppm $NO_x = 195$ ppm	$O_2 = 3.3\%$
Excess of air	1.18	1.32	1.083	1.1	1.154	1.189
Unburned C in fly ash	0.051	0.035	0.042	0.086	0.0828	0.047
Velocity in reactor	4.25 m.s^{-1}	4.50 m.s^{-1}	4.50 m.s^{-1}	3.30 m.s^{-1}	4.24 m.s^{-1}	4.73 m.s^{-1}

Table 2. Basic characteristics of the combustion tests

Sample		heat exchanger 85/15	Filter 85/15	Filter 0/100	Cyclone 0/100-wet	Cyclone 0/100-dry
As	result	11.6	11.4	9.1	7.6	10.3
	insecurity	0.6	0.6	0.5	0.4	0.6
Ba	result	900	1,120	1,480	780	920
	insecurity	60	70	90	50	60
Cd	result	8	13	13	6	7
	insecurity	5	7	7	3	4
Co	result	< 20	< 20	< 20	< 20	< 20
Cr	result	149	110	97	285	285
	insecurity	9	7	6	18	18
Cu	result	180	147	200	96	108
	insecurity	13	11	14	7	8
Hg	result	< 5	< 5	< 5	< 5	< 5
Mn	result	1,510	1,810	1,400	1,110	1,150
	insecurity	70	80	60	50	50
Mo	result	< 20	< 20	< 20	< 20	< 20
Ni	result	105	78	69	201	201
	insecurity	7	5	5	13	13
Pb	result	80	82	101	61	75
	insecurity	6	6	8	5	6
Sn	result	100	68	68	59	66
	insecurity	20	14	14	12	14
V	result	52	42	61	49	50
	insecurity	5	4	5	4	5
Zn	result	610	1,080	1,400	550	660
	insecurity	40	70	90	40	40
Loss of annealing	Result in %	16.8	8.47	5.55	6.25	0.57
	Insecurity in %	0.2	0.09	0.06	0.07	0.01

Table 3. Concentration of heavy metals in pilot plant - test No.5., 6 by X-ray fluorescence spectroscopy (mg.kg^{-1})

The model research verify if the alternative fuel produced from biomass and sewage sludge may be used as alternative energy source in respect of the EU legislation, and/or its other modifications (with additives, decontamination technologies) for suitable fuel, which would comply with emission limits or the proposed energy process optimizing the preparation of coal/sludge mixture for combustion in the existing power engineering equipment.

The limiting factor for sewage sludge utilization from WWTP (waste water treatment plant) in agriculture is the increased content of risk elements and also the occurrence of organic pollutants – primarily polyaromatic hydrocarbons, PCB (polyaromatic byfenyls) and AOX (adsorbable organic halid). Other alternative fuels have not these limiting conditions. The limiting factor for sludge combustion at incineration plants is water content. With regard to the fact that from 2005 the EU Directives EU expects to ban waste disposal sites with any material with content of organic substances above 10 %, it is apparent that the priority condition for sludge utilization is sludge decontamination or power engineering utilization (Loo & Kopperjan 2008). Results from tests may be evaluated as very good with the prerequisite for utilization, testing of investigated substances in real combustion units. On the basis of carried out laboratory and pilot tests one may expect good results from these real units (equipment with greater output), many of these experiments have been already performed. From the results of experiments and thermoanalytical studies it is clear that 15 % of alternative fuels – biofuels based on sludge and brown coal can be used in the large fluidized bed boilers located in the Czech Republic. The combined combustion will enable to fulfil the Czech Republic's pledge to the European Commission concerning the development of renewable energy resources by 2010.

3. Diagnostic methods for operating surveillance of large fluidized bed boilers

Once large units are put into operation during co-combustion of alternative fuels and coal, guaranteed-performance tests must be conducted. The aim of the guaranteed-performance tests is to verify design parameters. Guaranteed-performance figures are compared with reality. Apart from basic measurements there are a number of other similar measurements of specific equipment parts that might be initiated because: the manufacturer is interested in using the experience to improve or design new units and the operator is interested in both eliminating problems and improving the economics of the operation process - Table 4. (Čech 2006). This chapter reviews the development of verification methods and presents some equipment for the determination of all important and interesting measuring data. The conclusions might be useful to energy companies and operators that want to verify operation data of fluidized bed boilers, flue gases and air channels.

Diagnostic measurements at a particular unit basically cover:

- The measurement of fluidized bed temperatures, furnace temperatures, flue gases temperatures at ancillary heating surfaces up to the boiler.
- The measurement of flue gas velocity in the furnace chamber and exits of cyclones, in cyclones, at the cyclone exit to second pass as well as in the area of additional boiler surfaces, sampling of flue gas in the boiler.
- Sampling of characteristic solid ash particles including isokinetic sampling to determine solid particle concentrations.

Place	Year	Type	No	Tph	Producer	Fuel	System
Třinec	1995	CFB	1	160	Lurgi (SES Tlmače)	HC, BIO	ABB
	1997	CFB	1	160			Siemens
Poříčí	1996	CFB	1	250	Foster Wheeler (CNIM)	HC/BC	Siemens
	1998	CFB	1	250		BC, BIO	
Tisová	1996	CFB	1	350	EVT (Vítkovice)	BC, BIO	Valmet
	1998	CFB	1	350	LURGI (SES Tlmače)	BC, BIO	Valmet
Zlín	1996	CFB	1	160	Babcock ABB Alstom	HC, BIO	Honeywell
	2000	CFB	1	125	Lurgi (SES Tlmače)	HC	Honeywell
Komořany	1995 - 99	Fluidized bubble bed	10	125	Power International	HC, BIO	ABB Honeywell
Hodonín	1996	CFB	2	170	AEE Austria	Lignit, HC, BIO	Valmet
Ledvice	1998	CFB	1	350	ABB Alstom	BC	ABB
Olomouc	1998	CFB without hot cyclon	1	190	Foster Wheeler (FORTUM)	HC, BC, BIO	Valmet
Štětí	1998	CFB Retrofit	1	220	Foster Wheeler	BC, BIO	Valmet
Ml.Boleslav	1998	CFB	2	140	EVT (Vítkovice)	HC	ABB
Kladno	1999	CFB	2	375	ABB Alstom	HC	ABB
Plzeň	1999	CFB	1	180	ABB Alstom	BC	ABB

HC – hard coal, BC – brown coal, BIO – biomass , Tph – tons steam per hour - output

Table 4. Newly-built fluid boilers with circulating fluid layer in the Czech Republic

3.1 Flue gas elements

Sampling of flue gas elements from the entire boiler can be divided into three groups:

- Sampling of flue gases from the bottom part of the fluidized bed.
- Sampling of flue gases from the boiler second pass up to the exit to the chimney.
- Sampling of flue gases from the boiler furnace, cyclones and cyclone link channels.

To monitor the fluidized bed boiler operation process O_2, CO, CO_2, NO_X, and SO_2 measurements can be taken. Other elements are usually monitored up to the exit from the separator of solid particles in front of the chimney. Fig. 5. illustrates a cooled sampling probe that might be used to take flue gases samples. The probe has an identical construction to that used for temperature measuring. During exhaustion gas is rapidly cooled down (from 800 °C to approx. 30 °C in cooled probe) so that there is no reaction with any other flammable waste gases. Gas is then sampled to be analyzed in the mobile laboratory. It is always recommended that a cooled probe is used to take samples from the furnace, cyclone and cyclone linking channels – Fig. 6. Sampling of flue gases from the second pass of the boiler occurs at temperatures safely below 800 °C. Thanks to that temperature a larger part of the gaseous sample is not able to oxidize quickly and thus it is possible to use a sampling tube made of stainless steel or sintered corundum, Al_2O_3. To set concentration (e.g., SO_2), sampling channels must be heated up during the sampling operation so that no reaction with water occurs.

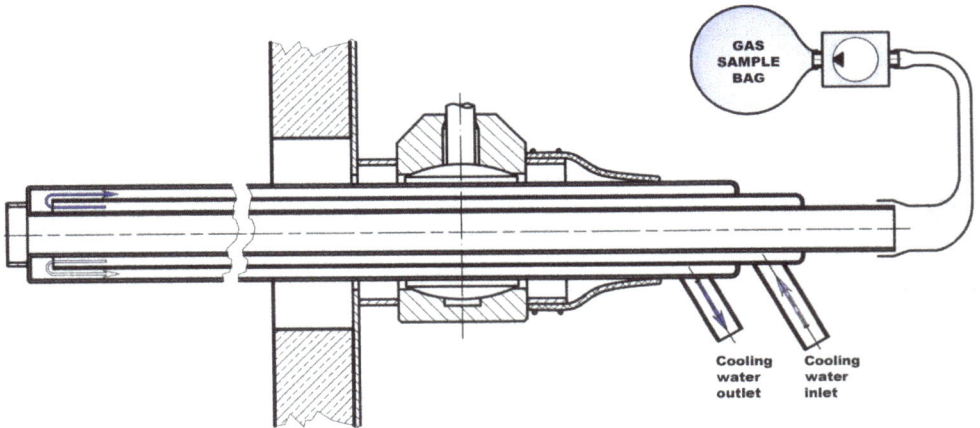

Fig. 5. Probe for flue gas sampling from fluid layer

Fig. 6. illustrates a sampling probe used in the detailed net measuring of O_2 concentration in the boiler combustion chamber with a steam output 125 t/h. (15 % biofuels and 90 % lignite coal).

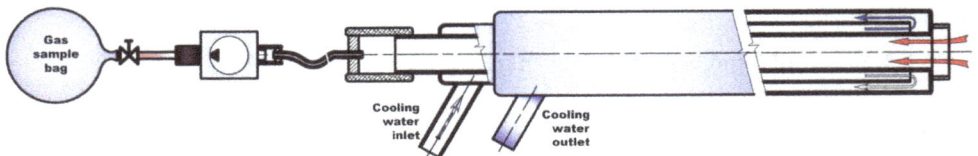

Fig. 6. Probe for flue gas sampling from furnace of boiler

A grid method of measuring O_2, CO and NO_X concentration was used. Measurements were taken using instrumentation openings in the middle of the side walls of the combustion chamber. The results of O_2 concentration measurements for 60 % and 90 % nominal output (350 t/h steam) are illustrated in Fig. 7. The results of CO and NO_X concentration measurements for 60 % nominal output are illustrated in Fig. 8.

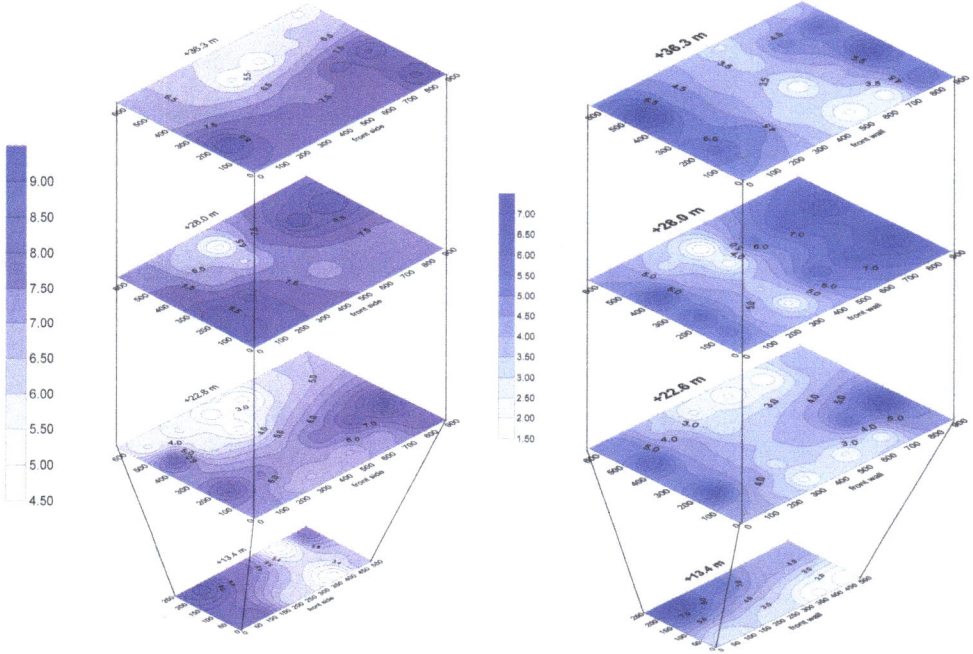

Fig. 7. Results of O_2 concentration measurements by water-cooled sampling probe in relation to height at 60 % and 90 % nominal output during co-combustion tests in power station Tisová.

The measurment results suggest that there is intensive suppression of NO_X formed increases in areas of secondary and tertiary air supply. CO concentration has developed as expected. The concentration decreases if the secondary air supply is gradual.

3.2 Solid particle concentration measurements

To determine the solid particle concentration in air flow, the Czech standard ČSN ISO 9096 needs to be observed. It is gravimetric determination of concentrations based on isokinetic sampling of solid particles from air flow. In the case of fluidized bed chambers the aim is to determine the solid particle concentration in the lower part of the fluidized bed layer and in the boiler furnace. For a bigger or smaller particle separation, the fluidized layer density varies depending on furnace height as well. The density in the lower part is in the range of 500 – 800 kg.m^{-3}, in the upper part of the furnace with the circulating layer the range is 0.1-0.5 kg.m^{-3}. Pressure in the fluidized layer is always measured for various height levels. The acquired pressure data are continuously monitored by operation measuring instruments. To determine the solid particles concentration in flue gases, the gravimetric method with solid particles isokinetic sampling can be used. Fig. 9. illustrates the measurement unit for solid particle isokinetic sampling.

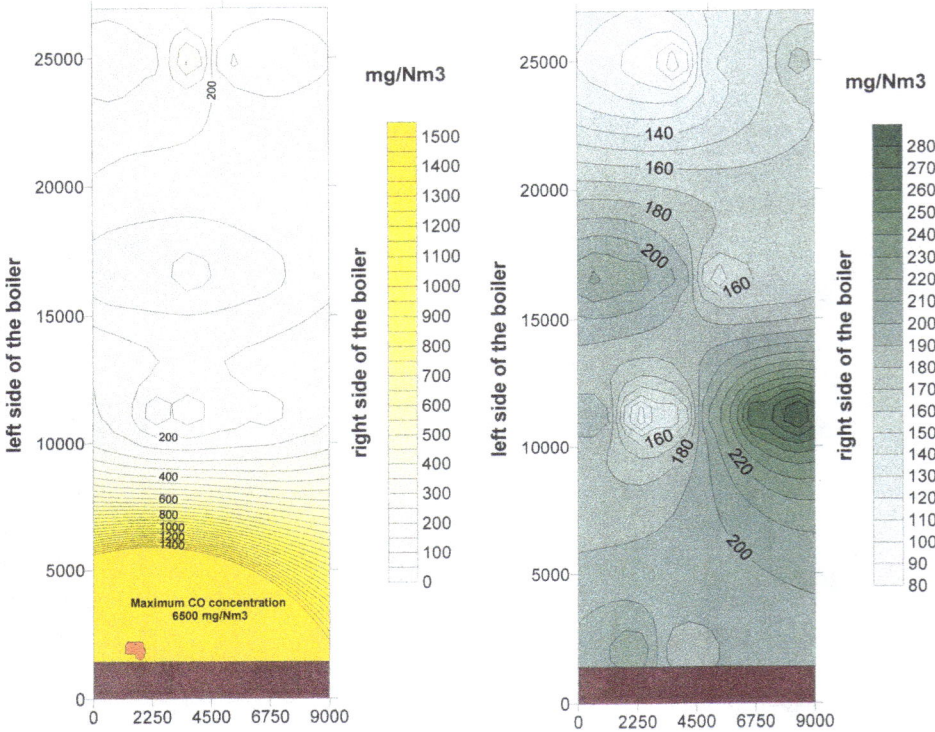

Fig. 8. Average CO and NO$_X$ concentration distribution along the height of the combustion chamber of the boiler at 60 % nominal output in Power station Tisová.

To determine the concentration at the measuring points, a disposable sampling probe is used. Another option would be a sampling probe with cooled support as illustrated in Fig. 10. This probe was developed to measure solid particles through an opening in the membrane wall flag. The sampling device touches the cooled parts only with smaller part to prevent gases from cooling below the dew point during the sampling process. The results of measurement of the solid particle concentration are illustrated in Table 5. for the 100 %, 70 % and 40 % nominal outputs for 15 % biofuel and 90 % lignite. (Power station Tisová 350 t/h steam, 9,42 MPa/ 505 ºC). The circulation number is quite high but it corresponds with the very low content of combustible particles measured in chutes under the cyclones. The results of cyclone efficiency measurements for various boiler outputs are from 96,5 % to 98,7 % for right and left cyclone. Generally we can say that hot cyclones of fluidized bed boilers have high separation ability.

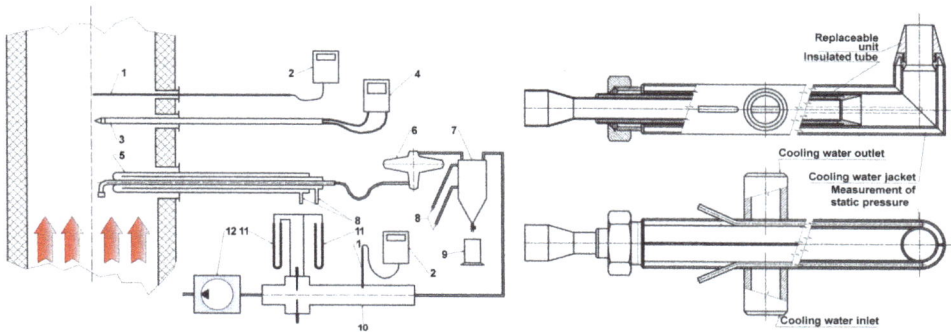

Thermocouple 2. Digital thermometer 3. Cylindrical probe 4. Digital pressure sensor 5. Disposable probe for isokinetic sampling 6. Separator 7. Steam condenser 8. Cooling water inlet and outlet 9. Bottle 10. Measuring equipment 11. U tube manometer 12. Air pump

Fig. 9. Diagram of measurement unit for solid particle isokinetic sampling.

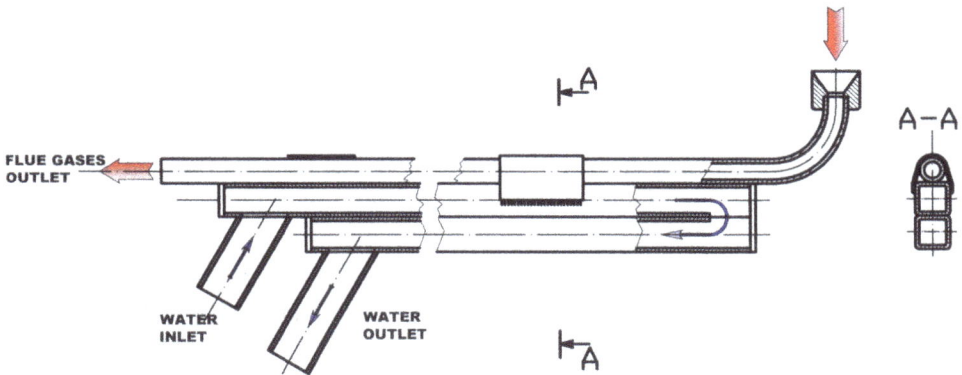

Fig. 10. Sampling water-cooled probe

Boiler output	40 %	70 %	100 %	
Medium concentration of solid particles – left side	841.9	3 308.5	4,427	[g.m^{-3}N]
Medium concentration of solid particles – right side	670.1	1,165.7	1,827	[g.m^{-3}N]
Mass flow of solid particles – left side	31.21	173.6	234.9	[kg.s^{-1}]
Mass flow of solid particles – right side	20.14	65.91	100.12	[kg.s^{-1}]
Total mass flow of solid particles	51.35	239.51	335.02	[kg.s^{-1}]
Lower heating value	11,860	11,950	11,870	[kg.s^{-1}]
Fuel consumption	10.041	16.72	23.283	[kg.s^{-1}]
Volume of ash and limestone supply to boiler	2.447	5.677	6.995	[kg.s^{-1}]
Circulation number of solid phase – left side	42	81	77	[1]
Circulation number of solid phase – right side	27	31	32	[1]
Circulation number of solid phase through cyclone	34.5	56	54,5	[1]
Distribution of bed ash and light ash	40:60	25:75	13:87	[%]

Table 5. Determination of the circulation number at different boiler outputs

4. Co-combustion of coal and solid waste fuels

Substitution of conventional fossil fuels (like bituminous coal or lignite) by low-carbon fuels for the energetic use is an efficient and cost-effective means of meeting the Kyoto Protocol establishing greenhouse emission targets for each of the participating developed countries (related to their 1990 emission levels). Considerable reductions of CO_2 emissions can be achieved by combustion of waste; therefore combustion of waste materials of various origins (industrial, agricultural etc.) or their co-combustion with fossil fuels in fluidized bed boilers became a legitimate alternative to conventional coal combustion. Another reason why particular attention is paid to energetic utilization of wastes is also elimination of waste and minimizing costs of waste deposition. (Loo & Kopperjan 2008).

But there are still challenges to be solved such as behaviour of the mineral matter during the wastes' combustion. Although the elemental behaviour during coal combustion has been studied and described in detail, the works dealing with redistribution of elements during waste combustion are quite rare, nevertheless, the conclusions described in these works are rather analogous – the application of the results obtained for the coal combustion on the combustion of wastes is not possible since the character of these materials is quite different. (Bartoňová at al., 2008). Another problem is that even if the waste materials differ from one another in their characteristics and content of toxic elements, most works only focus on wood and bark combustion.

This chapter intends to shed more light on the spectrum of alternative fuels used for energy production focusing on the evaluation of the effect of co-combustion of waste fuel and coal on the environment. In the circulating fluidized bed power station in Tisová - 350 t/h – Table 4., the waste alternative fuel (WF) containing plastics (1-20 %), fabric and carpets (45-75 %), rubber (5-15 %), paper (1-10 %) and wood (1-10 %) was co-combusted with the coal and the limestone. The samples of coal, limestone, bottom ash and fly ash were collected at regular time intervals and unburned carbon particles were separated from bottom ash by hand. Analysis of major, minor and trace elements was performed by X-ray fluorescence spectrometry (SPECTRO XEPOS) and mineral analysis was carried out using X-ray diffraction analysis (BRUKER D8 ADVANCE). Ash content of the samples was determined at 815°C. The distribution of macro pores was determined by means of mercury porozimetry (Micromeritics – AUTOPORE IV); SORPTOMATIC 1990 (Thermo Finnigan) equipment was used for the determination of specific surface area and mezopore-size distribution. Scanning electron microscope micrographs were taken by SEM PHILIPS XL – 30.

4.1 Mineral analyses

The X-ray diffraction patterns were obtained for the samples of unburned carbon (UC), bottom ash (BA) and fly ash (FA) Fig. 11. With the aid of elemental analyses of unburned carbon and ash samples the major mineral phases were established in the diffraction patterns and are marked with abbreviations explained in the figure caption. Coal has already been given in indicating the dominant occurrence of quartz and kaolinite. Diffuse area observed in the unburned carbon diffraction pattern (approximately from 25° to 31°) corresponds with semi-crystalline carbon phases. Somewhat lower crystallinity (broadened peaks) is evident also in case of magnetite and calcium hydroxide. Conversely, high-degree crystalline levels are represented e.g. by sharp peaks of quartz, lime or anatase. The comparison of the diffraction patterns revealed nearly the same mineral composition obtained for both unburned carbons – the dominant mineral phase in both samples was quartz and minor occurrence of anatase was identified as well. The both bottom ashes showed the similar mineral composition as well – it was lime there that was the most abundant mineral phase and also minor amount of quartz, anhydrite and anatase was identified in these samples. The similar mineral composition was obtained also for both fly ashes where quartz was the most dominant mineral and where the occurrence of lime, anhydrite, anatase and calcite was of minor significance. Hence, it can be concluded that the addition of solid waste fuel to coal during the combustion did not change the mineral composition of both unburned carbon and the ash samples. (Bartoňová at al., 2009).

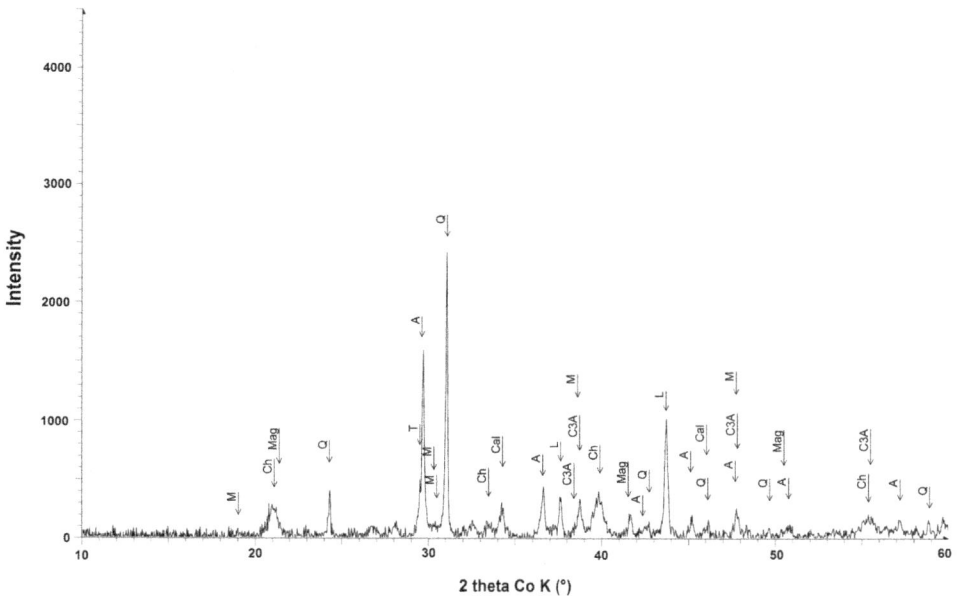

Fig. 11. X-ray diffraction patterns of fly ash (C). Q-quartz, L-lime, Cal – calcite, A – anhydrite, Mag – magnetite, C3A – tricalcium aluminate, Ch - calcium hydroxide, M - mullite T – anatase

4.2 Chemical analyses

By means of X-ray fluorescence spectrometry the contents of major, minor and trace elements were determined in coal (C), unburned carbon (UC), bottom ash (BA), fly ash (FA) and waste alternative fuel (WF). These results as well as the ash contents in these materials are given in Table 6. The porosity of the coal and bottom ash is rather low, whereas unburned carbon shows highly-developed system of ruptures, pores and cavities leading to high porosity of this material. That is why unburned carbon is being studied in relation to its adsorption properties.

4.3 Surface morphology and pore-size distribution

The morphology of coal, waste fuel, unburned carbon and bottom ash grains was studied using scanning electron microscopy with the secondary-electron beam method. The surface structure of coal and waste fuel was determinated and the texture of a typical grain of unburned carbon and bottom ash is shown in Fig. 12, 13. A general view (with magnification of 50x) and a surface detail (with magnification of 1500x) are shown for each material studied.

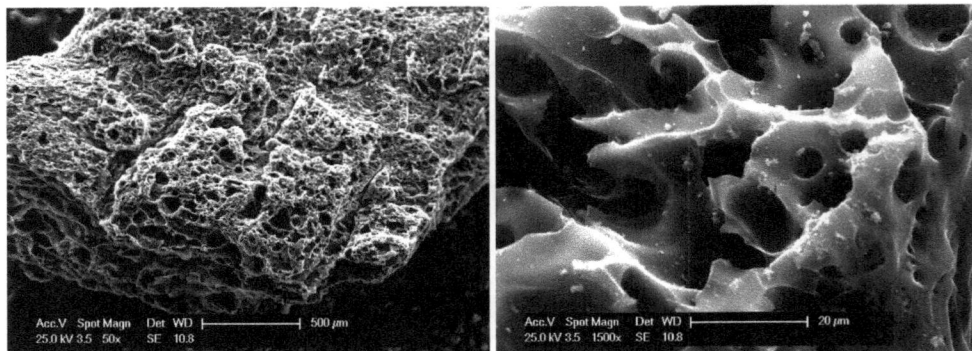

Fig. 12. SEM micrographs of unburned carbon particle

A) general view (magn. 50x)   B) surface detail (magn. 1500x)

Fig. 13. SEM micrographs of bottom ash

The surface texture shown in Fig. 12., 13. indicates that the porosity of unburned carbon collected at waste fuel co-combustion with coal is much better developed than that of unburned carbon when pure coal without waste fuel was combusted. But some caution is needed in such conclusions due to somewhat low representativity of one studied grain towards the average unburned carbon sample. Therefore pore-size distribution and specific surface area measurements were conducted in order to prevent misinterpretation when comparing adsorption properties of unburned carbon collected during pure coal combustion and during co-combustion of coal and waste fuel. Specific surface area of unburned carbon collected at pure coal combustion was 194 m^2/g, whereas during co-combustion of the same coal with waste fuel the specific surface area of unburned carbon reached 297 m^2/g, which is significantly higher value. This work was focused on the comparison of minor and trace elements behaviour during the co-combustion of coal and waste alternative fuels with the previous results regarding the combustion of the same pure coal in the same power station but without the added waste fuel. Elemental behaviour exactly in the combustion chamber did not change noticeably when waste alternative fuel was co-combusted with the coal. Even the most abundant elements in waste alternative fuel (related to coal) - Zn, Cl and Br - showed nearly the same behaviour. This observation can be explained through similar high volatility of these elements both in the coal and in the waste materials. (Bartoňová at al., 2009).

Element	Measured contents w_i				
	Sample				
	C	UC	BA	FA	WF
Ash (%)	23.4	67.7	98.0	98.8	5.6
Na_2O (%)	< 0.2	< 0.2	< 0.3	< 0.3	< 0.1
MgO (%)	< 0.1	< 0.1	0.3	0.6	< 0.02
Al_2O_3 (%)	5.9	21.4	11.6	17.6	0.6
SiO_2 (%)	11.8	37.3	21.0	30.9	3.9
P_2O_5 (%)	0.1	0.3	0.16	0.3	0.08
K_2O (%)	0.1	0.4	0.5	0.4	0.07
CaO (%)	0.7	1.5	44.0	29.8	2.4
TiO_2 (%)	1.5	5.0	3.0	4.9	0.14
MnO (%)	0.02	0.04	0.10	0.04	0.003
Fe_2O_3 (%)	1.5	4.2	6.1	5.4	0.1
S (%)	1.1	0.7	3.1	2.7	0.1
V (ppm)	62.4	270.0	72.0	233.0	< 6.4
Cl (ppm)	41.5	72.8	87.4	411.0	121.4
Ni (ppm)	12.5	33.0	23.2	75.5	26.4
Cu (ppm)	67.0	176.8	82.2	185.0	25.0
Zn (ppm)	26.6	43.1	111.0	370.3	1717.0
Ga (ppm)	14.4	30.0	21.7	40.1	< 1.0
Ge (ppm)	5.4	11.4	7.4	15.9	1.1
As (ppm)	39.0	27.0	66.0	97.8	< 0.7
Se (ppm)	1.2	2.1	1.6	8.3	0.5
Br (ppm)	1.8	1.2	2.1	12.2	49.1
Rb (ppm)	11.4	39.9	35.2	30.1	< 0.6
W (ppm)	19.1	52.5	21.8	60.2	< 8.1
Pb (ppm)	6.9	22.2	18.2	36.0	1.46
Th (ppm)	5.7	14.4	8.9	17.4	< 1.0

Table 6. Ash contents and concentrations of elements in coal (C), unburned carbon (UC), bottom ash (BA), fly ash (FA) and waste fuel (WF)

Comparison of elemental contents in bottom ash and fly ash was performed to describe further behaviour of elements when leaving the combustion chamber. It was established that when waste fuel was co-combusted with coal, a slight shift towards the higher enrichment of most elements in fly ash (vs. bottom ash) was observed. This trend is the most significant in case of Zn, Cl and Br which are the very elements that were the most abundant in waste fuel (when compared to coal). Therefore it can be concluded that the elements showing high concentrations in waste fuel tend to concentrate in fly ash. Specific surface area of unburned carbon collected at the test where waste fuel was co-combusted with the coal (297 m²/g) was significantly higher that that of unburned carbon from the combustion test without waste materials (194 m²/g). Comparison of pore-size distribution curves obtained for both unburned carbons revealed that unburned carbon collected during coal and wastes combustion contains larger amount of small pores, whereas macropores are more abundant in the unburned carbon form coal combustion without the waste alternative fuel. The unburned carbon collected at the co-combustion of the coal and wastes is undoubtedly of better adsorption properties.

5. Co-combustion of coal and waste wood

Biomass represents a lot of various materials, either waste materials or special energetic plants. Fuels based on wood biomass (sawdust, shavings, chips, tree-bark) can be used also for the production of high-quality biofuels, such as wooden briquettes and pellets, or can be co-combusted with coal. (Bartoňová at al., 2008). Average ash content of wood is about 1 – 2 % and calorific value ranges from 11 to 18 MJ.kg⁻¹. Straw is another advantageous energetic source and its calorific value ranges from 17.6 to 18 MJ.kg⁻¹, ash content is about 5.3 – 7.1 % and is often used e.g. in Sweden, Denmark or USA. (Loo & Kopperjan 2008). The disadvantages of this material are its huge volume and heat-exchanger fouling problems. There are also other biomass materials used for the energetic utilization – various agricultural residues (green wastes, hull, shells, pruning, rice straw, rape residues, corncobs and stems, sugar cane trash, cassava rhizome) as well as growing energetic plants. (Winter & Hofbauer, 1997). This chapter mainly evaluates the environmental impact of fluidized bed combustion of different fossil and biomass fuels. Particular attention was paid to the comparison in the release of environmentally most significant molecular species – amount of solid coal combustion products and their leaching behaviour or emissions of sulphur and carbon dioxide. For this work the samples from circulating fluidized bed power station in Štětí - Table 4. - were collected. In this power station coal combustion and co-combustion of coal / wastes tests were performed in circulating fluidized bed boiler at 870°C. Simplified diagram of the combustion facility is given in Fig. 14. In this power station usually lignite is co-combusted with the wood waste (coming from the cellulose production). Usual lignite / wood waste ratio is 10 :1.

5.1 Combustion tests

Three combustion tests were performed – Regime I, II and III. In Regime I lignite and limestone were combusted (in weight ratio of lignite/limestone = 10:1). In regime II lignite, limestone, sawdust and tree-bark were combusted in coal/wood waste ratio of 1:1.76. In regime III wood, sawdust and wood chips were combusted in ratio of 1:0.21:1. (This combustion test was rather unusual because no bottom ash was created and the only solid output flow was fly ash). Mass flows of input and output materials (BA – bottom ash, FA – fly ash, E,s – solid emission particles) and volume of gaseous emissions ($V_{E,g}$) are summarized in Table 8. The ash and water contents in these materials are given in Table 8. as well. Mass flows relate to undried samples. Proximate and ultimate analyses of input and output materials are given in Table 7.

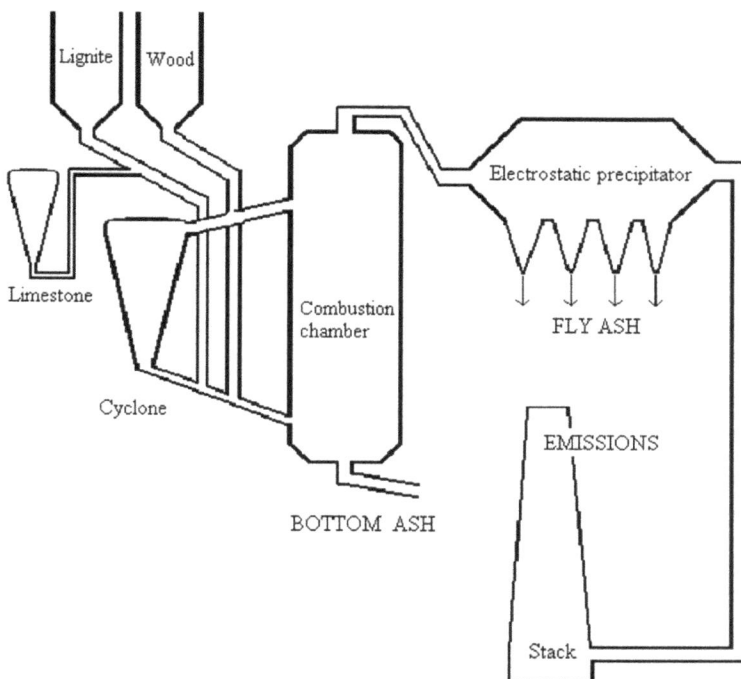

Fig. 14. Simplified diagram of the combustion facility

5.2 Analyses of emissions

Emissions from combustion unit were analysed and CO, NO_X and SO_2 were determined in flue gas, while As, Se, Cd, Hg and Pb were determined in solid particles captured on the filter in flue gas stream. The results of emissions analysis are given in Table 9. In the boiler mantle there are four holes into the combustion chamber. Using these holes as progressive sliping thermocouples to measure temperatures in the fluidized bed at different levels through the holes, where the proble was plugged, three samples of gaseous emissions and ash were collected directly from the fluidized bed. The sliping probe measured temperatures in the fluidized bed at the inlets. All combustion regimes were sampled from storage tanks of fuel and all four sections of the electrostatic precipitator. Furthermore, there was continuous measurement of emissions of NO_X, CO, SO_2 in the flue gases (see Fig. 5.). The balance of fuel and combustible waste, the mass flow, moisture content and ash, as well as the mass flow bed ash (BA) and fly ash (FA), the volume of gaseous emissions (VE, g), the quantity of solid emissions, can be evaluated in Table 10. The summary of calculated values shows a relation between the input (m_{inp}) and output (m_{out}) data. The difference between the weights of the input current m_{inp} and output current m_{out} under regime III can be explained by the fluid in the boiler not "running" the whole regime III - cleaning ash from coal combustion and therefore part of the ash has gone into the output stream and so its weight is greater than the output current m_{out}.

Regime	Input	Output
I	Lignite (C): m_C = 25,920 kg.hr^{-1} (W = 14.7 %, A = 18.3 %) Limestone (L) m_L = 2,630 kg.hr^{-1} (W = 0.45%, LOI = 34.1 %) LOI value in limestone is thought to be CO_2 released during the combustion	m_{BA} = 3,250 kg. hr^{-1} m_{FA} = 3,170 kg. hr^{-1}, $m_{E,s}$ = 0.42 kg. hr^{-1} $V_{E,g}$ = 201,130 Nm^3.hr^{-1}
II	Lignite (C): m_C = 11,840 kg.hr^{-1} (W = 16.4 %, A = 16.8 %) Limestone (L): m_L = 970 kg.hr^{-1} (W = 0.45 %, LOI = 34.1 %) Sawdust (S): m_S = 5,220 kg.hr^{-1} (W = 28.1 %, A = 0.64 %) Tree-bark (B): m_B = 5,620 kg.hr^{-1} (W = 26.7 %, A = 4.8 %)	m_{BA} = 2,020 kg. hr^{-1} m_{FA} = 1,360 kg. hr^{-1} $m_{E,s}$ = 0.57 kg. hr^{-1} $V_{E,g}$ = 234,830 Nm^3.hr^{-1}
III	Wood (W): m_W = 14,905 kg.hr^{-1} (W = 12.8 %, A = 0.25 %) Sawdust (S): m_S = 3,114 kg. hr^{-1} (W = 29.7 %, A = 0.47 %) Wood chips (Ch): m_{Ch} = 14,870 kg. hr^{-1} (W = 26.8 %, A = 1.04 %)	m_{BA} = 5 kg. hr^{-1} m_{FA} = 396 kg. hr^{-1} $m_{E,s}$ = 0.16 kg. hr^{-1} $V_{E,g}$ = 205,340 Nm^3.hr^{-1}

Table 7. Mass and volume flows of input and output materials

	Regime I	Regime II			Regime III		
	Lignite	Lignite	Sawdust	Tree-bark	Wood	Sawdust	Wood chips
C (%)	47.84	47.77	33.58	34.43	39.87	32.76	33.02
H (%)	3.98	4.05	4.14	4.26	4.91	3.92	4.03
N (%)	0.88	1.12	0.11	0.50	0.12	0.14	0.27
S_{total} (%)	0.59	0.84	< 0.01	< 0.01	< 0.01	< 0.01	< 0.01
O (%)	13.71	13.06	33.43	29.3	42.03	33.03	34.79
W^a (%)	14.7	16.41	28.12	26.70	12.83	29.71	26.84
A^a (%)	18.3	16.8	0.64	4.8	0.25	0.47	1.04

Table 8. Proximate analysis of lignite and wood wastes (related to undried samples) for regimes I, II, III.

Input mass flow of carbon converted to carbon dioxide (CO_2): I – 28 kg/h.GW and II-12 kg/h.GW, Table 11. - where the index corresponds to the C carbon in coal - are then calculated giving all the input flows of carbon-converted CO_2. For simplicity it is assumed that all the carbon is burned and transferred to the emissions in the form of CO_2.

	Regime I	Regime II	Regime III	Limits
CO [mg/Nm³]	17.0	19.9	7.50	250
NO_x [mg/Nm³]	308	286	150	400
SO_2 [mg/Nm³]	294	236	53.81	500
F- [mg/Nm³]	-	-	0.9	-
As [mg/Nm³]	1.29	1.92	1.76	-
Se [mg/Nm³]	0.10	0.15	0.13	-
Cd [mg/Nm³]	0.19	0.17	0.25	-
Pb [mg/Nm³]	0.79	0.85	1.06	-

Table 9. Analysis of emissions for regimes I, II , III and emission limits

	Input (t/h)							Output (t/h)		
	Coal	Lime stone	Saw dust	Bark	Wood	Chips	M_{inp}	FA	BA	M_{out}
I	4,743	1,733	-	-	-	-	6,47	3,21	3,210	6,420
II	1,989	0,640	0,033	0,75			3,41	1,36	2,020	3,380
III	-	-	0,015	-	0,037	0,155	0,20	0,38	0,004	0,390

Table 10. Mass flow of inorganic materials

	Input CO_2 (t/h)							$m_{CO2,out}/Q_p$ (t/h.GW)
	Carbon	Limestone	Sawdust	Bark	Wood	Chips	$m_{CO2,inp}$	
I	45,46	0,90	-	-	-	-	46,36	0,20
II	20,74	0,33	6,43	15,70	-	-	43,20	0,16
III	-	-	3,73	-	21,74	18,01	43,48	0,14

Table 11. Calculation of the incoming flow of inorganic materials for 1 GW of power boilers

	Output (kg/h)				% $S_{E,g}$	$m_{s,E}/Q_p$ (kg/h.GW)
	FA	BA	$m_{s,E}$	$m_{s,out}$		
I	102	59	29,6	190,6	15,5	0,13
II	39.0	58.8	27,3	125,5	21,8	0,10
III	13,9	0,1	5,5	19,5	28,2	0,018

Table 12. Output flows of sulphur (S)

The results confirm that burning wood emits less CO_2 to the atmosphere per unit of energy input than burning brown coal. The data listed in Table 12. show that the minimum content of SO_2 emissions (% $S_{E,g}$) is the combustion of coal with limestone (regime I). Absolute numbers of sulphur contained in the mass emissions ($m_{s, E}$), however, clearly demonstrate that by burning wood the amount of sulphur getting in the emissions into the atmosphere is about 10 times smaller than that of burning coal. This parameter is much more favourable for burning wood. The most significant results are summarized below:

- Mass balance calculations suggest that mass flow of inorganic matter produced per 1 GW of boiler output has dropped from 28 kg /hr.GW for lignite combustion to 0.7 kg /hr.GW when wood wastes were combusted.
- This observation is a source of many advantages relating to ash land-filling e.g. decreasing the amount of ashes produced during the combustion process will consequently result in decreased amount of toxic leachates, above all sulphates, and also the increase of pH (due to high amount of Ca-bearing minerals present in coal ash) will not be as significant).
- Mass flow of CO_2 produced during the combustion was related to 1 GW boiler output. 0.20 kg /hr.GW was obtained for lignite combustion and it has dropped to 0.14 kg /hr.GW released when wood wastes were combusted.
- Sulphur emissions were also recalculated to 1 GW boiler output - sulphur emission flow calculated for lignite combustion (0.13 kg /hr.GW) was considerably higher than that obtained for wood wastes combustion (0.01 kg /hr.GW).

In conclusion - the results described above unambiguously suggest that the waste wood combustion produces lower amount of environmentally-hazardous pollutants than fossil fuel combustion, even if combusted with Ca-bearing additives. (Klika 2010).

6. Co-combustion of coal and sewage sludge

The sewage sludge is a heterogenous mixture of organic elements (both live and lifeless microorganizm cells) and incorganic elements. The organic part of the sewage sludge is mainly represented by the proteins, sugars and lipids. The inorganic part susteins mainly of the compounds of silicon, ferrum, calcium and phosphorus. Morover the sludge consist of a wide range of harmful substances as well – heavy metals, persistent organic elements PCB, PCDD/F, PAU etc. and other organic harmful elements. The Table 13. illustrates the summary of the organic polutants in the sewage sludge dry residues taken from the Central Sewage Plant of Ostrava (CSPO) and it is evident that almost all limits of the monitored polutants are exceeded. Such high values prevent the sewage sludge from being used for agricultural purposes and land reclamation – necessitating the usage of both the underground and exterior storage. The biggest problem in this case is the high content of the polyaromatic hydrocarbons that is ten times higher than the allowed limit. It is probably because of the industrial waste-water disposal. The value of TOC (Total Organic Carbon) that does not fit can rather be considered a useful than limiting factor. The energetical content of the sewage sludge is based on the chemical energy of the organic components that are capable of oxidation. To be able to describe the sewage sludge as fuel a material that converts its primary energy into the thermic energy - the condition of being flamable must be met. To make the combustion process balanced it is necessary to achieve fuel efficiencies from dry sludge residues and other heat distributed to the furnace, making it possible to use the water vaporization heat contained in fuel, the heat needed for the superheating of the water vapours in the waste gases and the heat needed for the waste gases heating. The important criterion of keeping the combustion process balanced is thus the water ratio in the sludge. Thus a problem exists because water ratio of the mechanically drained sludge is high (cca. 60 – 80 %) for the relatively low fuel efficiency and therefore the sludge cannot be combusted by itself. The most important energy characteristic of each single fuel is its efficiency. The dry residue efficiency of the anaerobicly stabilized sewage sludge is in the range of 7 – 10 MJ.kg^{-1}. Fig.15. shows the sewage sludge structure.

indicator – sample from the CSPO		Limit value	Rated value
Benzen	[mg/kg] dry residue	0.1	0.135
BTEX	[mg/kg] dry residue	10	5.46
EOX (Cl)	[mg/kg] dry residue	10	11.7
NEL	[mg/kg] dry residue	200	4,840
ΣPAU (15)	[mg/kg] dry residue	10	103
PCB (summary of 6 kongerens)	[mg/kg] dry residue	0.2	0.3
TOC	[%]	20	25.3
Tetrachlorethen	[mg/kg] dry residue	0.5	< 0.030
Trichlorethen	[mg/kg] dry residue	1	0.233

Table 13. Organic polutants in sewage sludge

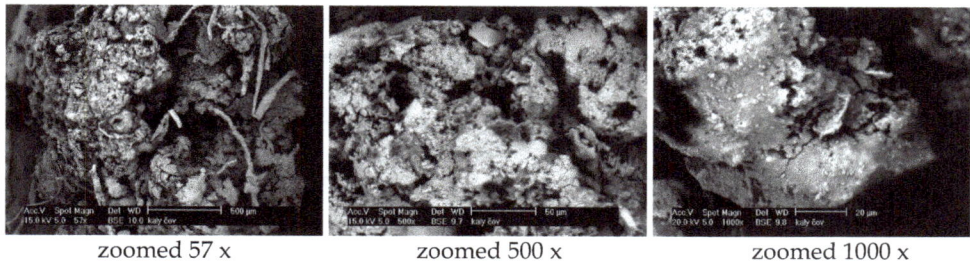

| zoomed 57 x | zoomed 500 x | zoomed 1000 x |

Fig. 15. The structure of the sewage sludge from CSPO

6.1 The combustion test description

The combustion test with the mechanically drained digested sewage sludge (the water proportion in the sludge was approx. 63 %) was carried out at circulating fluidized bed power station in Třinec with an output of 130 MW$_t$ - Table 4. The mixture of hard energy coal and the coal sludge of average efficiency Q_i^r = 19 MJ.kg^{-1}, water ratio W^r = 7,5 %, ash content A^r = 30 % was combusted at the fluid boiler. During the combustion test the fuel was distributed to the boiler in the ratio: 11 %$_{weight}$-sewage sludge from the Central Sewage Plant of Ostrava, 28 %$_{weight}$-energy coal and 61 %$_{weight}$-coal sludge. During the additional combustion of the sludge the mixture characteristics changed as follows: heating value Q_i^r = 17 MJ.kg^{-1}, water ratio w^r = 14,5 %, ash content A^r = 28 %. Based on the fact that the total heating value of the fuel mixture thus dropped by cca 2 MJ/kg during the additional combustion, the volume of the mixture must be enlarged by approx. 0,65 kg.s^{-1} to maintain the constant boiler output. However the total coal consumption does not raise and this fact is important. The description of the combusted fuel is illustrated in Table 14., 15., and 16. The glory-angle of the mixture was rapidly changed for the worse, compared to the hard coal. The chain feeders of the crude fuel worked reliably and had no failures. Thanks to the mixture passing through the chain feeder the big pieces of the sludge were crushed. The combustion test showed that 15 % of the sludge content in the mixture was the cut-off amount able to pass the swing-hammer crasher. The moisture was of a fundamental importance concerning the allowable amount of the sludge in the mixture.

Crude form	sample		Energy coal	Sewage sludge	Coal sludge
Water ratio	W^r	[%]	10.23	62.36	6.22
Ash content	A^r	[%]	24.77	19.52	31.78
Heating value	Q_i^r	[kJ·kg⁻¹]	18,854	1,476	18,962
Waterless	sample		Energy coal	Sewage sludge	Coal sludge
carbon	C^d	[%]	59.88	22.50	56.34
hydrogen	H^d	[%]	3.92	3.48	3.36
sulphur	S^d	[%]	0.35	0.63	0.22
nitrogen	N^d	[%]	1.30	2.40	1.21
oxygen	O^d	[%]	6.96	19.14	4.99
ash	A^d	[%]	27.59	51.85	33.88

Table 14. The fuel characteristics in crude and waterless form

			Energy coal	Sewage sludge	Coal sludge
carbon	C^h	[%]	82.70	46.72	85.21
hydrogen	H^h	[%]	5.42	7.22	5.08
sulphur	S^h	[%]	0.48	1.31	0.33
nitrogen	N^h	[%]	1.79	4.99	1.84
oxygen	O^h	[%]	9.61	39.75	7.55

Table 15. Fuel combustible composition

Sample	SiO_2 [%]	TiO_2 [%]	Al_2O_3 [%]	Fe_2O_3 [%]	MnO [%]	MgO [%]	CaO [%]	Na_2O [%]	K_2O [%]	SO_3 [%]	P_2O_5 [%]
Energy coal	16.02	0.28	7.2	1.96	0.04	<0.83	0.92	<2.02	0.86	1.6	0.22
Sewage sludge	13.78	0.25	2.49	10.24	0.11	<0.83	13.73	<2.02	0.38	3.35	6.44
Coal sludge	22.46	0.35	9.48	3.26	0.05	<0.83	1.07	<2.02	1.28	1.87	0.24

Table 16. The silicate analysis of the energy coal, sewage sludge and coal sludge (RTG-fluorescence method)

During the combustion test the sludge moisture was approx. 65 % compared to the hard coal moisture of 7, 5 %. The higher moisture makes the temperature drop behind the crasher and it results in sealing the crasher with the mixture of the wet mud.

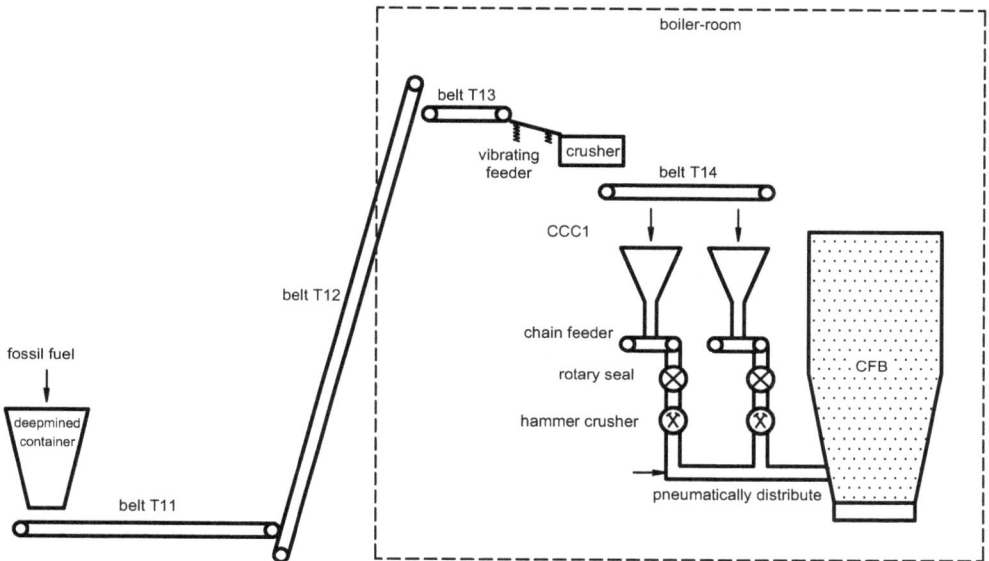

Fig. 16. The scheme of the distribution of the fuel to the CFB boiler

6.2 The boiler efficiency and its operational reliability

Based on the combusted coal and the sewage sludge of given ratio the approx. 0.3 % drop in efficiency of the boiler was observed - Table 17. The content of the combustible carbon in the combustion products corresponds with the fine hard coal combustion.

Fuel	η_k	C_{LP}	C_{UP}
	%	%	%
Energy coal and coal shed	89.1 – 90.5	0.25	6.5
Energy coal, coal shed and sludge	88.8 – 90.2	0.3	5.8

Table 17. The boiler efficiency η_k and the combustible matter content in the ash, where C_{LP} indicates the combustible matter content in the bedding ash and C_{UP} indicates the combustible matter content in the ash

If we focus on the operational efficiency of the boiler under the condition of the additional combustion of the sludge it it is advisable to monitor unwanted states like rust formation caused by high and low temperataure and silting the heat transfer surfaces and abration. In the case of boilers with the fluid bed and the additive desulphurisation, the marks of the chlorine rust pop up even if the chlorine content in the fuel is low. The ratio Cl/SO_2 has an impact on the high-temperature chlorine rust intensity. As for the chlorine content in the fuel sludge, it is evident that it does not exceed the volume found in the hard coal. The HCl concentration in waste gases influence the low-temperature rust intensity. In this particular case the rust rust in the recuperative air heater needs to be considered as well. Another characteristic of the operational efficiency of the combustion equipment is the silting of the heat transfer surfaces. It is important to notice the research on the thermoplastic characteristics of he ash. These characteristics are demonstrated by the following temperatures: t_A – softening point , t_B – melting point a t_C – pour point. The lowest temperature for the mixture of the sludge and coal is approx. 1220 °C for the half-reductive atmosphere.

6.3 The emissions

During the combustion test continuous measurements of the harmful gases like CO, NO_X, SO_2 and relative oxygen behind the boiler. were made by the Technical University of Ostrava. Furthermore single measurements of other emissions like cadmium, mercury, lead, arsen and their compounds, polychlorinated dibenzodioxines PCDD, polychlorinated dibenzofurans PCDF, polychlorinated bifenyls PCB, polycyclic aromatic hydrocarbons PAU, gaseous anorganic chlorine and fluor compounds, hard pollutants TZL, were also by the company TESO Ostrava. The measured values of the pollutants were re-counted to 6 % O_2. The pollutants are illustrated in Table 17. The table provides evidence that the components CO, NO_X, TZL, PCDD/F, HF, Hg meet the requirements of the public notice No. 354/2002 of the Codes of Law but the emissions of SO_2 and HCl do not. This can be attributed to the fact that the sorbent dosing to the boiler was put out of action during the combustion test and thus the process of the conversion of both SO_2 to $CaSO_4$ and HCl to $CaCl_2$ could not occur.

emission	unit	Measured concentration	Limit values *)	
NO_X	$mg.m^{-3}_N$	80	238	Meets the requirements
SO_2	$mg.m^{-3}_N$	560	333	Does not meet the requirements
TZL	$mg.m^{-3}_N$	11	26	Meets the requirements
HCl	$mg.m^{-3}_N$	17	10	Does not meet the requirements
HF	$mg.m^{-3}_N$	0.2	1	Meets the requirements
PCDD/F	$ng(TE).m^{-3}_N$	0.006	0,1	Meets the requirements
Hg	$mg.m^{-3}_N$	0.0013	0,05	Meets the requirements

*) ... limit values given by No. 354/2002 of The Codes Of Law.

Table 18. The chosen emissions of the pollutants (6 % O_2, 101,32 kPa, 0 °C)

Regarding the boilers and reducing their emissions of SO_2, the fine grounded lime stone $CaCO_3$ is continually added directly to the furnace. The lowerig emissions of SO_2 approx. by 100 $mg.m^{-3}_N$ was observed during the combustion test with the the shut-down desulphurisation. This reduction in SO_2 content in the waste gases is obtainable only under the condition of additional combustion of the sewage sludge. The additive gets into the sewage sludge during the process of the sludge hygienisation by the lime dosing at the sewage plant. Furthermore it is hydrated on $Ca(OH)_2$ by the sludge humidity. The lime hydrate is rid of CaO while entering the fluidized bed boiler. The CaO then reacts with the SO_2 and the $CaSO_4$. The amount of the additives in the sludge lowers the lime stone consumption as the primary source (Szeliga 2008). The analysis of heavy metals and microelements in the combusted fuels (the energy coal, the coal sludge, the sewage sludge) were carried out in the laboratories of The Technical University of Ostrava. The evaluation was made for the single coal combustion and then for the mixture of coal and sludge. The redistribution of heavy metals and microelements during the additional combustion of the sewage sludge to the combustion hard residues and the emissions are a matter of further research. The combustion test proves there are further opportunities for additional fuel combustion in the fluidized bed boilers. The advantages of this kind of sewage sludge usage are mainly in the reliable decomposition and oxidation of the organic harmful elements and significant sludge volume reduction. Another suitable way of using the sludge is to reduce its humidity, which improves fuel efficiency, transport and manipulation. The disadvantage of the thermic usage of the sewage sludge is higher concentration of heavy metals and microelements entering the combustion equipment. Co-combustion of coal and sewage is possible only if there is appropriate content of heavy me tals in the sewage sludge entering the combustion process. The monitoring and analyses of heavy metals in the sewage sludge are nessesary.

7. Findings

The most important findings from the research can be summarized as follows:

1. Stability of combustion depends on two factors: a) regular and uniform feed regulation of the fuel mixture, b) perfect homogenization of the fuel mixture. Otherwise, pulsation in the furnace can occur.

2. Experience with the combustion of sewage sludge showed that the highly volatile matter contents significantly affect the overall combustion process. Care must be taken to achieve complete combustion of the volatiles to ensure higher combustion efficiency and low emissions of CO, hydrocarbons and PAH (polyaromatic hydrocarbons).

3. During devolatilization the biomass undergoes a thermal decomposition with subsequent release of the volatiles and formation of tar and char. The results show that the quantities of char and gas formed depend on the type of carbonised material. Furthermore, increasing the pyrolysis temperature leads to a decrease in the quantity of char formed and an increase in the quantity of volatiles. Analyses of the compositions of the volatiles from straw and stover as well as from wood chips and sewage sludge show that CO, H_2, CO_2 and CH_4 are the main gaseous components. High moisture contents have been found to increase the devolatilisation time. For dry residues, in addition to the expected immediate ignition and the highly volatile matter contents, the volatiles consist mainly of combustibles – CO, H_2 and C_xH_Y.

4. The composition of the ashes from sewage sludge, coal, peat and wood influences melting point. It is known that the Na_2O contents of the residues are low and

comparable to those of sewage sludge, wood, peat and coal. The K_2O content of the fuel ashes on their melting points is well demonstrated.

5. Combination of low flow rate and high temperature causes the particles, which are coated with fuel ash, to contact each other and form weak physical bonds or to agglomerate. The formation of these weak bonds or agglomeration is due to the surface of the particles having a low eutectic point or ash softening temperature. This low value is caused by the high alkali content, specifically sodium and potassium compounds, formed during combustion of the boiler fuel. The agglomerated particles, subjected to high temperatures, then begin to sinter or stick together through bond densification thereby forming a strong physical and chemical bond.

6. Agglomeration begins when part of the fuel ash melts and causes adhesion of bed particles. Beginning of agglomeration in the fluidized bed is often indicated by occurrence of temperature differences in the bed and the presence of large fluctuation of bed pressure. When the feeding of the fuel continues it eventually leads to a de-fluidization of the whole bed.

7. To rate the propensity of fuels against fouling, the alkali index has been developed. This index relates the mass of alkali metal oxides $K_2O + Na_2O$ produced with ash to the GJ of energy generated thermally and may be used for biomass feedstock. Above 0,17 kg alkali/GJ fouling is likely and above 0,34 kg/GJ fouling is virtually certain to occur.

8. Ash deposition from biomass fuels which contain certain chemicals can also create corrosion and erosion of metals. Two most abundant inorganic elements are Si and K that form silicates with a low melting point. The combustion leads to the condensation of molten silicates which are likely to cause fouling and corrosion. Analyses showed that corrosive reactions occur between chemical compounds in the ash particles and the elements in the metal on un-cooled samples at gas temperatures near 650 °C.

9. Solutions for the problems resulting from the low melting points of the ash are: use of additives, use of alternative bed materials in the case of fluidized bed combustion and blending of biofuels with coals, lignite.

10. There are three routes of formation of NO_X during coal combustion, namely: thermal, prompt and fuel - NO_X. Biomass has high contents of volatile matter and low contents of fixed carbon, so that the effect of char on formation of NO_X and N_2O may be significant. However, the catalytic effect of the ash could be important for residues which have high CaO contents.

11. Concentrations of heavy metals are in compliance with environmental directive of EU2000/76/EG (including cancerous harmful components and benzopyren +Cd+Co+Cr+As). Combustion of alternative fuels and coal has no significant influence to leaching and Ph factor.

8. Conclusion

Since 1996 29 large fluidized bed boilers with desulphurization ability during combustion process have been launched in the Czech Republic. The differences in design and various concepts of these units have helped collect a lot of valuable data and gain a great deal of experience. The opportunity had not existed before these units were constructed.

Because the boilers for co-combustion of coal and alternative fuels in the Czech Republic were developed from know-how of foreign suppliers, it was not possible to get familiarized with the technical parameters until they started operation. The first operation hours of the most of these boilers were affected by the typical characteristics of Czech coal. Highly abrasive ash matter, high humidity, clay impurity of the fuel and higher content of other elements in the raw fuel (stone, wood, metal) made it necessary to modify fuel feed channels, crushers, separating plants and fuel intake to fluid channels. Many times before, these problems resulted in total unit reconstruction or even replacement. Frequent fuel supply discharges led to reduction of durability of heavy linings of the combustion chamber, especially cyclone bricking and chutes under the cyclones. Some problems were caused by ash extraction from fluidized layer, its cooling down, granulometrics finishing and further manipulation. Other problems occured in sintering fluidized particles when combustion temperatures were well below 900 °C. In spite of this mass sintering happens in various parts of the boiler. Last but not least, there is a trend to reduce desulphurization costs if the molar ratio Ca/S is in the range of / from 2.5 to 3, which means higher operation costs compared to wet tailings.

A quite new area of fluidized bed boilers is the combined combustion of coal and alternative fuels or the co-combustion of assorted fuels from renewable sources. Despite some slowdown in the expansion of activities in energetics, there are further projects in the area of applied research focused on operation process optimization, efficiency improvements and operation costs minimization. These are the areas where the information obtained from measurement results in various boiler types can be used.

9. Acknowledgments

The author would like to express their acknowledgements to the National Program of Research *ENET* – energy units for utilization of non-traditional energy sources cz.1.05/2.1.00/03.0069 for the financial support of this work.

10. References

Bartoňová, L. at al. (2008). Elemental enrichment in unburned carbon during coal combustion in fluidized bed power stations. the effect of the boiler output, In: *Chemical and Process Engineering*, Vol.29, (May 2008), pp. 493-503, ISSN 0208-6425

Bartoňová, L. at al. (2009). Evaluation of elemental volatility in a fluidized bed power station in term of unburned carbon study, In: *Chemical and Process Engineering*, Vol.30, (May 2009), pp. 495-506, ISSN 0208-6425

Čech, B. & Kadlec, Z. (2006). Diagnostic Methods of Combustion in the Fluid Bed Boiler, In: *Acta geodynamica et geomaterialia*, Vol. 3, No. 1, (May 2006), pp.13-41, ISSN 1214-9705

Klika, Z. at al.(2010). Comparison of elemental contents in unburned carbon, coal and ash during brown coal combustion in fluidized bed boiler. In: *Transcriptions of the VSB – Technical university of Ostrava, mechanical series*, Vol. LVI, No. 1, 2010, (August 2010), pp. 1-7, ISSN 1210-0471

Loo, S. & Kopperjan, J. (2008). *Handbook of Biomass Combustion and Co-firing*, (Vol. 2), Earthscan, ISBN 978-1-84407-249-1, London, UK

Szeliga, Z. at al.; (2008). The Potential of Alternative Sorbents for Desulphurization – from Laboratory Tests to the Real Combustion Unit, In: *Energy & Fuel*, Vol. 22 (5) (July 23, 2008), pp. (3080-3088), ISSN 0887-0624

Winter, F. & Hofbauer, H. (1997). Temperatures in a Fuel Particle Burning in a Fluidized Bed: The effect of Drying, Devolatilisation and Char Combustion, In: *Combustion and Flame*, Vol.108, 1997, pp. 302-314, ISSN: 0010-2180

6

Energy-Efficient Standalone Fossil-Fuel Based Hybrid Power Systems Employing Renewable Energy Sources

R. W. Wies, R. A. Johnson and A. N. Agrawal
University of Alaska Fairbanks
USA

1. Introduction

The cost and efficiency of fossil fuel based electric power and heat production in remote areas is an important topic, such as in Alaska with more than 250 remote villages, and developing countries such as Mexico, with approximately 85,000 villages, each with populations less than 1000 persons. The operating cost of fossil fuel based generators such as diesel electric generators (DEGs) is primarily influenced by the cost associated with the purchase, transportation, and storage of diesel fuel. It is very expensive to transport fuel for DEGs in some villages of Alaska (Denali Commission, 2003) due to the extreme remoteness of the site. Furthermore, there are issues associated with oil spills and storage of fuels (Drouhillet & Shirazi, 1997). As of the year 2010, the average subsidized cost of electricity (COE) for a remote Alaskan community is about 0.53 USD/kWh for the first 500 kWh per residential customer per month. The unsubsidized COEs are as high as 2.00 USD/kWh for some extremely isolated communities (Denali Commission, 2003). An extension of the main grid is not possible for such communities due to high cost and losses for the transmission lines.

Based on energy consumption studies compiled by the US Department of Energy, Alaska spends about 50% more (28.71 USD per million BTU) for electrical energy than the rest of the United States (19.37 USD per million BTU) (EIA, 2002). A Memorandum of Agreement (MOA) was signed between the Denali Commission, the Alaska Energy Authority, and the Regulatory Commission of Alaska to supply reliable and reasonably priced electricity to the rural communities of Alaska (Denali Commission, 2003). With the rising cost of fuel and the need for more efficient systems with higher reliability and lower emissions, integrating renewable energy sources and energy storage devices could prove to be more cost effective solutions for electrical power in remote communities (Fyfe, Powell, Hart, & Ratanasthien, 1993). Consequently, there is great need for energy-efficient standalone smart micro-grid systems in these remote communities that employ renewable power sources and energy storage devices.

Distributed power generation systems consisting of two or more generation and storage components, including solar PV arrays, WTGs, battery banks, DEGs, and microhydro, are widely used to supply energy needs. Renewable energy sources such as solar photovoltaics (PV) and wind turbine generators (WTGs) could be used in conjunction with DEGs to supply electricity in remote Alaskan communities and other remote regions (Denali Commission, 2003), (Drouhillet & Shirazi, 1997), (Dawson & Dewan, 1989), (Wies, et al., 2005a), (Wies, et al., 2005b), (Wies, et al., 2005c), & (Borowy, 1996). Besides reducing fuel consumption, the use of renewable energy sources has been shown to increase system efficiency and reliability, while reducing emissions (Drouhillet & Shirazi, 1997), (Dawson & Dewan, 1989), (Wies, et al., 2005a), (Wies, et al., 2005b), (Wies, et al., 2005c), & (Borowy, 1996). It has been predicted that by the year 2050, despite the increase in the demand for electric power, the global CO2 level which is the major greenhouse gas would be reduced to 75% of its 1985 level due to the increase in the use of renewable energy sources for energy production (Johansson, et al., 1993).

This remainder of this chapter presents an economic and environmental model for standalone fossil fuel based micro-grid systems employing renewable energy sources based on an existing diesel-electric power generation systems in remote arctic communities. A simulator called the Hybrid Arctic Remote Power Simulator (HARPSim) was developed using MATLAB® Simulink® to estimate the reduction in fuel consumption of DEGs and the minimization in the cost of producing electricity in remote locations by integrating solar PV and WTGs into the system. HARPSim is used to predict the long-term economic and environmental performance of the system with and without the use of renewable sources in combination with the diesel electric power generation system. A battery bank is also included in the system to serve as a backup and a buffer/storage interface between the DEGs and the variable sources of power from solar PV and wind.

The economic part of the model calculates the fuel consumed, the kilowatt-hours (kWhrs) obtained per liter (gallon) of fuel supplied, and the total cost of fuel. The environmental part of the model calculates the CO_2, particulate matter (PM), and the NOx emitted to the atmosphere. The Life Cycle Cost (LCC), net present value (NPV), efficiency, and air emissions results of the Simulink® model are compared with those predicted by the Hybrid Optimization Model for Electric Renewables (HOMER) software developed at the National Renewable Energy Laboratory (NREL) (NREL HOMER, 2007). A sensitivity analysis of fuel cost and investment rate on the COE is also performed to illustrate the impact of rising fuel costs on the long-term system economics.

2. Distributed generation system

Distributed generation systems like the one described here are currently used in many parts of the world. While this work focuses on modeling a distributed electric power generation system for the remote arctic community in Alaska, the general model can be applied to any distributed generation system containing these components, but can also be extended to include other energy technologies.

2.1 General block diagram

A simple block diagram of a standalone distributed (hybrid) power system is shown in Fig. 1. The sources of electric power in this system consist of a DEG, a battery bank, a WTG, and a PV array. The output of the diesel generator is regulated AC voltage, which supplies the load directly through the main distribution transformer. The connection of the battery bank, the WTG, and the PV array are through a DC bus. The control unit regulates the flow of power to and/or from the sources, depending on the load. The load in the hybrid power system can be an AC load, a DC load, a heating load (resistive load bank), or a hybrid load.

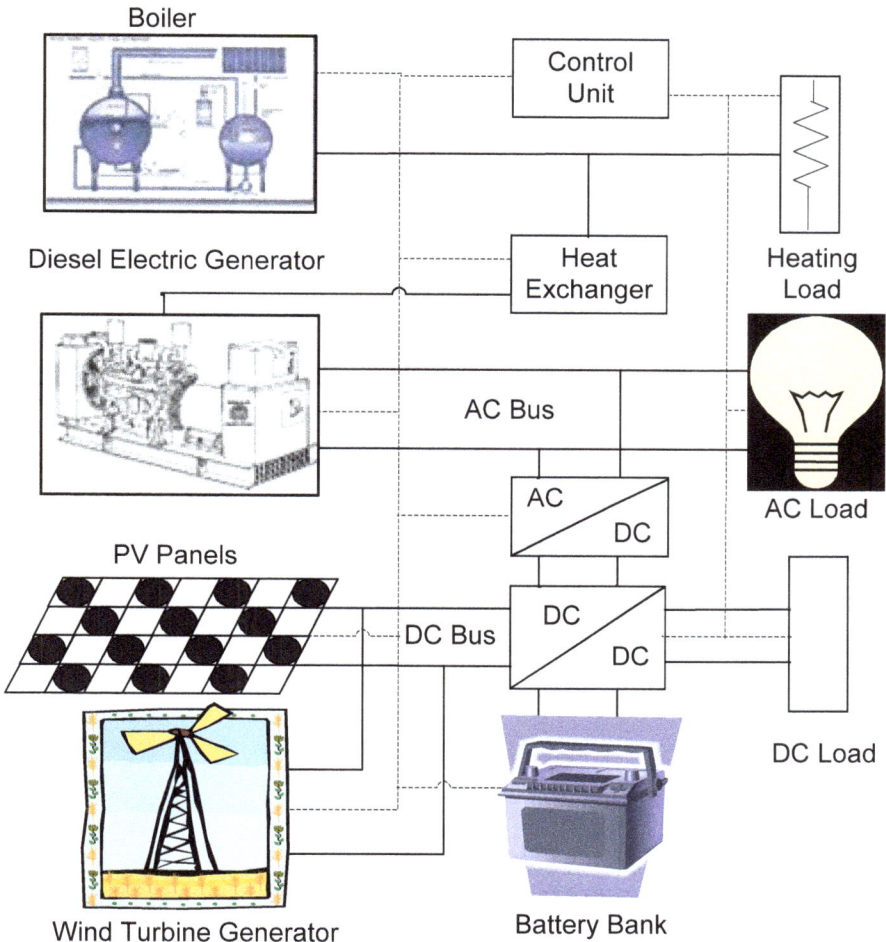

Fig. 1. General distributed (hybrid) power generation system.

2.2 Sample standalone village power system

The sample standalone distributed (hybrid) electric power system used in this analysis consists of four DEGs rated at 235 kW, 190 kW, 190 kW, and 140 kW. The average electrical load is 95 kW with a minimum of 45 kW and a maximum of 150 kW. One DEG is sufficient to supply the village load. Currently, a PV array and a WTG are not installed in the system. In order to analyze the long-term performance of the system while integrating a PV array, a WTG, and a battery bank, simulations were performed using HARPSim for a PV-diesel-battery system, a wind-diesel-battery system, and a PV-wind-diesel-battery system. The simulation results were compared with those predicted by the HOMER software. The system performance is analyzed by incorporating a 100 kWh absolyte IIP battery bank, a 12 kW PV array, a 65 kW 15/50 AOC WTG, and a 100 kVA bi-directional power converter.

3. Simulation model

A general model block diagram for the wind-PV-diesel-battery hybrid power system is shown in Fig. 2. The model is based on previous work with a PV-diesel-battery system (Wies, et al., 2005a) & (Wies, et al., 2005b), and a wind-diesel battery system (Wies, et al., 2005c). The basic model blocks in Fig. 2 and their subsystems are described in detail in Chapter 2 of (Agrawal, 2006). The model consists of nine different subsystems contained in blocks. The electrical energy sources in the model include DEGs, subsystems are described in detail in Chapter 2 of (Agrawal, 2006). The model consists of nine different subsystems contained in blocks. The electrical energy sources in the model include DEGs, WTGs, a PV array, and a battery bank. Currently, the Simulink® model performs a long term performance analysis including the environmental impact calculations of the hybrid power system under consideration. The different inputs required include the annual load and power factor profile, the annual wind speed for the WTGs, the annual insolation profile for the PV array, the annual ambient air temperature in which the power system is operating, the kW ratings of the generators, and the kW rating of the battery bank.

Some basic information about the DEG, Fuel Consumption, Wind, PV, and Battery subsystem models are provided in the following sections.

3.1 DEG and fuel consumption model

The DEG consists of two parts: the electric generator and the diesel engine. The electric generator model consists of the efficiency curve that describes the relationship between the electrical efficiency and the electrical load on the generator. Fig. 3 shows a typical electrical efficiency curve for an electric generator. The fuel curve for a diesel engine describes the amount of fuel consumed depending on the engine load. A typical diesel engine fuel curve is a linear plot of load versus fuel consumption as shown in Fig. 4.

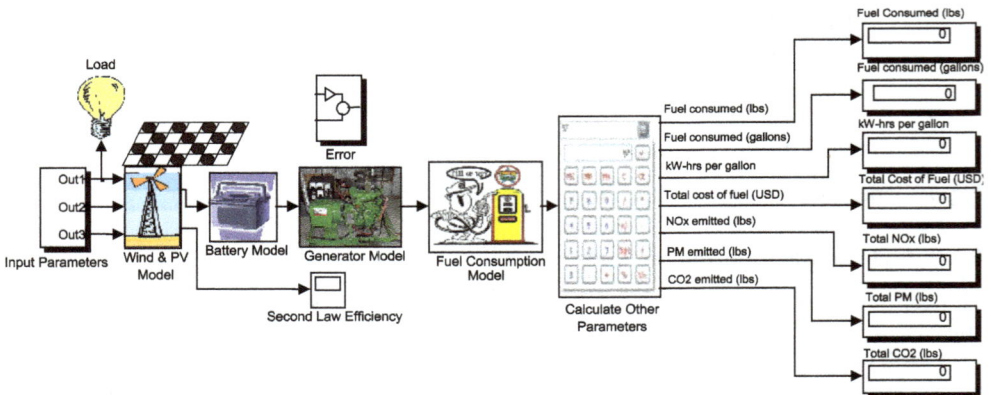

Fig. 2. PV-wind-diesel-battery hybrid power system model.

A fourth order polynomial fit for the electrical efficiency curve as a function of the generator electrical load 'L_{gen}' at unity 'η_{e1}' and 0.8 'η_{e2}' power factor is used. The actual load on the electric generator is converted to its percentage value by dividing the actual load by the electric generator rating and multiplying by 100. This operation is performed so that the same efficiency equations are independent of the rating of the electric generators. The values for the electrical efficiency η_{el} of the generator and the mechanical load 'L_{eng}' on the engine for any given power factor 'pf' are determined using linear interpolation as follows:

$$\eta_{el} = \eta_{el2} + \left(\frac{(\eta_{el1} - \eta_{el2})}{0.2} * (pf - 0.8) \right) \tag{1}$$

$$L_{eng} = \frac{L_{gen}}{\eta_{el}} \tag{2}$$

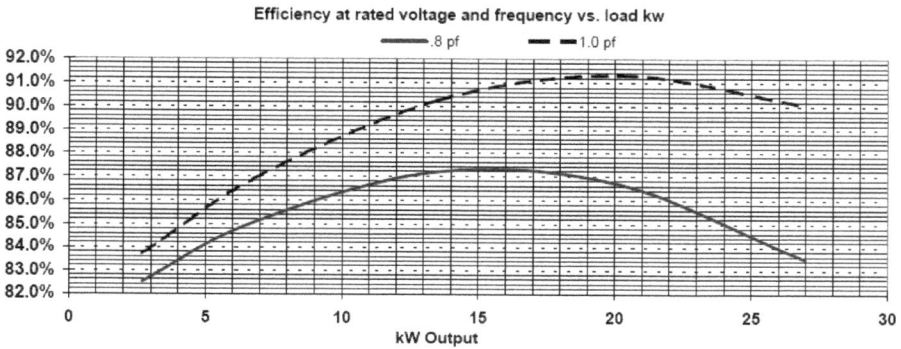

Fig. 3. Typical efficiency for an electric generator.

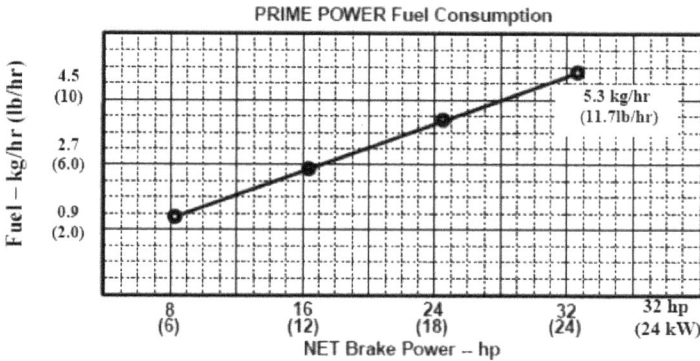

Fig. 4. Typical fuel consumption curve for DEG.

The linear fit for the diesel engine fuel curve is given as

$$\dot{F}_c = 0.5 * \left(L_{eng} * \frac{kW\_A}{100} \right) - 0.44 \tag{3}$$

and

$$\text{Total } F_c = \int_0^T \dot{F}_c \, dt \qquad (4)$$

where '$\dot{F}_c$' is the fuel consumption rate in kg/hr (lbs/hr), 'L_{eng}' is the percentage load on the engine, 'kW_A' is the rating of the electric generator, 'F_c' is the total fuel consumed in kg (lbs), 'dt' is the simulation time-step, and 'T' is the simulation period. The fuel consumed in kg (lbs) is obtained by multiplying the fuel consumption rate of kg/hr (lbs/hr) by the simulation time-step 'dt' (given in hours), and the total fuel consumption in kg (lbs) is obtained by integrating the term '$\dot{F}_c$ dt' over the period of the simulation.

When two or more DEGs supply the load, it is important that the DEGs operate optimally. The following steps are performed to find the optimal point of operation for DEG 2.

1. The electrical generator (Fig. 3) and diesel engine (Fig. 4) performance curves are used to determine overall fuel consumption for the given load profile.
2. The load on the DEGs is varied from 0 to 100%.
3. The fuel consumption for each DEG is noted at different load points.
4. The point of intersection of the two curves is the optimal point of operation for DEG 2. Beyond this point DEG 1 is more efficient than DEG 2.
5. If the two curves do not intersect, the optimal point is taken as 0. This situation implies that DEG 1 is efficient throughout the operating range of the load.

Fig. 5 shows the overall fuel consumption curves for the two DEGs and the optimal point of operation for DEG 2. In order to avoid premature mechanical failures, it is important that DEGs operate above a particular load (generally 40% of rated). The long-term operation of DEGs at light loads leads to hydrocarbon built-up in the engine, resulting in high maintenance cost and reduced engine life (Malosh & Johnson, 1985). If the optimal point is less than 40% load, it is adjusted so that DEG 2 operates at or over 40% load.

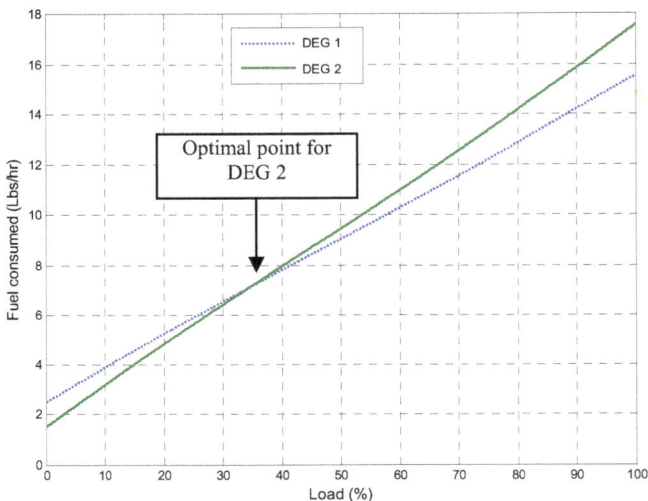

Fig. 5. Optimal point of operation for DEG 2.

3.2 Wind model

The *Wind Model* block calculates the total power available from the wind turbines based on the power curve. The power curve gives the value of the electrical power based on the wind speed. The wind turbine used in this simulation is the 15/50 Atlantic Orient Corporation (AOC). Fig. 6 shows the power curve for the 15/50 AOC wind turbine generator (AOC, 2007). This block calculates the power available from the WTGs depending on the speed of wind based on a look-up table (Table 1).

The wind model block also calculates the second law efficiency of the WTG. The second law efficiency of the WTG is given as

$$\eta_{second\,law} = \frac{actual\,power}{max\,possible\,power} \tag{5}$$

where 'η_{second_law}' is the second law efficiency of the WTG, 'actual_power' is the actual power output from the WTG and 'max_possible_power' is the maximum possible power output from the WTG.

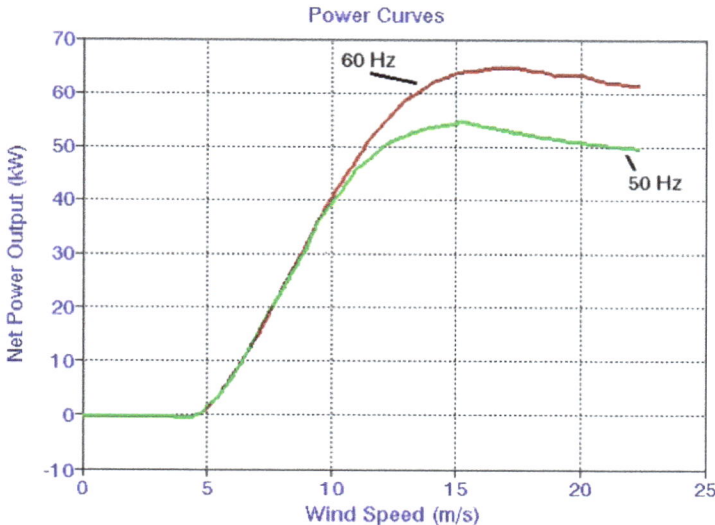

Fig. 6. Power curve for 15/50 Atlantic Oriental Corporation WTG [13].

The actual power of the wind turbine is obtained from the manufacturer's power curve and the maximum possible power is obtained from the Betz formula described in (Patel, 1999) as

$$P_{max} = \frac{1}{2}\rho AV^3 \times (0.59)P_{max} \tag{6}$$

where 'P_{max}' is the maximum possible power, 'ρ' is the density of air taken as 1.225 kg/m^3 (0.076 lb/ft^3) at sea level, 1 atmospheric pressure i.e. 101.325 kPa (14.7 psi), and a temperature of 15.55°C (60°F), 'A' is the rotor swept area in m^2 (ft^2), 'V' is the velocity of wind in m/s (miles/hour), and the factor '0.59' is the theoretical maximum value of power coefficient of the rotor (Cp) or theoretical maximum rotor efficiency which is the fraction of the upstream wind power that is captured by the rotor blade. It should be noted from (6) that the wind power varies with the cube of the air velocity. Therefore, a slight change in wind speed results in a large change in the wind power.

Sr. No.	Wind speed		Net power output (kW)
	Meters/second	Miles/hour	
1	0	0	0
2	5	11.18468	2
3	10	22.37	40
4	11.5	25.725	50
5	13.5	30.1986	60
6	15	33.554	63
7	17	38.028	65
8	19	42.5	63
9	21	46.975	62
10	22.5	50.331	61

Table 1. Look-Up Table for the 15/50 AOC Wind Turbines

The air density 'ρ' can be corrected for the site specific temperature and pressure in accordance with the gas law

$$\rho = \frac{p}{RT} \tag{7}$$

where 'ρ' is the density of air, 'p' is the air pressure, 'R' is the gas constant, and 'T' is the temperature.

3.3 PV model

The PV model block calculates the PV power (kW) and the total PV energy (kWh) supplied by the PV array using the following equations.

$$P_{PV} = \eta_{pv} *ins*A*PV \tag{8}$$

and

$$E_{PV} = \int_0^T P_{PV}.dt \,, \qquad (9)$$

where 'P_{PV}' is the power obtained from the PV array (kW), 'η_{pv}' is the efficiency of the solar collector, 'ins' is the solar insolation (kWh/m²/day), 'A' is the area of the solar collector/kW, 'PV' is the rating of the PV array (kW), and E_{PV} is the total energy obtained from the PV array.

The efficiency of the solar collector is obtained from the manufacturer. The data sheets for the solar panels manufactured by Siemens and BP are available in Appendix 4 of (Agrawal, 2006). The solar insolation values are available from the site data or can be obtained by using the solar maps from the National Renewable Energy Laboratory website (NREL GIS Solar Maps, 2007). The area of the solar collector depends on the number of PV modules and the dimensions of each module. The number of PV modules depends on the installed capacity of the PV array and the dimensions of each PV module are obtained from the manufacturer's data sheet.

3.4 Battery model

In the Simulink® model, the battery bank is modeled so that it acts as a source of power, rather than back-up power. The battery model block controls the flow of power to and from the battery bank. A roundtrip efficiency of 90% is assumed for the battery charge and discharge cycle. The battery model incorporates the effect of ambient temperature as described in (Winsor & Butt, 1978) into the hybrid power system model. Therefore, the model can be used for cold region applications.

The life of the battery bank depends on the depth of discharge and the number of charge discharge cycles. In the Simulink® model the battery bank is modeled so that it acts as a source of power rather than back-up power. Therefore, the depth of discharge of the battery-bank is assumed between 95% and 20% of the rated capacity. This higher depth of discharge reduces the number of battery operating cycles for the same energy output. It should be noted that the number of battery cycles plays a more significant role in the life of the battery bank.

3.5 Fuel consumption and emissions

The *Calculate Other Parameters* block calculates parameters like the total kWhrs/gallon supplied by the generator, the fuel consumed in lbs, the fuel consumed in gallons, the total cost of fuel in USD, the amount of CO_2 emissions, the amount of particulate matter (PM$_{10}$) emissions, and the amount of NO$_x$ emissions. For example, the kWhrs/gallon supplied by the generator and the total cost of fuel in USD are calculated as

$$\text{kWhrs/gallon} = \frac{\text{kWhr}_{Gen}}{F_C} \qquad (10)$$

and

$$\text{Total cost (USD)} = F_C{}^*\text{cost/gallon} \,, \qquad (11)$$

where $kWhr_{Gen}$ is the total kWhr supplied by the diesel generator and F_C is the total fuel consumed in gallons. The quantity cost/gallon is the cost of fuel (USD) per gallon and varies for different locations.

The total CO_2 emissions were estimated based on the equation for the combustion of diesel fuel. For example, one empirical formula for light diesel $C_nH_{1.8n}$ is given in (Cengel & Boles, 2002). For this empirical formula, with 0 % excess air the combustion reaction is given as

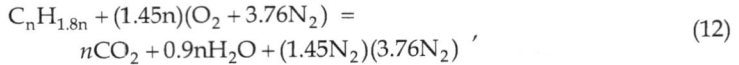

$$C_nH_{1.8n} + (1.45n)(O_2 + 3.76N_2) = nCO_2 + 0.9nH_2O + (1.45N_2)(3.76N_2) \quad , \tag{12}$$

where n is the number of atoms. For any n, the mass in kg (lb) of CO_2 per unit mass in kg (lb) of fuel = $44/(12 + 1.8) = 3.19$. For example, to get the emissions per unit electrical energy output, the above is combined with an engine efficiency of 3.17 kWh/liter (12 kWhr/gallon) and a fuel density of 0.804 kg/liter (6.7 lb/gallon). Doing this results in specific CO_2 emissions of $3.1*(0.804/3.17) = 0.786$ kg (1.73 lb) of CO_2 per kWh of electricity which agrees closely with 0.794 kg/kWh (1.75 lb/kWh) obtained from the DEG manufacturer.

The annual CO_2 amount was calculated from the lb CO_2/kWh and the annual kWh produced and is given as follows:

$$\text{Total pollutant in kg (lb)} = \frac{\text{pollutant}}{\text{kWh}} * kWh_{Gen} , \tag{13}$$

where kWh_{Gen} is the total kWh supplied by the diesel generator during the simulation period. The corresponding values for PM_{10} and NO_x emissions can be obtained from the manufacturer using relations similar to (13).

3.6 Overall model operation and algorithm flow

Fig. 7 shows the algorithm flow chart for the PV-wind-diesel-battery hybrid power system. In the PV-wind-diesel-battery system, the PV array and the WTGs have the highest priority to supply the load. If the load is not met by the PV array and WTGs, the battery bank is used to supply the required load, and if the battery bank is less than 20% charged, the controller sends a signal to the diesel generator to turn "on" and the diesel generator is then used to supply the desired load and charge the batteries at the same time. On the other hand if there is excess power available from the PV array and WTGs, the excess power is sent to a resistive/dump load which can be used for space heating purposes. It should be noted that there is a high demand for heating load during the long winter months in remote communities of Alaska.

Various output parameters from the model include: the second law efficiency of the WTGs (%), the power supplied by the WTGs (kW), the power supplied by the PV array (kW), total fuel consumed in liters (gallons), total fuel cost (USD), total CO_2 emitted (metric tons), total NOx emitted in kg (pounds), and total PM_{10} emitted in kg (pounds). These output parameters are used to calculate the life cycle cost (LCC) and net present value (NPV), the cost of electricity (COE), the payback period for the PV array and the WTGs, and the avoided cost of pollutants.

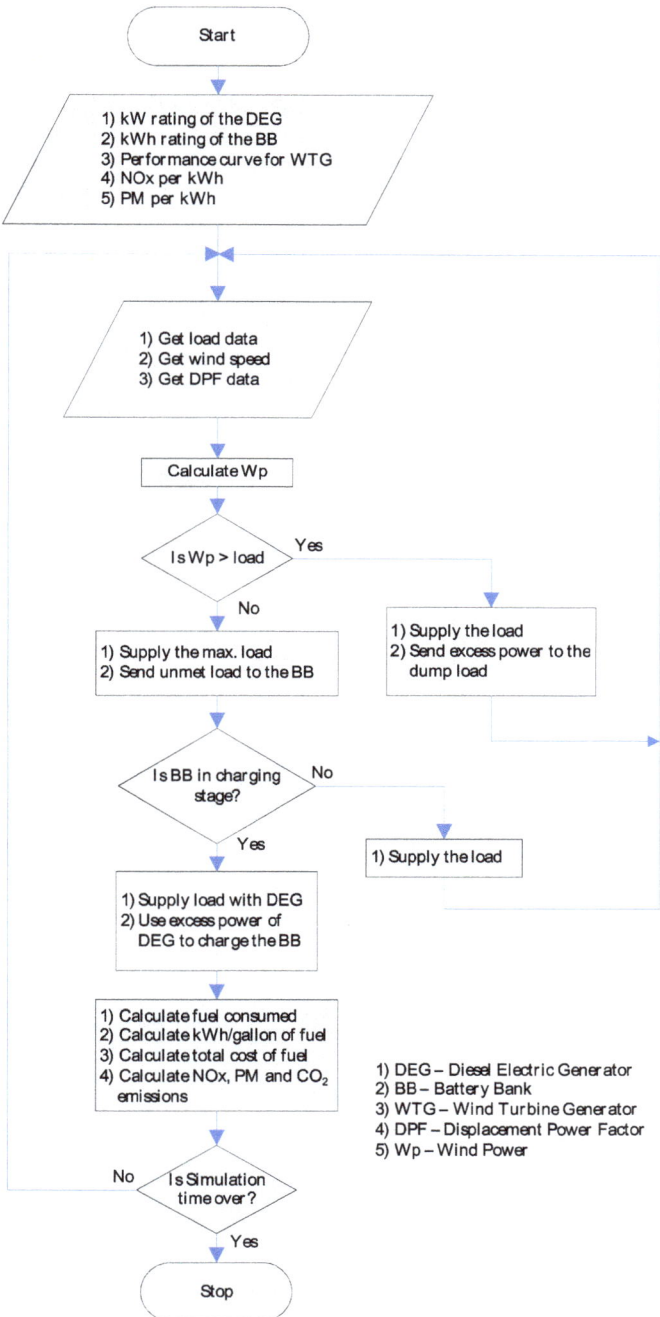

Fig. 7. Flow chart for model algorithm.

4. Simulation cases, results and discussion

4.1 System load, wind, and solar flux profile

The annual synthetic load profile from January 1st, 2003 to December 31st, 2003 with one hour average samples, the annual synthetic wind speed profile, and the annual solar flux profile used for analyzing the performance of a sample village power system are shown in Fig. 8, Fig. 9, and Fig. 10, respectively. It can be observed from Fig. 8 that the maximum load

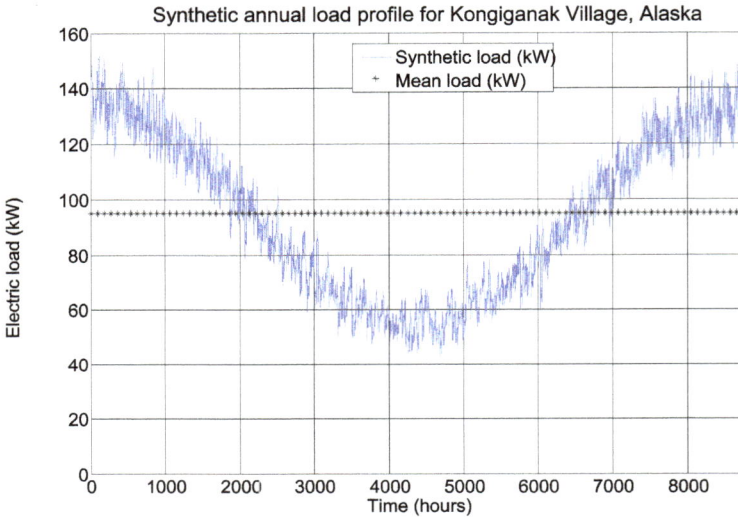

Fig. 8. Synthetic annual load profile for sample village electric power system.

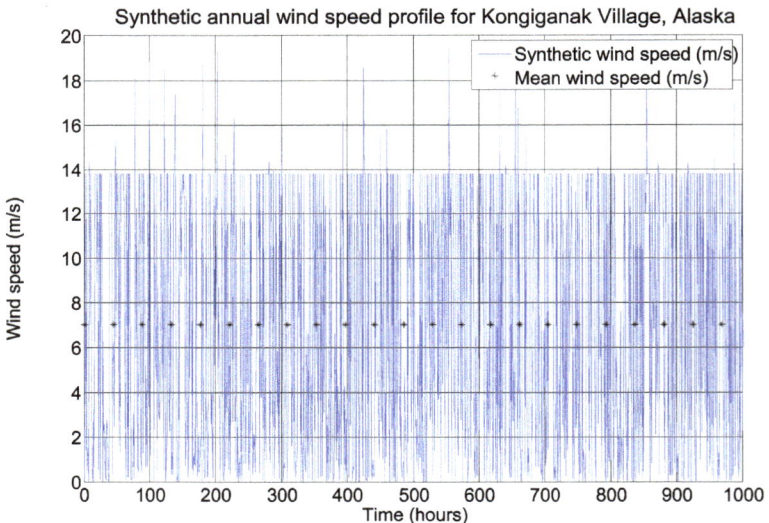

Fig. 9. Synthetic annual wind speed profile for Kongiganak Village, Alaska.

of the system is about 150 kW, the minimum load is about 45 kW and the average load is about 95 kW. From Fig. 9 it can be observed that the annual average wind speed is about 7 m/s (15.66 miles/hr). From Fig. 01 it can be observed that the village has low solar flux during winter months and high solar flux during summer months. The clearness index data for the solar insolation profile is obtained using the solar maps developed by NREL (NREL Solar Radiation Resource, 2007).

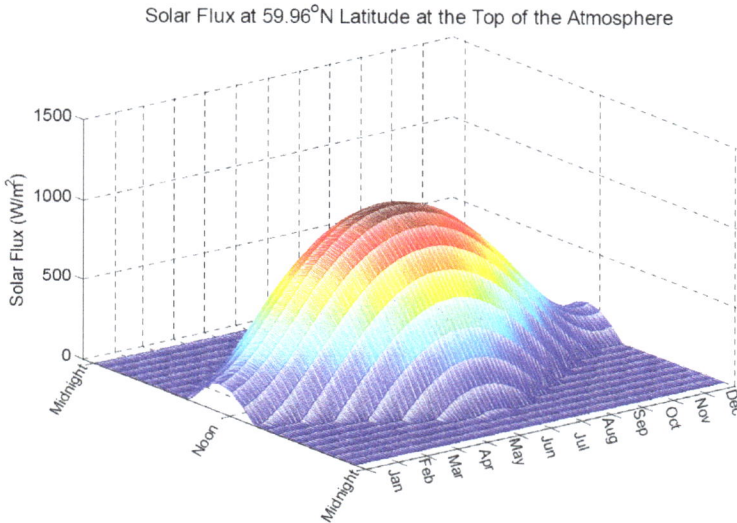

Fig. 10. Annual solar flux for Kongiganak Village, Alaska.

4.2 Simulation cases and results

Simulations were performed for the standalone hybrid power system using the annual load profile for four systems: (i) diesel-battery system, (ii) PV-diesel-battery system, (iii) wind-diesel-battery system, and (iv) PV-wind-diesel-battery system.

The following assumptions were used for the Kongiganak Village simulations:

1. Interest rate i = 7%.
2. Fuel cost of 0.80 USD/liter (3.00 USD/gallon).
3. Life cycle period for PV and WTG (n) = 20 years.
4. Life cycle period for diesel-battery system = 5 years.
5. Life cycle period for diesel-battery system when operating with PV and WTG = 5.5 years.

Table 2 shows the installation costs (USD) for different components for the hybrid electric power system. The post simulation results obtained from the HARPSim model were compared with those obtained from the HOMER software. Table 3 shows the comparison of results from the HARPSim model with HOMER for the hybrid electric power system. It can be observed from the table that the wind-diesel-battery system is the most cost effective system with the lowest NPV, COE, and payback period. This is because of the high energy available from the WTG. The WTG penetration level is observed as 28%. Due to its location, the solar flux available in this region is low resulting in low energy penetration from the PV array. The payback period of the WTG is obtained a little over a year and the payback period for the PV array and the WTG for the PV-wind-diesel-battery system is obtained as a little over two years. It can also be observed that the NPV of the wind-diesel-battery system using HARPSim is less than HOMER. This is because in HARPSim the battery bank charges and discharges while supplying the load. Therefore, the DEGs operate more efficiently resulting in fuel savings while emitting less pollutant. However, this fuel savings is achieved at the expense of the battery life.

Item	Cost per unit (USD)	No of units	Diesel-only system (USD)	Diesel-battery system (USD)	PV-diesel-battery system (USD)	Wind-diesel-battery system (USD)	PV-wind-diesel-battery system (USD)	2 wind-diesel-battery system (USD)
140 kW diesel generator	40,000	1	40,000	40,000	40,000	40,000	40,000	40,000
190 kW diesel generator	45,000	1	45,000	45,000	45,000	45,000	45,000	45,000
Switch gear to automate control of the system	16,000	1	16,000	18,000	20,000	20,000	22,000	30,000
Rectification/ Inversion	18,000	1	0	18,000	18,000	18,000	18,000	28,000
New Absolyte IIP 6-90A13 battery bank	2,143	16	0	34,288	34,288	34,288	34,288	68,576
AOC 15/50 wind turbine generator	55,000	1	0	0	0	55,000	55,000	110,000
Siemens M55 solar panels	262	180	0	0	47,160	0	47,160	0
Engineering		1	3,000	3,500	4,000	4,000	4,500	6,000
Commissioning, Installation, freight, travel, miscellaneous		1	13,000	14,000	16,000	18,000	20,000	30,000
		TOTAL	117,000	172,788	224,448	234,288	285,948	357,576

Table 2. Installation Costs for Different Components.

Since the wind-diesel-battery system was observed to be the most cost effective system, further work was carried out to study the effect of installing another WTG into the wind-diesel-battery system. The addition of a second WTG required an increase in the capacity of the battery bank to accommodate more energy storage. Therefore, the battery bank capacity and the inverter rating were increased from 100 kW and 100 kVA to 200 kW and 200 kVA, respectively.

Item	Diesel-battery system		PV-diesel-battery system		Wind-diesel-battery system		PV-wind-diesel-battery system	
	HARPSim	HOMER	HARPSim	HOMER	HARPSim	HOMER	HARPSim	HOMER
System cost (USD)	172,788	172,788	224,448	224,450	234,288	234,288	285,948	285,950
Engine efficiency (%)	29.3	28.63	29.3	28.51	29.3	27.03	29.3	26.88
kWh/liter (kWh/gallon) for the engine	3.11 (11.75)	3.04 (11.48)	3.11 (11.75)	3.02 (11.43)	2.87 (11.75)	2.87 (10.84)	3.11 (11.75)	2.85 (10.78)
Fuel consumed in liters (gallons)	267,662 (70,810)	273,910 (72,463)	264,834 (70,062)	272,568 (72,108)	193,249 (51,124)	216,027 (57,150)	190,837 (50,486)	214,776 (56,819)
Total cost of fuel (USD)	212,429	217,390	210,185	216,325	153,373	171,451	151,458	170,456
Energy supplied								
(a) Diesel engine (kWh)	832,152	832,205	823,368	823,422	597145	619,504	588,362	612,287
(b) WTG (kWh)	-	-	-	-	235,007	238,000	235,007	238,000
(c) PV array (kWh)	-	-	8,784	8,783	-	-	8,784	8,783
Energy supplied to load (kWh)	832,152	832,205	832,152	832,205	832,152	832,205	832,152	832,205
Operational life								
(a) Generator (years)	5	1.87	5	1.87	5	1.8	5	1.8
(b) Battery bank (years)	5	12	5.5	12	5.5	12	6	12
Net present value (USD) with i = 7% and n = 20 years	-	1,992,488	2,545,084	2,945,502	1,954,127	2,383,766	1,974,389	2,421,502
Cost of Electricity (USD/kWh)	0.301	22.6	0.304	0.334	0.237	0.27	0.24	0.275
Payback period for renewable (years)	-	-	Never	-	1.07	-	2.12	-
Emissions								
(a) CO_2 in metric tons (US tons)	660 (728)	703 (775)	653 (720)	700 (772)	477 (526)	555 (612)	471 (519)	552 (608)
(b) NO_x in kg (lbs)	7,322 (16,143)	-	7,245 (15,972)	-	5,288 (11,657)	-	5,222 (11,512)	-
(c) PM_{10} in kg (lbs)	308 (679)	-	305 (672)	-	222 (490)	-	220 (484)	-

Table 3. Comparison of Results from HARPSim with HOMER.

Table 4 shows the comparison of results from the HARPSim model with HOMER for the two wind-diesel-battery hybrid power system. It can be observed that the addition of the second WTG into the wind-diesel-battery hybrid power system resulted in the further reduction in the NPV and the COE, while the payback period with the two WTGs increased slightly. The WTG penetration level increases to 50% for this case. The payback period of the WTGs has increased to 1.56 years due to the extra cost involved in the addition of the second WTG.

Item	Two wind-diesel-battery system	
	HARPSim	HOMER
System cost (USD)	357,576	357,576
Engine efficiency (%)	29.3	26.6
kWh/liter (kWh/gallon) for the engine	3.11 (11.75)	2.78 (10.53)
Fuel consumed in liters (gallons)	151,252 (39,961)	201,444 (53,222)
Total cost of fuel (USD)	119,883	159,876
Energy supplied		
(a) Diesel engine (kWh)	469,542	561,741
(b) WTG (kWh)	470,015	475,999
Energy supplied to load (kWh)	832,152	832,205
Operational life		
(a) Generator (years)	5	1.8
(b) Battery bank (years)	5.5	12
Net present value (USD) with i = 7% and n = 20 years	1,748,988	2,407,895
Cost of Electricity (USD/kWh)	0.22	0.273
Payback period for WTG (years)	1.56	-
Emissions		
(a) CO_2 in metric tons (US ton)	367 (405)	517 (570)
(b) NOx in kg (lbs)	4,068 (9,112)	-
(c) PM_{10} in kg (lbs)	171 (383)	-

Table 4. Comparison of Results from HARPSim with HOMER for Two Wind-Diesel-Battery Hybrid Power System.

4.3 Life cycle cost and net present value analysis

The life cycle cost (LCC) is the total cost of the system over the period of its life cycle including the cost of installation, operation, maintenance, replacement, and the fuel cost. The life cycle cost also includes the interest paid on the money borrowed from the bank or other financial institutes to start the project. The life cycle cost of the project can be calculated as follows:

$$LCC = C + M + E + R - S \qquad (14)$$

where 'LCC' is the life cycle cost, 'C' is the installation cost (capital cost), 'M' is the overhead and maintenance cost, 'E' is the energy cost (fuel cost), 'R' is the replacement and repair costs, and 'S' is the salvage value of the project.

The net present value (NPV) is the money that will be spent in the future discounted to today's money. The NPV plays an important role in deciding the type of the system to be installed. The NPV of a system is used to calculate the total spending on the installation, maintenance, replacement, and fuel cost for the type of system over the life-cycle of the project. Knowing the NPV of different systems, the user can install a system with minimum NPV. The relationships used in the calculation of NPV are given as follows:

$$P = \frac{F}{(1+I)^N} \tag{15}$$

and

$$P = \frac{A[1-(1+I)^{-N}]}{I}, \tag{16}$$

where 'P' is the present worth, 'F' is the money that will be spent in the future, 'I' is the discount rate, 'N' is the year in which the money will be spent, and 'A' is the annual sum of money.

Fig. 11 and Fig. 12 show the LCC analysis of the PV-wind-diesel-battery hybrid power system using HARPSim and HOMER, respectively. It can be seen that in HARPSim, the cost of DEGs is 4% less while the cost of the battery bank is 2% more than in HOMER. This is because in HARPSim, the battery bank acts as a source of power rather than as the backup

20-year LCC analysis of the Kongiganak Village hybrid power system using HARPSim

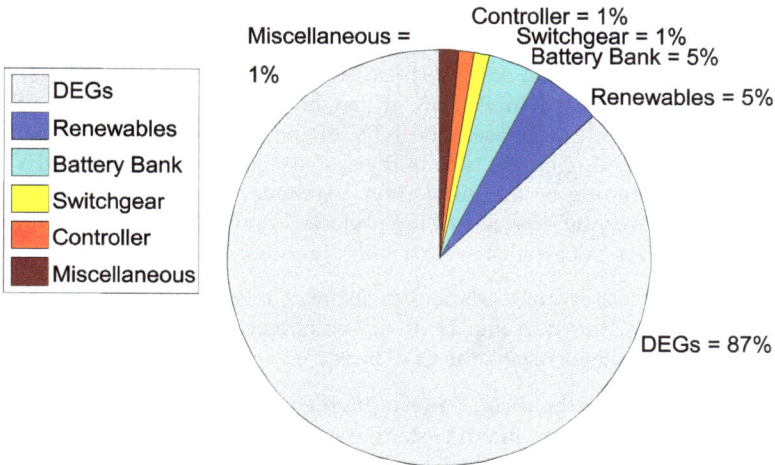

The NPV of the system, with i = 7% and fuel cost = 0.79 USD per liter (3.0 USD per gallon), is 1,974,389 USD

Fig. 11. 20-year LCC analysis of the hybrid power system using HARPSim.

20-year LCC analysis of the Kongiganak Village hybrid power system using HOMER

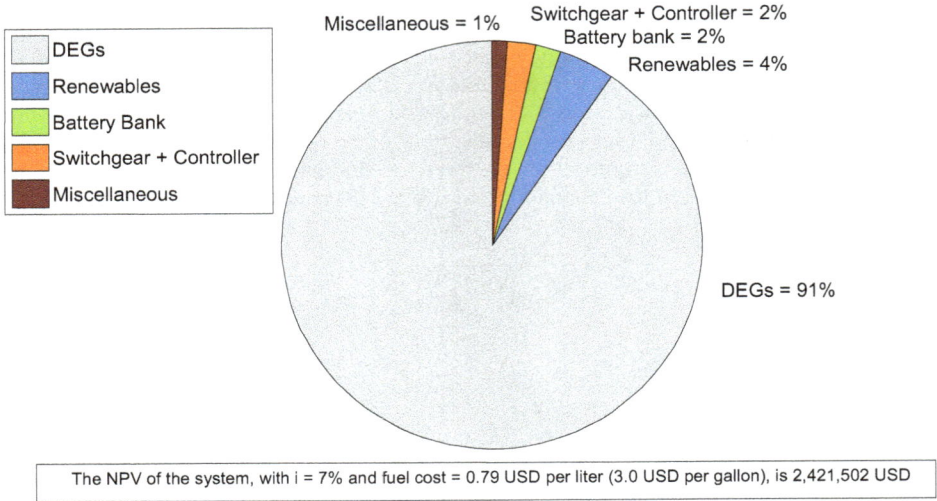

Fig. 12. 20-year LCC analysis of the hybrid power system using HOMER.

power source used in HOMER. Therefore, the life of the battery bank is less in HARPSim due to the annual increase in charge/discharge cycles. This results in more efficient operation of the DEGs while reducing the fuel consumption and saving in the cost of the DEGs. Overall, the LCC analysis shows a lower NPV in HARPSim than in HOMER.

4.4 Sensitivity analysis: Fuel cost on NPV, cost of energy, and payback period

The plot for sensitivity analysis of fuel costs and investment rate on the NPV for the PV-wind-diesel-battery system is shown in Fig. 13. It can be seen that as the cost of fuel increases and the investment rate decreases, the NPV of the system increases. The NPV plays an important role in deciding on the type of the system to be installed. The NPV of a system includes the total spending on the installation, maintenance, replacement, and fuel cost for the type of system over the life-cycle of the project. Knowing the NPV for different system configurations, the user can install a system with minimum NPV.

The plot for sensitivity analysis of fuel costs and investment rate on the COE for the PV-wind-diesel-battery system is shown in Fig. 14. It can be observed that as the cost of fuel increases and the investment rate increases, the COE increases.

In order to calculate the COE for the diesel-battery (high emissions plant) system and the PV-wind-diesel-battery (low emissions plant) system, it is necessary to know the A/P ratio for the system, where 'A' is the annual payment on a loan whose principal is 'P' at an interest rate 'i' for a given period of 'n' years (Sandia, 1995).

The ratio A/P is given as follows:

$$\frac{A}{P} = \frac{i(1+i)^n}{(1+i)^n - 1} \tag{17}$$

The annual COE for different systems given a fuel price in USD per liter (4.00 USD per gallon) and an investment rate (%) is calculated as follows:

$$COE_L = \left(\frac{A}{P}\right)_L (C_{PV\text{-}wind} - C_{DB}) + \left(\frac{A}{P}\right)_H (C_{DB}) + C_F \tag{18}$$

and

$$COE_H = \left(\frac{A}{P}\right)_H (C_{DB}) + C_F, \tag{19}$$

where $C_{PV\text{-}wind}$ is the cost of the PV-wind-diesel-battery system from Table 2, C_{DB} is the cost of the diesel-battery system from Table 2 and C_F is the annual cost of fuel from Table 3.

The plot for sensitivity analysis of fuel costs and investment rate on the payback period for the PV-wind-diesel-battery system is shown in Fig. 15. It can be seen that the payback period of the PV array decreases as a function of a fifth order polynomial with the increase in the cost of fuel.

The simple payback period (SPBT) for the PV array and WTG is calculated using data from Table 2 and Table 3 as

$$SPBT = \frac{\text{Extra cost of PV system}}{\text{rate of saving per year}}. \tag{20}$$

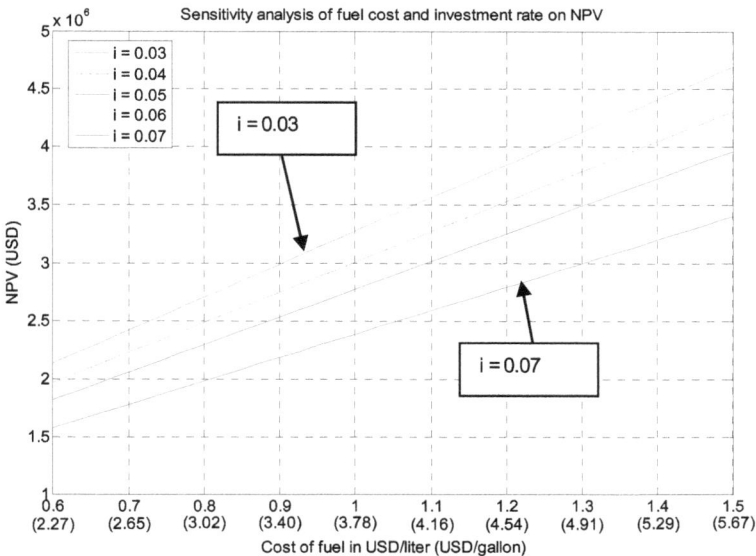

Fig. 13. Sensitivity analysis of fuel cost and investment rate on the NPV.

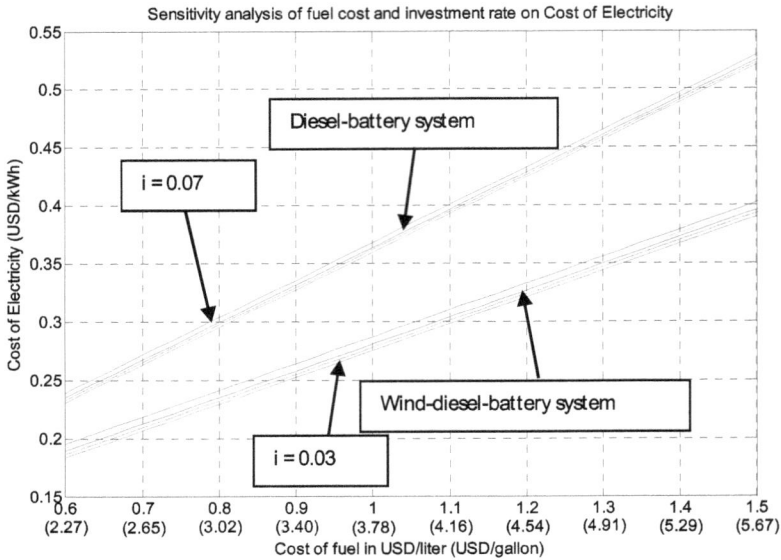

Fig. 14. Sensitivity analysis of fuel cost and investment rate on the COE.

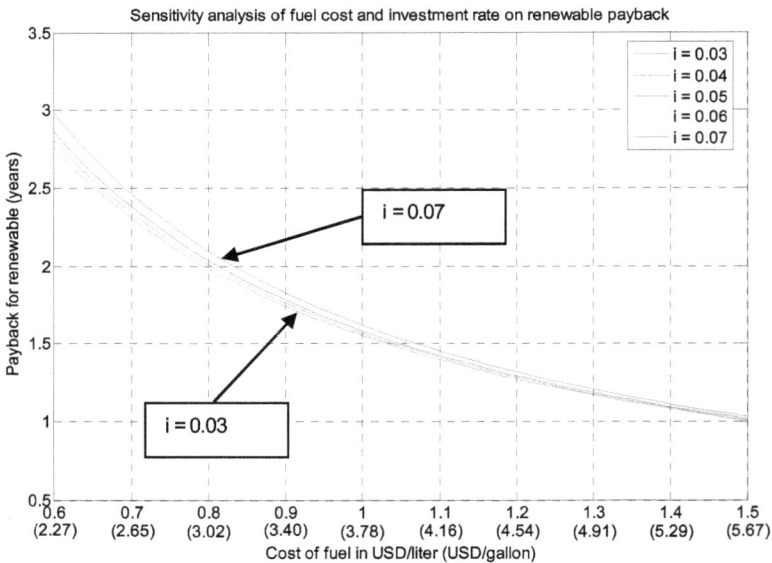

Fig. 15. Sensitivity analysis of fuel cost and investment rate on the payback period.

The extra cost of the PV array and WTG in the system is obtained as the difference between the system cost of the PV-wind-diesel-battery system and the diesel-battery system from Table 2 and the rate of savings per year is obtained from the savings in the cost of fuel per year as given in Table 3.

5. Conclusion

A model called HARPSim was developed in MATLAB® Simulink® to demonstrate that the integration of WTGs and PV arrays into stand-alone hybrid electric power systems using DEGs in remote arctic villages improves the overall performance of the system. Improved performance results from increasing the overall electrical efficiency, while reducing the total fuel consumption of the DEG, the energy costs, and emissions.

The LCC cost analysis and the percentage annualized cost from the Simulink® model were comparable to those predicted by HOMER. The Simulink® model calculates the CO_2, NOx and the PM_{10} emitted to the atmosphere over the period of one year. These results can also be utilized to calculate the avoided costs of emissions.

Distributed or hybrid energy systems which result in more economical and efficient generation of electrical energy could not only improve the lifetime and reliability of the diesel-electric generation systems in remote communities, but could also help to extend the future of non-renewable energy sources.

6. Acknowledgment

The authors would like to thank Peter Crimp of the Alaska Energy Authority and Dennis Meiners of Intelligent Energy Systems for providing the power system information and data for the sample village electric power system. The authors would also like to thank Siemens, Entegrity Wind Systems (formerly Atlantic Orient Corporation), Caterpillar (Detroit Diesel) and GNB Industrial for providing the design specifications for the PV panels, wind turbines, diesel-electric generator and battery bank, respectively.

7. References

Agrawal, A. (2006). "Hybrid Electric Power Systems in Remote Villages: Economic and Environmental Analysis for Monitoring, Optimization, and Control," *Ph.D. Dissertation*, Dept. of Elect. and Comp. Eng., Univ. of Alaska, Fairbanks.

Atlantic Orient 15/50 Brochure (2005). *Official Website of Atlantic Orient Canada Inc.*, accessed Aug 8th, 2011, Available from: http://www.atlanticorientcanada.ca/pdfs/AOCI-SalesSheet2005-1.pdf.

Borowy, B. (1996). "Design and Performance of a Stand Alone Wind/Photovoltaic Hybrid System," *Ph.D. Dissertation*, Dept. of Elect. Eng., Univ. of Massachusetts, Lowell.

Cengel, Y. & Boles, M. (2002). "Engineering Thermodynamics", *McGraw Hill Publications*, 4th ed.

Dawson, F. & Dewan, S. (September 1989). "Remote Diesel Generator with Photovoltaic Cogeneration", *Solar'89*, pp. 269-274.

Denali Commission (August 2003). Memorandum of Agreement Re Sustainability of Rural Power Systems, *A Memorandum of Agreement between the Denali Commission, the Alaska Energy Authority and the Regulatory Commission of Alaska*, August 2003.

Drouilhet, S. & Shirazi, M. (September 1997) "Performance and Economic Analysis of the Addition of Wind Power to the Diesel Electric Generating Plant at Wales, Alaska", *a report prepared by the National Renewable Energy Laboratory*.

US Energy Consumption Database (2002). *US Department of Energy Energy Infomation Authority Website*, accessed on July 9, 2002, Available from: http://www.eia.doe.gov/emeu/sep/ak/frame.html.

Fyfe, W.; Powell, M., Hart, B. & Ratanasthien, B. (1993) "A Global Crisis: Energy in the Future", *Nonrenewable Resources*, pp. 187-195.

HOMER Software, *Official Website of the National Renewable Energy Laboratory*, accessed August 8th, 2007, Available from:, http://www.nrel.gov/homer/.

Johansson, T.; Kelly, H., Reddy, A. & Williams, R. (1993). Renewable Energy Sources for Fuels and Electricity, *Island Press*, Washington, D.C.

Malosh, J. & Johnson, R. (June 1985). "Part-Load Economy of Diesel-Electric Generators", *a report prepared for Department of Transportation and Public Facilities*, report # AK-RD-86-01.

NREL GIS Solar Data and Maps (2007). *Official Website of the National Renewable Energy Laboratory*, accessed August 8th, 2011, Available from: http://www.nrel.gov/gis/solar.html,.

NREL Solar Radiation Resource Information (2007). *Official Website of National Renewable Energy Laboratory Renewable Resource Data Center*, accessed August 8th, 2011, Available from: http://rredc.nrel.gov/solar.

Patel, M. (1999). "Wind and Solar Power Systems", *Florida: CRC Press LLC*, 1st ed.

Sandia National Laboratories (March 1995). "Stand-Alone Photovoltaic Systems - A Handbook of Recommended Design Practices", *a report prepared by Sandia National Laboratories*, report # SAND87-7023.

Wies, R.; Johnson, R., Agrawal, A. & Chubb, T. (2005a). "Simulink Model for Economic Analysis and Environmental Impacts of a PV with Diesel-Battery System for Remote Villages," *IEEE Transactions on Power Systems*, vol. 20, no. 2, pp. 692-700.

Wies, R.; Agrawal, A. & Chubb, T. (2005b). "Optimization of a PV with Diesel-Battery System for Remote Villages," *International Energy Journal*, vol. 6, no.1, part 3, pp. 107-118.

Wies, R.; Johnson, R. & Agrawal, A. (2005c). "Life Cycle Cost Analysis and Environmental Impacts of Integrating Wind-Turbine Generators (WTGs) into Standalone Hybrid Power Systems," *WSEAS Transaction on Systems*, iss. 9, vol. 4, pp. 1383-1393.

Winsor, W. & Butt, K. (September 1978). "Selection of Battery Power Supplies for Cold Temperature Application", *Technical report, C-CORE Publication* No. 78-13.

7

Estimating Oil Reserves: History and Methods

Nuno Luis Madureira
ISCTE-IUL, CEHC
Portugal

1. Introduction

When human societies became aware that an increasing proportion of their power, heat and light were produced from fossil fuels and that fossil fuels were an exhaustible resource, there was no way back to the pre-industrial world. Drawing on the history and methodologies of estimating petroleum reserves this chapter explains how geologists, politicians, engineers, managers and the public at large have come to perceive the finite nature of energy sources.

On the technical side, the need to assess petroleum reserves fostered scientific advances in the domain of the stratigraphic study of rock reservoirs, in terms of the geological understanding of petroleum origins and formation, in the domain of statistical forecasting based on data from producing wells as well as in geophysical measurements. However, the path from measurement technique capacities towards a set of final aggregate figures proved anything but linear. From the outset, the estimation of reserves took place within the framework of a web of political stances, business and social interests and economic organizational realities. What sway did these forces hold over the course of events? What came to determine the core choices about the classification of reserves and the assessments of undiscovered petroleum? Furthermore, how relevant did the contribution of science and technology prove? To answer these questions the ensuing pages sketch the state of the art in geological surveying at the dawn of the twentieth century before examining the technology available for oil discovery and closing with a comparative view of the institutional scenario prevailing over the classification of oil reserves.

2. Volumetric methods and statistical methods: The first oil survey

The completion of the first oil survey unleashed generalized fears of imminent depletion in the United States (U.S.). Much in line with what had previously happened with coal surveys in Britain, key advances in the knowledge on fossil fuel stocks spread alarm about the finiteness of non-renewable resources. As early as 1909, Americans realized that oil wells might rapidly dry up and all the more so when the boom in automobiles could only worsen the situation.

2.1.1 Volumetric methods and statistical methods

The first national estimate of oil resources was commissioned from the U.S. Geological survey and integrated into the national inventory of mineral wealth conducted under the supervision of a new National Conservation Commission, chaired by the head of forestry Gifford Pinchot. The call for a systematic inventory of America's mineral wealth bore President Theodore Roosevelt's personal stamp and thus proved both a scientific project and a political venture. For the U.S. Geological Survey, as an institution, this represented an enormous challenge as oil expertise was concentrated in regional field-surveys and there were no means available to undertake extensive and accurate estimations on a national scale. To overcome tight deadlines and surmount the practical difficulties, the geologist in charge, David T. Day was compelled to resort to indirect forecasting methods that took advantage of the undisputed achievements of geological knowledge. One such undisputed area was the study of oil reservoirs as traps. By the time of this national survey, it had effectively become clear that petroleum needed a particular "trap configuration" to exist. This trap was made up of:

1. a source rock of shale or limestone where a type of organic matter that gives rise to petroleum could once have been deposited;
2. a layer or formation of rock, generally a sandstone or a limestone, both porous and permeable to allow for the formation of a reservoir in the pore spaces, cavities or fissures;
3. a cap rock or "cover", commonly a strata of shales, clays or marls, located above the reservoir to retain the petroleum and block any possible surface outflow;
4. a structural fault in the strata or a geological "unconformity" through which the reservoir was sealed while in this fold trap "gravitational factors" would impel oil to rise above water and gas to rise above oil.

In spite of this common ground, experts did not completely agree on issues including the origin of petroleum (point 1) or the theory of oil accumulation (point 2). Whilst some geologists, for instance, maintained that the oil and gas found in a porous reservoir originated somewhere in that reservoir, others claimed that most oil found its way there from beyond the reservoir's own extent (Johnson & Huntley, 1916). Nevertheless, irrespective of these differences, the trap-configuration approach was solid enough to ensure the understanding that each reservoir enclosed a measurable space from which a commercially relevant amount of oil could be drawn. This enabled the oil reserves in place to be statistically inferred from the analysis of volumetric parameters.

David T. Day resorted to the volumetric method in regions where no other accurate information was available. In order to complete the survey's blanks, he made a simplified calculation based on the average porosity of oil-bearing sands and the amount of crude oil obtained per cubic foot of pay sands. He then extrapolated this cubic foot gauge into a rough forecast of entire oil-pools and applied a value for the recovery factor, setting a ratio between the total oil found underground and the oil that could actually be raised to the surface under the technical conditions prevailing. Curiously, the volumetric method came to be tested on the regions David Day knew best from previous geological field work: Pennsylvania, New York, West Virginia, Kentucky, Tennessee, Ohio, and Indiana (Day, 1909).

The process turned out rather differently when the geologist gained access to primary sources containing oil-well production statistics. In this case, the future recovery could be estimated from the past yield, since the record of the actual output of a well was interpreted as an index to the quantity of recoverable oil. Texas, Louisiana, Oklahoma and California were the regions selected to apply this method of statistical forecasting. Actually, the method was very simple and dubbed the "per cent decline curve". As its very name indicates, the per cent graph depicted the decline in the production of a well, of a property, or an oil-field by expressing each year's production as a percentage of the first year of output. Under the assumption that nearby wells will return similar "curves", the amount of ultimately recovered oil could be estimated for entire oil-pools.

Barely on the radar before the assessment undertaken by the U.S. Geological Survey, the "per cent decline curve" subsequently rose to prominence largely on account of its simplicity and utility, particularly for commercial and financial transactions. In one fell swoop, something that was designed with a view to guiding public policy became a decentralized instrument for private calculus applied to such diverse facets as assessing the rate of return on capital; evaluating the amount a company might afford to pay for the rights to a certain oil-land; future property values; and the distribution of quotas or taxes among producers (Requa, 1918; Beal, 1919: 80-89).

Pressed by the political agenda and the deadlines set by the Conservation Commission, the U.S. Geological Survey had to come up with a solution able to transpose the expertise held by regional oil-pool surveys into a national forecast. The mixed bag strategy of resorting in some circumstances to the volumetric approach and in others to the cutting-edge approach of production history statistics became the means of circumventing the practical difficulties.

The final figures portrayed bleak prospects for the future. David Day's assessment pointed to 10 to 24 billion barrels of oil left underground, with a general estimation of 15 billion recoverable barrels of oil (15×10^9). This would last for a minimum of 80 years should the 1909 consumption level be somehow stabilized or less than 25 years should the recent upward trend continue (Day, 1909). In short, oil might be about to end quite suddenly.

Although the author recognized the conjectural and indirect nature of the assessment, this did not prevent him from proposing clear-cut policies. For David Day, it was the federal government's responsibility to divert oil from power stations, railways, automobile drivers and exports and channel its usage to vital needs like lubrication and the military for which there were no substitutes. "Waste" was the produce of erroneous choices stirred by market prices, abundant resources and unsuitable lifestyles (Day, 1909). In this vein, depletion ceased to be a demand-side problem like that of uncontrolled wood-cutting in forests and became a moral and social question, a true challenge to citizenship impacting on the everyday life of Americans living off perceivably unsustainable resources. With depletion looming in the next quarter century, mechanisms other than price and individual choice in the allocation of resources seemed all the more justified. Moreover, the case of oil provided the ultimate confirmation of conservationist warnings as it demonstrated that those who had buttressed the moral, patriotic, democratic and ecological character of natural resources had been right all along.

2.1.2 Forest conservation and oil conservation

Much like a cause that becomes part of its effects, the conservationist administration of Theodore Roosevelt campaigned for an inventory of American mineral wealth, whose conclusions only underscored the need for conservation. Not only did the conclusions agree with the premises but the conclusions were, to a certain extent, also part of the premises. Identified by the catch-phrase "progressivism", Roosevelt's presidency (1901-1909) prompted a period of soul searching for an American identity deeply rooted in the President's personal fondness for the "true" American values and way of life such as the open country, breathtaking scenery, the wild and everything associated with the outdoor life: hunting, fishing, horse riding and wood-chopping. The great and fair America envisioned by progressivism required an active stance at the federal level on environmental issues, sometimes stretching presidential powers well beyond the limits set by the constitution (Cooper, 1990). Roosevelt's personal inclinations were further reshaped by increasingly closer contact with the Forest Bureau Director, Gifford Pinchot, appointed to advise Roosevelt on forestry matters. However, the mutual friendship and admiration that ensued brought Pinchot into the inner circle of the White House, making him into one of the most influential advisers on political questions with a say over all the critical points on the presidential agenda (Steen, 2001: 133-141). Pinchot's political weight meant the conservation policy became attuned to a commercially oriented perspective based on competitive bidding, the regulation of big business and a call for citizenship. The very language of foresters, with their emphasis on "repairing" forests destroyed by settlers or "restoring" virgin lands, had clear affinities with Roosevelt's message of "regeneration". Culturally, both stressed a return to America's roots. In accordance with key principles such as generational responsibility, social justice and industrial liberty, the government claimed natural resources should be handed back to the average citizen. In Roosevelt's words, the function of "Government is to ensure to all its citizens, now and hereafter, their rights to life, liberty, and the pursuit of happiness. If we of this generation destroy the resources from which our children would otherwise derive their livelihood, we reduce the capacity of our land to support a population and degrade the standard of living"(Roosevelt, 1909: 3). Emboldened by the idea of change and reform, the Federal state pushed the institutions designed for the conservation of natural resources to the forefront of national policy and into the headlines of the press.

With the President's second term drawing to an end and the survey's discovery that oil would run out in 25 years, the eyes of the administration fastened on petroleum with enhanced drama and urgency. Whereas forest conservation had been the hallmark of Theodore Roosevelt's time in office, oil conservation would prove its legacy. Continuity meant that much of the oil related public policy was outlined by the agenda, the knowledge and the legal instruments previously applied to forestry. Furthermore, this also meant that oil geologists, after the foresters, joined the rank and file in the protection of national resources in siding with the Federal State. It is important to note that progressivism consolidated a group of highly trained and qualified civil servants imbued with a sense of mission and whose careers depended on the ability to wield preservation as a key political issue for public opinion. The very program of American regeneration was anchored on a powerful social network of university friendships and further cemented through highly personalized and faithful administrative bureaucracies (Schulman, 2005). Because corporate actions were sometimes "illegitimate", "under cover" and lacking in "industrial democracy", they ought to be counteracted by equidistant public powers exercised by leading experts in transportation, agriculture, geology, utilities and public health (Miller, 2009). The best antidote to "the relentless exercise of unregulated control of the means of production" was joint action by government, scientific expertise and citizenship. This triple alliance formed the core message of progressivism. Indeed, the Director of the U.S. Geological Service had no problems in putting this down in black and white: "I have come to think of geology more as a phase of citizenship than as merely a branch of science" (Smith, 1920).

Once the case of impending depletion spilled over into public opinion and the "oil fraternity" at large, two areas of federal and state intervention surfaced with paramount urgency: one was precautionary action so as to guarantee the fundamentals of national security; the other involved rationalizing measures to tackle all sources of waste and find means to economize on oil consumption. On both fronts, petroleum geologists filled an important role by placing themselves as interpreters of the national interest. Unsurprisingly, the very first measures devised to face possible exhaustion mirrored the very same that had been applied in forest conservation. Security concerns prompted a strong position over oil-land ownership with rationalization efforts triggering a debate over tighter regulation. Globally, the 1909 oil survey proved instrumental in tilting the balance of power towards the "permanent public good" and against the "merely temporary private gain" (Roosevelt, 1909:4).

2.1.3 Shortage fears, precautionary action and regulation

The idea of ring-fencing strategic natural resources away from commercial usage by placing them under the management of the federal state had been enacted by the Amendment to the Land Revision Act of 1891. Though this, the U.S. Senate recognized the President's authority to establish forest reserves by proclamation, leading to the constitution of a non-market sector removed from lands that would otherwise be available. However, it was only during Theodore Roosevelt's presidency that the purchase of Western and even Eastern lands shot up and leading to the founding of a Federal reserve of 150.8 million acres embracing 159 forests and extending over 27% of the U.S. forested extent (Hays, 1959; *Brown, 1919:3-7*). Nevertheless, the more Roosevelt's policy moved towards public ownership, the more Congress responded with hostility towards conservation and bowed to local commercial pressures for quicker private sector exploitation. The clash came to a head in 1907 when Congress forbade the creation of more forest reserves in Western states in arguing that the creation of public property and environmental concerns were just a means to expand presidential authority (Penick, 1968).

Under these circumstances, it would be mightily hard to transpose the recipe for public forest ownership onto the oil realm. All the more so when the geological warning about forthcoming depletion was clearly offset by the intuitive evidence of prices, driven down to an average of $70 cents for the 42-gallon barrel (Williamson, 1968:38-39). Inasmuch as the current evidence ran counter to the geological forecast, there was a kind of pathological split in public opinion. However, against all odds, a small but influential group of oil geologists backed by the director of the U.S. Geological Service, George Otis Smith, was able to circumvent opposition and carve a federal oil reserve through the withdrawal of Californian public lands, and their subsequent conversion to supply the navy. Working behind the scenes and taking advantage of consolidated intra-governmental networks, geologists were able to seize the opportunity that came about in California, firstly by withdrawing prime oil land from the agricultural register, and then by switching its usage to public property. As a result, in September 1909, roughly three million acres of oil-rich lands became the future Naval Petroleum Reserves (Shulman, 2003). This pro-active course of events was particularly significant because when the breakthrough decision was made, the Navy remained undecided over which battle ship types should be fully converted to fuel oil. Equally, the usurpation of private ownership arrayed the opposition of the most significant sectors of the petroleum industry (Olien & Olien, 1993:49-50).

Aside from the shift in natural resource property rights, the first oil survey also set in motion unprecedented appeals for regulation, on behalf of saving the threatened liquid fossil fuel reserves. Geologists and public authorities repeatedly asserted that consumption should be constrained. This was particularly true in the righteous domain of automobile driving, which should be restricted to unavoidable work-related activities such as making deliveries, transporting doctors, transporting children, guaranteeing public order and easing the life of isolated farmers. Beyond this array of utilitarian functions stood nothing less than the hedonistic usage of cars "for pleasure", a social behavior increasingly targeted not only by conservationist writings but also by articles published in specialist magazines like "Motor", "The Motor Age" or the "Oil & Gas Journal". As the chief geologist of the U.S. Geological Service put it in an interview that addressed the danger of exhaustion: "the use of pleasure cars is growing beyond comprehension" (White, 1919; see also McCarthy, 2001).

More telling was the attempt to regulate the industry from the supply side, where sizable wastes were deemed to occur. Scarcity implied the rationalization of exploration. By the close of the nineteenth century, experts in scientific forestry had discovered the potential clash between the needs of common management of natural resources and the economic system of separate ownership and private appropriation. The discovery of the problem of ecological commonalities (indivisibility, interdependency, sustainability) had the effect of pushing the state into previously excluded areas asserting the importance of expert knowledge, norms and regulations. In fact, it might be said that the ascension of professional scientific groups came with the territory.

Although the oil industry long knew about the tremendous waste involved in legendary oil rushes and town-lot developments like the Sindletop oil-field of Texas or the Breman oil-field of Ohio, the explanation for the harmful outcome boiled down to the reckless behavior of adventurers, real-estate speculators and wildcatters. It was only by the close of the First World War that the issue of waste came to be perceived less in terms of economic greed and more in terms of the geologic preservation of reservoir indivisibility, interdependency and sustainability. What transformed the reservoir into an ecological commonality was the understanding of the bond between gas pressure and oil recovery: since the pressure of gas forced oil out of the rocks into the wells, pressure turns into a matter of great economic interest. One of the most influential manuals on oil geology summed up petroleum extraction as a two-step process in which the bore becomes gradually filled with oil, accumulating gas below, until the pressure is sufficient to cause the oil to overflow. Then, as the oil flows to the casing head, pressure is relieved allowing the gas to expand suddenly and to rise up the column with force (Emmons, 1921:184). Doubts still remained as to whether any increase in temperature, in accordance with the deeper burial of organic sediments, would linearly accelerate the process of gas formation leading to an overall increase in pressure. As a general principle, it was accepted that deeper reservoirs would return higher pressures; but geologists also pointed out, cautiously, that "rock pressure" was driven by an array of factors that acted over time: hydrostatic pressure; weight of superincumbent strata; rock movements; deep-seated thermal conditions; long-continued formation of natural gases; and the resistance to fluid movements through the strata. (Mills & Wells, 1919: 30). This would later be summed up in the principle that the level of maturation of the organic matter, through which lighter gas hydrocarbons are generated (methane, ethane, propane, butanes), increases exponentially with temperature and linearly over time.

In addition to natural conditions, human-made intervention plays no less a role in the reservoir pressure level. It was particularly the spacing of wells and the pace of oil extraction that captured the attention of geologists and public authorities. Industrial practices once tolerated henceforth came under close scrutiny (Requa, 1918). The threat of depletion and the harbingered scarcity brought into the spotlight practices like allowing gas to issue freely from open wells or gas flaring, which both contributed to abnormal decreases in reservoir pressure, and subsequently forbidden in several states. The legal framework for property rights known as the "rule of capture" was also criticized as a source of waste. Applied to oil fields, the rule of capture meant the owners of land atop a common pool could take as much oil as they wanted even when unduly draining the pool and reducing the output of nearby wells. This process meant everyone was in harsh competition to extract as much and as swiftly as they could. The ensuing haste and dense well spacing led to a steady drop in reservoir pressure with an inherent reduction in the volume of oil able to be brought to the surface during the exploration life-cycle. Contemporaries estimated that as much as half of the petroleum in U.S. reservoirs remained underground after the fields ceased to yield. This value was later corrected to between 75% and 65% and finally settling on 60%, and interpreted as a regrettable dissipation of national resources. (Emmons, 1921:184; McLaughin,1939:127; Schurr & Netschert, 1977:357-358). Ultimately, the amount of oil reserves could grow simply by improving the oil-recovery factor. However, to achieve this goal, production had to be more efficient, more science based and more regulated.

Stirred by eager conservationist exposés, petroleum shortage forecasts made good copy in popular periodicals and provided appealing headlines. Soon, the basic principles of conservation made their way into Oklahoma state through the institution of pro-rationing among the wells, control over storage and transportation facilities and restrictions over the subsurface waste that caused pressure depletion (1913 and 1915, albeit with few practical consequences). Later on, Texas and Kansas followed in the footsteps of the Oklahoma regulations. With American participation in World War I, conservationism was temporarily diverted from its inward regulatory drive and oriented towards government-business cooperation. The rapprochement stems from the idea that U.S. shortages would have to be met by acquiring foreign oil lands and by taking a more aggressive stance in support of corporate interests abroad (*Nordhauser*, 1979; Clark, 1987).

To conclude this section, the first oil survey was framed by the necessity to present data on the conservation of natural resources so that the final figures released matched the pattern of a readily quantifiable total. Owing to the usage of expedient methodologies, tested in scattered oil-pools, inferences from volumetric parameters and inferences from historical record of production were aggregated into national forecasts. In this manner, the amount determined, 15 billion oil barrels left in "reserve" represented a menace to the future. Geologists looked at the glass as if half empty rather than half full. It suffices to point out that if the amount accrued by new discoveries (plus revision of the previously found and now recoverable petroleum), exceeded the amount of oil extracted from existing fields, the results would have shown a net increase in the volume of reserves. Indeed, as long as this situation lasted, the deadline for depletion would be extended rather than shortened. It was precisely this type of knowledge deriving from the business dynamic of discovery-exhaustion-new discovery that the 1909 geological survey failed to take into account (Olien & Olien, 1993).

Important as this logical viewpoint may be, the fact is the rate of discovery did slow considerably just after the survey's publication. The 1910s was a decade of "dry" wells, rising prices and new discoveries falling short of replacement needs. National Petroleum News, the journal of the independent oilmen, reported in 1913 that "during the past years the prospector has gone over the country with the drill, selecting the most favorable locations, and has not in a single instance been rewarded with a barrel of commercial oil, outside of what is generally accepted as the proven area" (Dunham,1913). Because compensation for growing demand barely occurred, the balance between withdrawals and additions to petroleum reserves moved the countdown of the time elapsing until exhaustion from 17 years (1918-1919 surveys), 13-15 years (1921-1922 surveys) and 6-10 years (1923-1925 surveys) (McLaughin,1939: 128-129; Clark, 1987:148).

This course of events definitely leaned towards the interests and the views of conservationists and geologists.

2.2 Oil discovery: Surface indicators and geophysical surveys

The first concept of oil reserves closely embraced the amounts of oil available in tapped reservoirs. Forecasting techniques like the volumetric and historical-statistical methods required that some successful drilling had already been carried out. In this sense, reserves resulted in ex-post measurements, with geologists tracking a path previously opened by wildcatters and oil-companies. Considering the epochal criteria, two omissions stand out as particularly relevant:

-   The first is the failure to account for enhanced recovery practices implemented in pools that had long since passed their maturity, such as New York and Pennsylvania. In these regions, continued production was maintained chiefly by cleaning and deepening old wells or by obtaining oil from shallow sands which had been were thought too insignificant when the wells were first drilled (Bacon & Hamor, 1916:69). Thanks to these recovery methods, new oil from exhausted fields could be added to the reserves. While only small amounts were at stake in 1910s, the importance of enhanced recovery methods (ERH) would attain new heights in the 1920s with the injection of gas, the injection of compressed air and flooding water flooding into reservoirs on the verge of exhaustion (Miller & Lindsly, 1934). Fostered by a string of technological improvements, recovery techniques rebounded again in the 1950s and 1960s with steam injection, the injection of water-solutions with polymers, surfactants or caustic chemicals, in situ combustion and electric hydraulic shocks.
-   The second was the omission of prospective and untapped reserves. To put the 1909 oil-survey into perspective, it is worth recalling that for nearly 50 years geological coal surveys had followed the practice of ascertaining the recoverable coal left behind in pits plus assessment of seams with "hidden coal". Referred to as existent, probable and possible reserves, this assessment was quantitative in nature (Madureira, 2012).

Insofar as oil reserves were equated as fixed assets, the depletion of reservoirs could somehow be thought of as the depletion of a non-renewable forest. The geological survey thus became a contentious issue that carved a trench between the business view of a drifting amount determined by new discoveries and the official view of a fixed amount determined by the already confirmed oil reservoirs. Hence, the stage was set for a public confrontation between those who claimed "a petroleum famine is imminent" and those who countervailed with "there will always be enough petroleum to meet demand" (Garfias & Whetsel, 1936:213).

Henceforth, the surveys were clouded by the suspicion that the conservative nature of the forecasts set the tone for those who argued in favor of government interference through regulation, pro-rationing, production controls, waste-disposal or even – the rumors persisted - partial nationalization. In an attempt to calm these troubled waters, in 1922 the U.S. Geological Survey (USGS) mobilized ten geologists representing the American Association of Petroleum Geologists and six from the USGS for a comprehensive and accurate study aimed at once and for all stemming the controversies and bringing the debate back to indisputably geological grounds. For the first time, the distinction between known fields and undiscovered reservoirs was acknowledged. The oilman's view of exhaustion-discovery cycles was translated into probabilistic categorizations that accounted for "prospective" and "possible" oil. The concluding estimate identified 5 billion (5 x10^9) barrels of crude "in sight" and an additional 4 billion barrels as "prospective" and "possible". The former was judged "reasonably reliable" with the latter deemed absolutely "speculative and hazardous". In the end, neither the enhanced accuracy of petroleum in sight nor the acknowledgment of "speculative" discoveries reassured the industry. On the contrary, the enduring politicization of the geological survey opened the door to the institutionalization of competing reports on petroleum reserves sponsored by the government, by specialized reviews (*Oil & Gas Journal, Oil Weekly*), and by the American Petroleum Institute business association. From 1922 onwards, this pluralism of estimates became the rule: each vested interest, each major institution produced its own forecasts. Maybe the surprising issue in this evolution towards customized surveys is that there were hardly any discrepancies in the final figures of proven reserves, although that did not halt public and private bickering between institutions (Dennis, 1985).

The crux of the matter was naturally the amount of oil still undiscovered. In this regard, the uncertainty could hardly be solved. There were bold stands on the subject but little means to figure out a reasonable and acceptable forecast. As regards finding oil, geological knowledge had limited utility: it could forecast where oil was not supposed to be found (for instance in rocks dating from the Jurassic, Permian and Silurian Eras) and it could provide some advice on defining areas worth exploring (areas of extensive limestone dolomization, salt domes or beds of porous sandstone lying within shales) (Johnson & Huntley, 1916; Bacon & Hamor, 1916). Nevertheless, the only way to be certain about oil reserves was to drill; as an experienced field-worker reported: "geologists have gone deeply into the matter and in a way seem to be able to select oil producing territory. But they are not infallible. A hole in the ground seems to be the only sure test" (Horlacher, 1929:24).

Up to the First World War, all geological knowledge was in fact exclusively based on surface indicators providing a vague clue as to the location of underground reservoirs. Throughout the U.S., the most reliable signal for the oil prospector was the localization of natural eruptions like oil seepages or springs, natural gas springs, outcrops of sands impregnated with petroleum or bitumen, bituminous dikes and bituminous lakes. These "eruptions" were the first feature to look for as they demonstrated that at least some oil existed in the vicinity and was able to migrate to the surface. Other sedimentary formations such as sands, sandstones, shales and limestones were also potential, though less certain, clues. For field-working American geologists, this hint was nonetheless of limited relevance since the few unveiled seepages were quickly drilled by wildcatters. On the contrary, seepage search did prove very productive in countries like Mexico and Azerbaijan- Russia where oil and gas leaked copiously from source rocks. So abundant was this type of primary surface indicators that the methodology for the second comprehensive Mexican oil survey relied chiefly on inventorying "chapopoteras" (seepages) scattered all over the country, and complemented by a geological description of the underlying sedimentary rock structure (Villarello, 1908). Before the 1910 revolution, the country had consolidated a hub of national oil geology expertise centered in the "small but highly respected organization" of the *Instituto Geológico de Mexico*, which kept in close contact with their North American colleagues (Owen, 1975: 246-256). In Azerbaijan, on the other hand, far-reaching seepages made the tapping of oil from surface wells a remunerative business for local tribes and an ecological nightmare once every amateur, adventurer and speculator began drilling at random during the oil rush of the 1880s. In truth, drilling appeared to be the single talent required to find oil (Leeuw, 2000).

Finally, in the absence of any such clear-cut indicators, geological advice could do no better than recommending searching for the usual landscape fold bed surfacing in an upwards convex form, with the oldest geological beds at its core. Unlike the former empirical guidelines, this particular suggestion was grounded on a theory of oil occurrence – in fact, the most accepted epochal-theory within the scientific community: the anticlinal theory of oil accumulation (Arnold, 1923; USGS, 1934). This convex salience identified by the observer was likely to match a geologic structure called an anticline. Anticlines are rock formations bent into an undulating pattern by a tectonic process and whose fold traps form an excellent reservoir for hydrocarbons, particularly when container reservoir-like rocks at their core and impermeable seals on the outer layers (Figure 1). The hypothesis that an extended "nose" at the surface could become an underground petroleum-bearing fold aroused interest in the systematic exploration of the American countryside, bringing topography back into the arms of geology. From the common perspective, this was summed-up in the unwarranted idea that "all oil is found in folds" (Johnson & Huntley, 1916:50).

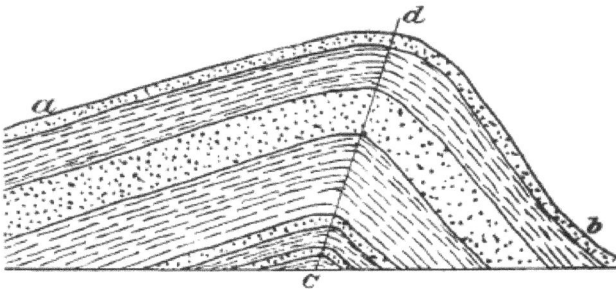

Source: Decker, Charles, E. (1920). *Studies in Minor Folds*, University of Chicago Press, Chicago-Illinois, 6.

Fig. 1. Diagram of a symmetrical anticline.

However, perhaps the most important contribution of the anticline theory to petroleum discovery lay in the technical innovations that accompanied it, especially the systematic observation of rocks altitudes and the representation of anticlines by contour-line subsurface maps. Invented for a geological survey undertaken in Trenton, Ohio (1889), topographic contour lines represented lines in depth below sea-level so that the highest points on the map were labeled with the lowest values. By disclosing the topographical relationship between the observable landscape and concealed petroleum reservoirs, the maps triggered debate about the whereabouts of gas and oil deposits. Above all, this new scientific "gadget" proved extremely useful to impress the value of geological prospecting on both the public and on companies. As expected, geologists endeavored to play their trump card by every feasible means.

The anticline theory gained momentum as more oil was found in anticlines with oil traps as theoretically predicted. West Virginia and South-western Pennsylvania in this respect offered the best supportive evidence; conversely Ohio, Indiana and Illinois cast serious reservations on the global validity of the theory. We know today that most of the world's oil was in effect discovered in anticline structures (Downey, 2009:98). However, this fact, per se, did not significantly raise the earlier probability of actually finding oil. Even when selecting anticlines as their main target, geologists of the 1920s could not single out precise location criteria. Surface indicators said little about whether or not anticlines might contain oil and gas, the amount of hydrocarbons in place, where the accumulation occurred, or the configuration of structural and stratigraphic traps. Ultimately, they could miss the spot simply because the oil was not at the top of a pronounced anticline or because the trap had an unexpected stratigraphic configuration. Furthermore, since oil was found in a great variety of structural positions, the basic anticline hypothesis underwent many vicissitudes (Hager, 1923; Hubbert, 1966).

The work with surface indicators required a sizable and labor-intensive organization. Nowhere as in the prospecting of foreign lands was this feature so remarkable. One may even say that an era of geologically-inspired "invasions" began with the dawn of the twentieth century sometimes involving the overseas relocation of battalions of forty to two hundred men. This stream was fostered by planned investments made by the largest oil companies and reflected the pressure to find untapped sources of supply in the face of increasingly global competition. Mesopotamia (1904 and 1908) Trinidad and the British West Indies (1908), Argentina (1908), Ecuador (1909) Egypt (1911), Algeria (1914) Venezuela (1917) were the most eminent cases of success in finding oil abroad. A geological expedition to China and Formosa (1914-1916) commissioned by the Standard Oil Company of New York also suggested there was a likelihood of discovering good reservoirs but the advance towards the production phase stalled for political reasons. In addition to the new production regions, multinational oil companies further reinforced their presence in Canada and in Peru, leading to a new cycle of discoveries, notably in Peru. So overwhelming was this trend that even firms long skeptical about geological endeavors ended up recruiting 10, 18, 26 geologists (Persia, Anglo-Persian, 1919-1924). Given the higher costs of oil prospecting in the international arena, the massification of discovery had to be spearheaded by some new institutional form of doing business: the multinational holding company was precisely the organizational structure able to finance a multiform presence in oil fields around the world.

After the First World War, the strategic commitment of these large corporations to get hold of secure supplies by constituting buffers of private reserves intensified the scrambling for oil and for leases. Soaring prices further increased the pay-offs for each dollar invested in prospection. The more active stand in geological affairs prompted a phase of swift technological innovation with a bet on every technique that might disclose the sedimentary structures lying beyond anticline's surfaces. Between 1919 and 1929 the core of geophysical technologies, as we currently know them, were experimented for the first time, improved and put to good usage.

Gravity surveys, magnetic surveys and seismic surveys derived from the idea that variations in rock density could be mapped by measuring the way they conveyed some signal. Hence, experiments with the torsion balance, a scientific instrument devised by the Hungarian Baron von Eoetvoes, relied on the assumption that the gravitational force exerted by very light rocks found close to the surface is less than those of very heavy rocks. By the same token, the electrical current sent by a magnetometer depicted a different magnetic "anomaly" when encountering minor magnetic sedimentary rocks and when coming across highly magnetic igneous rocks, thus enabling the identification of the former where oil was more likely to be found. Last of all, a concussive sound produced at the surface, in such a way that as much of its energy as possible was directed downwards, was then partially refracted backwards with greater or lesser velocity depending on the density or compactness of the geological formations encountered. In this echo-sounding technology, a picture could be formed by registering the way in which the velocity of vibrations changed with depth. The time taken for the sound wave to reach a seismic detector located on the surface was recorded on a strip of photographic paper. Owing to the fact that the speed of transmission was proportional to the density or compactness of the geological formation, the technique was firstly used to detect salt domes, which returned a high velocity of propagation. Later on, seismic refraction methods were improved and applied for the mapping of other rock strata (Forbes & O'Beirne, 1957:120-122).

Conceived for general scientific research in geodesy and geophysics (the gravitational method), for iron ore prospecting (the magnetic method) and for the location of enemy artillery firing positions (the seismic method), these technologies had to be further adapted to the particularities of oil surveying. As Bowker (1994:22) pointed out, during the first phase of learning and adjustment, the data produced by prospecting instruments could be correlated with underground structures and those structures could sometimes be correlated with the presence oil. Nevertheless, as of the 1920s, no link in this chain had been firmly established. It was only through further research and practical tests, financed by oil companies like Amerada Petroleum Company, Royal Dutch Shell and Shell's affiliate Roxana, Gulf Oil and its subsidiaries, Louisiana Land & Exploration, Calcasieu Oil, Standard Oil of New York, Humble, Pure and Louisiana, Aguila and Burmah Oil, that fundamental improvements were brought about. Within a short period of time, these investments paid off and paid off handsomely. Successful discoveries of new reservoirs in southern Texas, U.S., Mexico and Hungary with the use of gravitational methods; discoveries in the nearby counties of Texas, in Louisiana, U.S. and Mexico by means of seismic refraction methods; and new finds in Texas, Venezuela and Rumania by means of magnetic surveys and electric logs imparted an aura of buoyancy to geophysical techniques (Williams, 1928; USGS 1934; Forbes & O'Beirne, 1957; Owen, 1975; Bowker, 1994; Robertson, 2000, Petty, n.d.).

Afterwards, the effectiveness of these gravitational and magnetic methods became increasingly associated with reconnaissance surveys and efforts to measure sediment thickness. The seismic method additionally broadened its scope and seized the general purpose geophysical exploration market outside Texas, largely on account of its reliability, cost-benefit advantages and enhanced opportunity "for securing preferred acreage over mapped structures" (Bignell, 1934). The trend that turned seismic methods into the bedrock of core oil prospection activities was further reinforced by two international developments: first, the boom in offshore exploration that began in the late 1950s and was chiefly based on marine seismic surveys; in this respect the production of waterproof microphones (hydrophones) deployed along a cable or a steamer proved to be, far and away, the cheapest and most efficient technology; second, the interface with computing power which led to 3-D seismic surveys and the revolution in "the process of exploration and production, since the early 1990s" Among other aspects, 3-D surveys had the advantage of easing the identification of the optimal drilling point (Downey, 2009: 100-101).

A final piece in this puzzle may be called luck, coincidence or the unexpected coincidence of different series of events: in 1926 and 1927, a series of discoveries in Oklahoma, Texas and New Mexico hit some of the largest oil concentrations in the world, adding almost overnight 5 billion barrels (5 x10^9) to the proven reserves of the United States. The frenzied oil boom that ensued flooded the markets and drove prices down, silencing the "famine", "shortage" and "exhaustion" thesis for fifty years (i.e. up to the Club of Rome warnings). Institutions that had been founded to deal with scarcity and to fight "waste" were subsequently reshuffled to enforce conservation through the self-regulation of the industry. Overall, the rise of geophysical exploration played a minor role in this spurt (circumscribed to part of East Texas) as most of the discoveries resulted from wildcat drilling practices and surface indicator insights. Hence, the urgency that turned geophysical exploration into a key science for the future of humanity became less momentous. Conservationist ideas were also hit. The oil being endlessly pumped out of the earth washed away the bleak predictions of the early 1910s.

2.3 Classifying oil reserves: The U.S. and the Soviet Union

Geophysical surveys and particularly the promising branch of seismic refraction and reflection surveys changed the meaning of geological observations and mapping technologies. As aforementioned, the surface topography could henceforth be related with underground strata and, sometimes, with oil accumulations. For all these reasons, the stage seemed set for qualified assessments of prospective resources, at least of (untested) probable resources identified by geological or geophysical means. However, surprising as it may seem, the United States institutional evolution headed off in the opposite direction.

After 1925, geological uncertainty was removed from the very activity of measurement and substituted by the narrowest gauge of assessing "only the amount of crude oil which may be extracted by present known methods from fields completed developed or drilled or sufficiently drilled and explored to permit of reasonably accurate calculations" (API, 1938, as cited in Miser, Richardson & Dane, 1939: 289). This criterion pervaded the surveys of the American Petroleum Institute (API), the oil industry's trade association founded after the war by several oil companies. Over and over again, a special API committee called the Committee of Eleven, followed by the Committee on Petroleum Reserves, reasserted its pledge not to evaluate unproven reserves, for such estimates would only be "guess-work" and the "the committee refuses to indulge in speculation". Even when criticism mounted, the API stuck to its predefined policy guidelines (Pew, 1944).

The entrenchment around a narrow definition of "proven reserves" meant that the criterion of economic and technical feasibility overtook the criterion of geological probability. To put it differently, the business-oriented view of oil reserves outlived the scientific-administrative view. This conclusion does not imply a single-sided rule of data gathering and data analysis. It is worth recalling that an era of pluralism of estimates had just dawned and whose key institutional actors were the most reputable oil journals, the U.S. Geological Survey (USGC) and the API. During the interwar period, for instance, the USGC drew on its recognized regional expertise to map the probability of discovering oil (figure 2, possible and unfavorable categories).

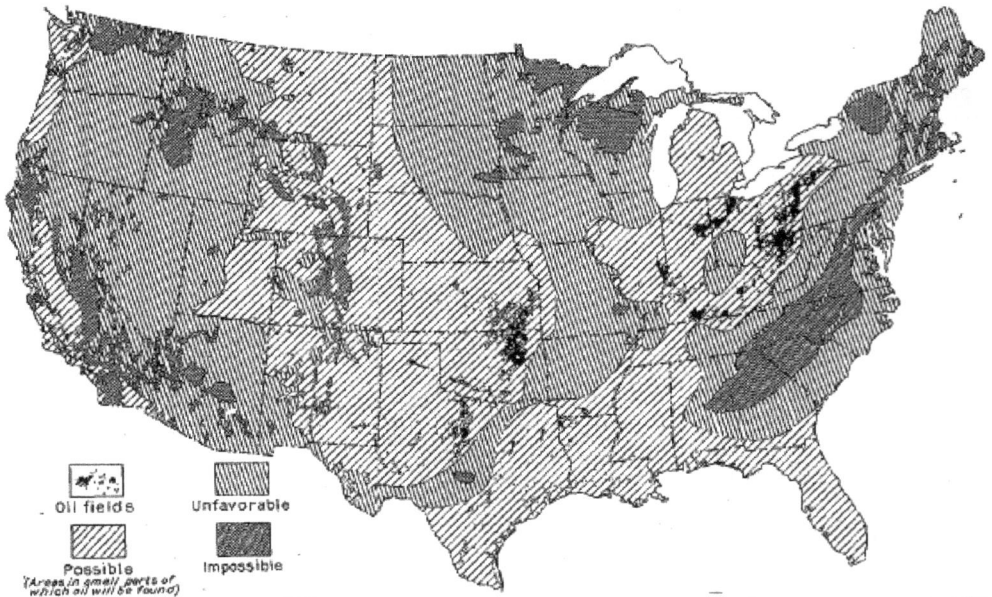

Credit: U.S. Geological Survey; Department of the Interior/USGS
Source: Miser, H.D., Richardson, G.B. & Dane, C.H. (1939) Petroleum Reserves, In: Energy Resources and National Policy, National Resources Committee (Ed.), 293, Government Printing Office, Washington.

Fig. 2. Map of the United States showing the classification of oil regions (1938).

However, beyond the ongoing pluralism, the API criterion broadly held sway in providing the standard upon which resource measurements were based. History and politics turned an instrumental categorization of proven reserves into an encompassing definition. From the outset, the business association was bent on producing its own trade statistics to protect the independence of the industry and counteract the arguments raised by conservationists, by antitrust reformers such as the La Follette Committee and by state and federal regulatory institutions. Drawing on the long-standing informal network of "scouts", traditionally employed by producers to collect confidential information on the activities of rivals, the first API secretary-general, Robert Welch, set up an overarching system of data-field collection, based on weekly reports telegraphed to the headquarters. However, this move was anything but pacific. Suspicions loomed across the affiliates that the information thereby provided might end up in the wrong hands, paving the way for antitrust prosecutions or, even worse, to harmful competition by direct business rivals. According to Joseph Pratt, the dispute over the API's proper role in the area of statistics climaxed in a public confrontation between Robert Welch and Robert Stewart, the President of Standard Oil of Indiana, with the two men almost coming to blows during an API board meeting (Pratt, 1980:78). To ease the concerns of members, Welch reassured that all information submitted would be treated as strictly confidential; that no information on prices would ever be subject to enquiry; and that no strategic information concerning future oil-exploration plans would be collected. Hence, the narrow definition of "proven reserves" was a natural consequence of institutional arrangements inside the API. The assembled information left aside data on private development strategies for as long as the reservoirs remained unexplored (falling into the category of insufficiently drilled fields) they were beyond the survey's scope. This meant that proven reserves was a narrower concept than the vaguer "oil in sight" measurement and also narrower than the standard settled for proven coal reserves (see McInnes, Dowling & Leach, 1913).

Soon API's simplicity and pointedness paid off. The deployment of operational economic criteria enhanced the accuracy, reliability and comparability of statistics. It was this grounding of surveys on proprietary operational information, furthered by Robert Welch's keen leadership and expertise, which established a first order reputation for the API's publications. Henceforth, they became the most current and the most accurate estimates of U.S. oil reserves. On the flip side, such a reality pushed the main government agencies - the U.S. Geological Survey, the Bureau of Mines, and the Bureau of Foreign and Domestic Commerce – into a secondary and complementary role in data collection (Pratt, 1980). The federal state acknowledged that government institutions could not compete with API's resourceful inroads into the industrial milieu. Accordingly, the oil-business association continued to issue reports until 1979, when the oil reserve estimation function was officially taken over by the U.S. Department of Energy. Along with that, the State department further embraced the in-built corporate concept of proven reserves. Finally, in 1983, the World Petroleum Congresses issued expanded definitions for categories ranging from "proven" to "speculative" reserves thereby reopening the door to probabilistic accounts of oil resources (Porter, 1995).

Due to its very history, the North American system of classifications bears the marks of private property, of individual rights and collective corporate action. To get a broader outlook of the ways in which social and economic systems imbue technical classifications, it is worth concisely considering the economic organization that took shape in the Soviet Union. The fact Russia only completed its systematic oil-surveys in the 1930s, when a central command economy was already well underway, eased the creation of a brand new system of oil-reserve classification.

Changes were nonetheless slow. The Bolshevik revolution and the Soviet nationalization of the oil industry did not alter the imperial tradition of oil-geology centered on Moscow with scarce field work outside the Caspian. Lenin's objective of boosting oil production to obtain hard currency from exports during the "New Economic Policy" (1921-1924) phase hinged on directing investments strictly towards productive activities so as to permit the recovery of the mature Baku-Azerbaijan oil fields. It was only with Stalin's drive towards an effective central–planned economy that large-scale use of petroleum-geology entered onto the agenda of the Soviet leadership. The (re)organization of the Chief Administration for Geology and Geodesy of the Supreme Council of the Soviet Union (1929) paved the way for encompassing geological surveys, the foremost of which were an estimate of the Baku oil reserves based on data from individual reservoirs collected by D. Golubyatnikov, the exploration of untapped Soviet basins in the Volga-Ural oil region by the academic V. Wassilieff and the systematic survey of proven Soviet Union oil reserves by the leading figure at the State Petroleum Research Institute, the academic I. M. Gubkin. In addition, innovative geophysical techniques, and particularly the gravitational method and the electric method, made headway in geological prospecting (Hassman, 1953; Maximov & Vinnikovski, 1983). By the time of completion, most of the operative oilfields were still concentrated in the Caspian zone (figure 3).

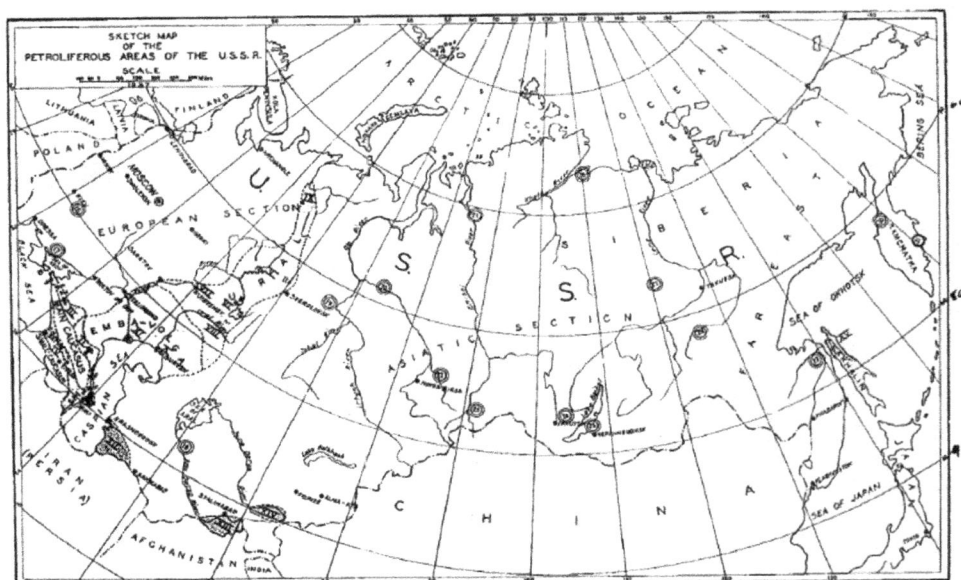

Productive oil fields shown as shaded areas

Caucasus region: I. Azerbaidjan—1, Apsheron Peninsula; 2, Kabristan; 3, Kura Plain: 4, Caspian. II. Northeast Caucasus—5, Grozny; 6,Daghestan III. Kuban and Azov-Black Sea—7, Maikop; 8, Kuban; 9, Taman Peninsula; 10, Crimea. IV. Georgia. Ural region: V. Permian Prikamye; 11, Verkhne-Chussovoskye-Gorodki; 12, Krasnokamsk. VI. Bashkira—13, Ishimbaevo; 14, Sterlitamak, Syzran. 15, Ryazanovka. VII. Samarskaya Luka. VIII. Orenburg-Aktyubinski. IX. Ukhtinski. Emba-Volga region: X. Emba. XI. Lower Volga and Kalmyk-Salsk Steppes. Central Asia region: XII. Turkmen. XIII. S. Usbek and S. Tadjik. XIV. Ferghana Valley. Far East region: XV. Sakhalin. Other areas: 16, Romanski; 17, Melitopol; 18, Western Uzbek; 19, River Tavdy; 20, River Yugan; 21, Estuary of Enisei; 22, Kuznetsk Basin; 23, Minusinsk; 24, Western Pribaikal; 25, Lake Baikal; 26, Khatanga; 27, Lena-Vilyui; 28, River Aldan; 29, Amur; 30, Voyampol; 31, Bogachevsk.

Source: Kemnitzer, W. Oil fields of the U.S.R.R. A descriptive outline of their geography, geology and relative productiveness. The Oil & Gas Journal, (December 30 1937), pp. 71-73.

Fig. 3. Productive oil fields in the U.S.S.R., 1937.

For the Soviets, the description of oil reservoirs had to be customized to the needs of a fully nationalized economy with the collective property of the means of production secured by the state. Categories thus had to convey good information to "distant" decision-makers enabling these central planners not only to allocate investments and resources but also to anticipate the conduct of operations with minimal margins of error. In practice, coordination proved far more difficult than expected.

Furubotn and Pejovich (1972) and Furubotn and Richter (1997:148-156) reveal how in a socialist economy the interests of decentralized managers seldom match the interests of central managers and politicians. In the case of oil discoveries, the interests of regional institutions, encharged with exploratory operations such as the Soviet industrial trusts and the geological services from the various republics, was to secure future investments and shelter their own organization by listing the maximum of reserves that could satisfy the industrial standards and technological requirements for production. On the contrary, the interest of central managers consisted of ascribing investments only to the most remunerative projects whose feasibility was fully established. Thus, the situation could be equated in terms of a principal-agent relationship in which the information is asymmetric and the agent's action on information cannot be observed directly by the principal. Under the conditions of the Soviet Union in the 1930s, the decentralized institutions (agents) had incentives and the means to overestimate the deposit sizes. This was particularly the case when the justifications for arriving at likely figures for untapped petroleum were based on volumetric method assessments resorting to a string of variables – volume of reservoir rock, porosity, oil saturation, recovery coefficient – that could be quietly manipulated (Campbell,1968:62-63). Central planners learned the lesson the hard way on disclosing the premature and wasteful nature of many investments, which after all proved to be over dimensioned given the reality of the petroleum reserves effectively obtainable. In an attempt to overcome this asymmetry of information, the political powers stipulated an intermediate level of certification and supervision, positioned between the central command level of the ministries and the decentralized level of oil exploration. Recognized by the acronym GKZ (Gosudarstvennaia Komissiia po Zapasam poleznykh iskopaemykh), this state commission tightened its grip on local organizations and imposed a system of oil-reserve certification with higher standards of geological evidence (1940). But despite all these endeavors, the principal-agent imbalance continued and attritions rumbled on down the years (Campbell, 1968: 62-68).

Just as the institutional configuration sought to enforce rules for efficient information control by distant decision-makers, so the design of categories sought the same purpose. The goal of monitoring was achieved by breaking up established international categories into minor markers in order to single out, in each marker, data on oil reservoirs and additional meta-data on how the reservoirs were estimated. The emphasis on meta-data loomed therefore as the distinctive feature of the Soviet central command economy.

Reserves were classified by six different tags, labeled A, B, C1, C2, D1, D2, so that central planners could confidently calculate the total quantity of oil extractable over the next five years, controlling for the reliability of the geological forecast. No equivalence whatsoever existed between these six category-tags and the American classification system. The API concept of proven reserves, for instance, was "lost in translation". According to Krylov (Krylov et. al. 1998; see also Poroskun et. al., 2004), American proven reserves seem to correspond to three splits in the Soviet taxonomy:

-    The A Category: reserves of a pool, or part of a pool outlined by wells with proven production. Meta-data requirements involve matters such as the lithological characteristics of the reservoir, quality of oil, well-logging indications, type of drive, pressure, permeability, position of the oil-gas contact and others.

- The B Category: reserves of a pool, or part of a pool outlined by at least two wells with proven production of commercial flows of oil or gas. Meta-data requirements: equal to category A, though some features may have been less fully studied.
- 15% to 20% of the C1 Category: deposits for which favorable conditions were revealed by geological or geophysical data and a commercial flow of oil or gas has been obtained from at least one well. Meta-data requirements: permeability and porosity (based on 1960 descriptions as cited in Campbell, 1968:60-61).

One may conclude that "proven", in the Russian vocabulary, entails three levels of certification. In fact, the system of classification appears to have somehow been transformed into a system of certification.

From the perspective of "distant" decision-makers facing an asymmetry of information vis-à-vis the people in the field, more refined classifications could only mean enhanced control and lesser flaws. Ideally, there could be a quasi-perfect match between reserve categories and the successive stages of exploration. One may imagine the whole upstream sector of the nationalized oil industry functioning like an assembly line: each upgrade of petroleum reserves pushed oil reserves upwards in such a manner that the removal of pools from a lower category becomes the input for the next category (geological studies of category C2 let to them passing onto category C1; if confirmed, exploration and production of pools in category C1 made them pass to B, and so on). Because the whole process is supposed to be driven by adjustments in quantities, the sequence may theoretically be thought of as a linear enchainment (in normal circumstances no C1 pool is allowed to jump directly to A). The reader may remark, instead, how the current American tripartite classification of proved, probable and possible reserves supposes price-quantity adjustments and constitutes the basis of risk assessment and risk management by private investment companies (Poroskun et. al., 2004). Naturally, prices phase out the linearity of classificatory schemes.

For a better understanding of the entire framework, Figure 4 depicts Soviet classifications within the space of a rectangle named the McKelvey box. The McKelvey box (McKelvey,1973) generates an insight into the relative position of each classification according to its location along the horizontal axis (geological certainty increasing from right to left) and according to its location along the vertical axis (technical end economic feasibility increasing from bottom to top). Hence, while the more feasible and more economic oil stands in the northwest corner of the box - in this case the petroleum that has been produced over the years -, the less feasible and undiscovered petroleum lies in the southeast corner. The Soviet categories come somewhere in between these two extremes (figure 4).

The fundamental division in the Soviet scheme is between reserves and resources. The identification of reserves of commercial interest, through the discovery of a field, draws a boundary between categories A+B+C1 and the prospective resources of C2+D1+D2. In fact, the communist leadership only began paying serious attention to the latter group of prospective resources at the end of the 1950s. This means that for a long time category C2 was not tied in very precisely with the planning and conduct of explorations. Moreover, D categories only received closer consideration in the 1960s (Campbell, 1968:64-65).

		(C2 Category) Resources in the planning phase; part of the pool identified by geological and geophysical indicators.	Unmapped resources
Historical commercial petroleum; accumulated production	(A Category) Wells with proven production entirely certified by geological and geophysical indicators.		
	(B Category) Some wells with proven production; reasonably certified by geological and geophysical indicators.	(D1 Category) Resources in rock-stratigraphic units whose commercial production has been proven.	
	(C1 Category) At least one well with proven production; in part certified by geological and geophysical indicators.	(D2 Category) Resources in rock-stratigraphic units whose commercial production has not been proven.	
Reserves lying deeper than 5000 meters			

↑ Increasing economic and technical feasibility

← Increasing geologogic certainty

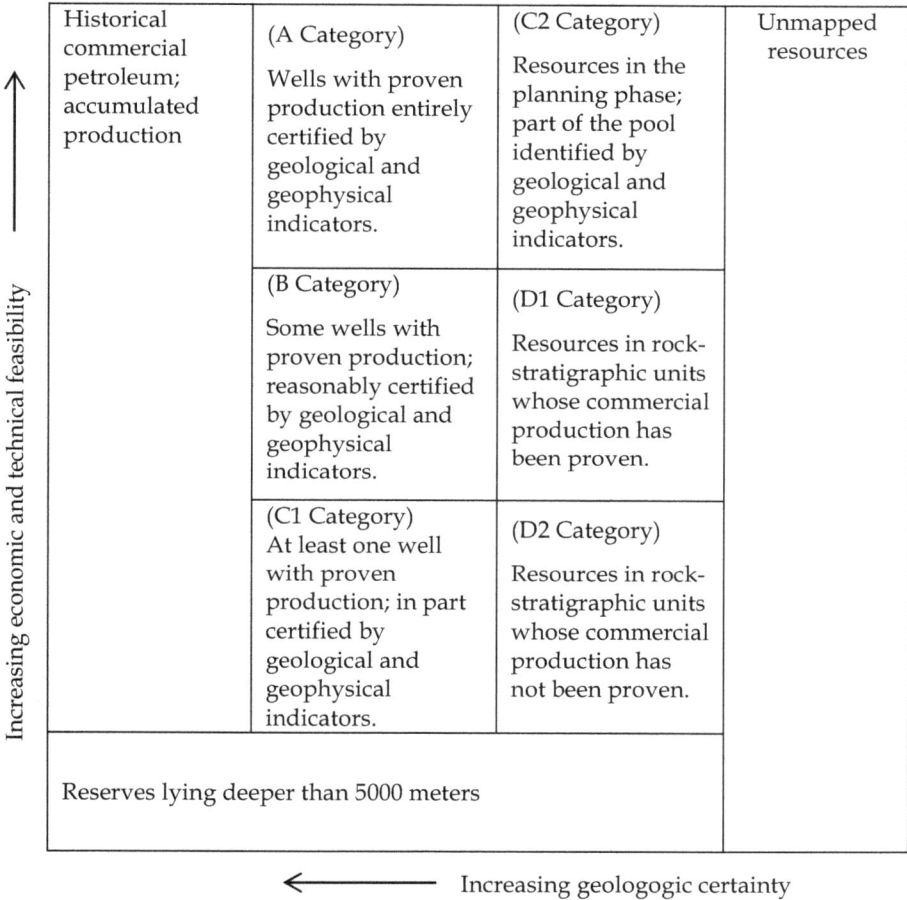

Fig. 4. Modified McKelvey Box with oil reserve and resource categories according to the Soviet Union's classification scheme (1960).

The Khrushchev era brought geology into full blossom. The continuing effort to establish a firmer base for estimating category D triggered a boom of discoveries in Western Siberia, in the Pechora-Timan basin of the Komi Republic and in the Pricaspian Basin, Kazakhstan Republic. Large reservoirs were hit provoking an unparalleled rise in USSR exploration efficiency (measured in tones of recoverable reserves per meter of exploratory drilling). The reasons for such success lay not only in boosting investments by the Ministry of Geology and Conservation of Natural Resources but also in a selective goal-oriented policy: surveys looked primarily for oil in anticline structural traps; in stratigraphic formations with depth limits of 5,000 meters; and looked solely to onshore areas that had previously been studied (Campbell, 1968:64-66; Elliot, 1974:90-106; Krylov, 1998). Furthermore, much of the mapping of D categories was accomplished with the usage of low-cost geological analogue methods, which, as the name indicates, draw inferences from the evidence on known fields recording, for example, its real extent, thickness and porosity, to extrapolate them to similar stratigraphic areas where oil and gas accumulation is expected to be found (Maximov & Vinnikovski, 1983).

To sum up, this chapter highlighted how the core concept of proven reserves has grown out of the willingness to strengthen the role of the business association, the American Petroleum Industry- API, in the turbulent waters of the 1920s. Just when the development of geophysical methods opened up the way for more accurate probabilistic assessments of oil reserves, North American institutions refrained from this path, preferring instead to draw their estimates from business-reports and surveys, therefore anchoring the concept on the solid ground of observable actions undertaken by corporations. In quite the opposite camp, the Soviet regime used oil reserve categories as instruments of control and certification thereby shortening the distance between decentralized operative exploration and central management. Although with less success, the regime also tried to use classifications to ease economic transactions by fitting the categories into the bureaucratic sequence of exploration, certification and production.

3. Conclusion

The history of estimating oil reserves is a history of long lasting misunderstandings. Although American geologists combined volumetric and statistical methods specifically tailored to the realities of the petroleum industry, the final figures from the first oil survey, released in 1909, came to be interpreted by analogy with the forest conservation practices and policies. The discovery of 15 billion barrels of oil left in reservoirs was regarded as a sort of opening shot in a race against the clock of depletion. "Reserves" were understood as a stock; a finite stock that had to be economized, held back and set aside for future uses or contingencies. By adopting terminologies with familiar nontechnical meanings furthermore colored by the moving debate on presidential powers and federal forest "reserves", geologists ascribed the meaning of the concept to an observable fixed asset. Furthermore, given there was, after all, not so much of it left underground, they conveyed the idea that America was approaching its resource supply potential. What ensued was a sort of pathological split between the "scarcity" and the "overflowing" stances, a split sturdily entrenched in discourses, social networks, newspapers, journals and institutions.

To counteract looming claims over the need for regulation, the American Petroleum Association set up its own survey, based on preferential access to oilfields and business records, thereby building a spotless reputation for data gathering. Its aim was to replace geological uncertainties by a narrow but accurate appraisal of oil reserves. Between 1925 and 1935, all open possibilities were locked-into the concept of "proven reserves", grounded on technical and economic feasibility. Such closure ran against the grain of current technological improvements as it excluded probabilistic methods of oil finding by the geophysical sciences and neglected the novel enhanced oil-recovery practices. This means that, in the end, economic-political factors superseded the technical and scientific factors. Comparatively, the Soviet system of oil reserve classification attempted to encapsulate technical feasibility and geological certainty in each category. Indeed, its major originality lay in carving up very analytical classifications with data imbued in meta-data. The fundamental asymmetry between decentralized exploration and central allocation of resources was expected to be overcome by this means.

Overall, the conservation debate continues today with the dispute over whether or not oil production is approaching its maximum level and going to enter into decline soon afterwards, following the downward slope of a logistics curve. Dubbed the "peak oil problem" or "Hubbert's peak" (for an overview see Deffeyes, 2006), its basic assumptions reverse the terms of the debate: whilst in the 1920s it was thought that the oil already found set a ceiling on the likelihood of any new discoveries, the peak oil theory argues the ability to find oil is dominated by the fraction of oil that remains undiscovered.

4. Acknowledgment

The author wishes to thank The Oil & Gas Journal for allowing the reproduction of copyrighted figures inserted in this chapter. Our appreciation is extensive to the University of Chicago Press and the United States Geologic Survey for their reply acknowledging that the figures inserted in this chapter are currently in the public domain.

5. References

Arnold, R. (1923). Two decades of petroleum geology 1903-1922. *Bulletin of the American Association of Petroleum Geologists*, Vol. 7, No. 6 (Nov-December 1923), pp.603-624, ISSN: 0883-9247.

Bacon, W.F. & Hamor, W.A. (1916). *The American Petroleum Industry*. McGraw-Hill, ISBN 9781144838599, New York, Vol.1.

Beal, C. H. (1919). *The Decline and Ultimate Production of Oil Wells, with Notes on the Valuation of Oil Properties*, U. S. Bureau Mines- Government Printing office, ISBN 9781146574471, Washington.

Bignell, L.G.(1934). Geophysical prospecting activity in all areas indicates need for additional oil reserves. *The Oil & Gas Journal*, (February 15 1934), pp. 16, 52, ISSN 0030-1388.

Bowker G. C. (1994). *Science on the Run Information Management and Industrial Geophysics at Schlumberger, 1920-1940*. The MIT Press, ISBN 0262023679, Cambridge-Massachusetts.

Brown, N.C. (1919). *Forest products and Their Manufacture and Use*, John Wiley & Sons, ISBN, 9781177904148, New York.

Campbell , R. W. (1968). *The Economics of Soviet oil and gas*. Johns Hopkins University Press, ISBN 9780801801051, Baltimore.

Clark, J.G. (1987). *Energy and the Federal Government. Fossil fuel policies 1900-1946*. University of Illinois Press, ISBN 9780252012952, Urbana and Chicago, USA.

Cooper, J.M. (1990). *Pivotal decades: The United States 1900-1920*, W.W. Norton, ISBN: 9780393956559, New York.

Day, D. T. (1909). The Petroleum Resources of the United States, In: *United States Geological Survey Bulletin 394*, USGS, pp.30-50, Government Printing office, ISBN 9780217915557, Washington.

Decker, C. E. (1920). *Studies in Minor Folds*, University of Chicago Press, Chicago-Illinois, ISBN 9781141569311.

Deffeyes, K. S. (2006). *Beyond Oil. The view from Hubbert's peak*. Hill and Wang, ISBN 9780374707026, New York.

Dennis, M.A.(1985). Drilling for Dollars: The Making of US Petroleum Reserve Estimates, 1921-25. *Social Studies of Science*, Vol. 15, No. 2 (May, 1985), pp. 241-265, ISSN:0306-3127.

Downey, M. (2009). *Oil 101*. Wooden Table Press, ISBN 9780982039205, New York.

Dunham, W.M. (1913). Foresees Famine in Light crude oil. *The National Petroleum News*, No.3 (May 1913), pp. 1-2, ISSN 0149-5267.

Elliot, I. F. (1974). *The Soviet energy balance. Natural gas, other fossil fuels and alternative power sources*. Praeger Publishers, ISBN 9780275089306, New York- London.

Emmons, H.E. (1921). *Geology of petroleum*. McGraw-Hill, ISBN 9781406707960 New-York/London.

Forbes, R. J. & O'Beirne, D. R. (1957). *The technical development of the Royal Dutch/ Shell 1890-1940*. E.J. Brill, ISBN 9781157320234, Leiden. Netherlands.

Furubotn E. G. & Pejovich, S. (1972). Property Rights and Economic Theory: A Survey of Recent Literature. *Journal of Economic Literature*, Vol. 10, No. 4 (December 1972,) pp. 1137-1162, ISSN 0022-0515.

Furubotn E. G. & Richter, R. (1997). *Institutions and economic theory. The contribution of the new institutional economics*. University of Michigan Press, ISBN 9780472108176, Ann Arbor, USA.

Garfias V.R. & Whetsel, R.V. (1936). Proven oil reserves. *Transactions of the American Institute of Mining and Metallurgical Engineers*, Vol. 118, AIMME, ISBN 9781150194221, New York, pp.211-214.

Hager, D. (1923). Hager on Geology and the geologists. *The Oil & Gas Journal*, (January 18 1923), pp. 91-93, , ISSN 0030-1388.

Hassman, H. (1953). *Oil in the Soviet Union: history, geography, problems*. Princeton University Press , Princeton.

Hays, S. P. (1959). *Conservation and the Gospel of Efficiency: The Progressive Conservation Movement, 1890-1920*. Harvard University Press, Cambridge–Massachusetts.

Horlacher, J.L. (1929). *A Year in the Oil Fields*. The Kentucky Kernel Press, ISBN 0714843229, Lexington, USA.

Hubbert, M.K. (1966). History of petroleum geology and its bearing upon present and future exploration. *American Association of Petroleum Geologists Bulletin* (December 1966) Vol. 50, No. 12, pp. 2504-2518, ISSN 0883-9247.

Johnson, R.H. & Huntley, L.G. (1916). *Principles of Oil and Gas Production*, John Wiley and Sons, ISBN: 1148318283, New York.

Kemnitzer, W. (1937). Oil fields of the U.S.R.R. A descriptive outline of their geography, geology and relative productiveness. The Oil & Gas Journal, (December 30 1937), pp. 71-73.

Krylov, N.A.. et.al, (1998). The oil reserves and resource base of Russia, , In: *The oil industry of the former Soviet Union*, N.A. Krylov, A.A. Bokserman & E.R. Stavrovsky (Ed.), 1-68, Gordon and Breach Science Publishers, ISBN 9789056990626, Amsterdam, Netherlands.

Leeuw, C. (2000). *Oil and gas in the Caucasus & Caspian*. St. Martin Press, ISBN 9780312232542, New York.

Madureira, N.L. (2012).The anxiety of abundance: Stanley Jevons and coal scarcity in the nineteenth century. *Environment and History*, Vol. 18, (forthcoming), ISSN 0967-3407.

Maximov, S.P. & Vinnikovski, S.A. (1983). Development of methods for the quantitative evaluation of Petroleum potential in the USSR. *Journal of Petroleum Geology*, Vol.5 No.3 (January 1983), pp. 309-314, ISSN 0141-6421.

McCarthy, T. (2001). The Coming Wonder? Foresight and Early Concerns about the Automobile. *Environmental History*, Vol. 6 (January 2001), pp. 46–74, ISSN: 1084-5453.

McInnes, W.,Dowling D.B. & Leach W.W. (1913). *The Coal Resources of the World: an enquiry made upon the initiative of the Executive Committee of the Twelfth International Geological Congress.* Morang & Co. and Dulau & Co, ISBN 9781149318232, Toronto-London, Canada-UK.

McKelvey, V. (1973). Mineral Estimates and Public Policy, In: United States Mineral Resources, D. Brobst & W. Pratt (Eds.), 9-19, U.S. Geological Survey Professional Paper 820, United States Printing Office, Washington.

McLaughlin, G.L. (1939) The economic significance of oil and gas, In: *Energy Resources and National Policy*, National Resources Committee (Ed.), 123-236, Government Printing Office, ISBN 9781153473552, Washington.

Miller, C. (2009). The Once and Future Forest Service: Land-Management Policies and Politics in Contemporary America. · *Journal of Policy History*, Vol. 21, No. 1, (Winter 2009), pp. 89-104, ISSN: 1528-4190.

Miller, H.C.& Lindsly,B.E. (1934). Report on petroleum production and development, In: *Hearings before a Subcommittee of the Committee on Interstate and Foreign Commerce*, USGS (Ed.), 1087-1222, Government Printing Office, ISBN: 1153526220, Washington, USA, Part 2.

Mills, R. & Wells, R.C. (1919). The evaporation and concentration of waters associated with petroleum and natural gas. In: *United States Geological Survey Bulletin 693*, USGS, pp.1-31, Government Printing Office, Washington.

Miser, H.D., Richardson, G.B. & Dane, C.H. (1939) Petroleum Reserves, In: *Energy Resources and National Policy*, National Resources Committee (Ed.), 286-297, Government Printing Office, ISBN 9781153473552, Washington.

Nordhauser, N. (1979). *The Quest for Stability: Domestic Oil Regulation 1917-1935.* Garland Publishing Co., ISBN 9780824036386, New York.

Olien, D.D. & Olien, R.M. (1993). Running Out of Oil: Discourse and Public Policy, 1909-1929. *Business and Economic History*, Vol. 22, No. 2, (Winter 1993), pp. 36-66, ISSN: 0358-5522.

Owen, E.W. (1975). *Trek of the Oil finders: A history of Exploration for Petroleum.* American Association of Petroleum Geologists, Tulsa-Oklahoma.

Penick, J.L. (1968). *Progressive politics and conservation: the Ballinger-Pinchot affair,* University of Chicago Press, ISBN 9780226654713, Chicago.

Pew. J.E. (1944). Statement of J. Edgar Pew on A.P.I. Reserves Committee Report on Crude Oil Reserves in the United States as of December 31, 1943. *Chemical & Engineering News*, Vol. 22, No. 8 (April 1944), pp.642, ISSN 0009- 2347.

Poroskun, V.I. et.al. (2004). Reserves/ Resource classification schemes used in Russia and Western countries: a review and comparison. *Journal of Petroleum Geology*, Vol. 27, No.1 (January 2004), pp. 85-94, ISSN 0141-6421.

Porter, E.D. (1995). Are We Running Out of Oil? In: *American Petroleum Institute-Policy Analysis And Strategic Planning Department*, Discussion Paper #081. 02.09.2011 Available from http://gisceu.net/PDF/U30.pdf.

Pratt, J.A. (1980). Organizing Information about the Modern Oil Industry in the Formative Years of the American Petroleum Institute, In: *Business and Economic History On-Line*, Vol.9, 26.09.2011. Available from http://h-net.org/~business/bhcweb/publications/BEHprint/toc91980.html.

Requa, M..L. (1918). Methods of Valuing Oil Lands. *Bulletin of the American Institute of Mining Engineers*, Vol. 134 (February 1918), pp. 409-428, ISSN 0096-7289.

Petty, O. S. (n.d.). Oil exploration, In: *Handbook of Texas Online* published by the Texas State Historical Association, 12.09.2011, Available from http://www.tshaonline.org/handbook/online/articles/doo15.

Robertson. H. (2000). A historic correspondence regarding the introduction of the torsion balance to the United States, *The Leading Edge*, Vol. 19, No. 6 (June 2000), pp. 652-654, ISSN 1938-3789.

Roosevelt, T. (1909). Special message from the President of the United States transmitting a report of the National Conservation commission, In: *Report of the National Conservation Commission*, H. Gannett (Ed.),1-9, Government Printing Office, ISBN 1150477636, Washington, USA.

Schulman, B.J. (2005). Governing Nature, Nurturing Government: Resource Management and the Development of the American State: 1900-1912. *Journal of Policy History*, Vol. 17, No 4, (Fall 2005), pp. 375-403, ISSN 1528-4190.

Schurr, S. H. & Netschert, B. C. (1977). *Energy in the American Economy 1850-1975: an economic study of its history and prospects*, Greenwood Press, ISBN 9780837194714, Westport Connecticut, 2nd edition.

Shulman, P.A. (2003). Science can never demobilize: the United States Navy and the Petroleum Geology. *History and Technology*, Vol. 19, No. 4, (Summer 2003), pp. 365-385. ISSN 1477-2620.

Smith, G.O. (1920). Opportunity for the oil geologists. *The Oil & Gas Journal*, (March 26 1920), pp. 36, 74, , ISSN 0030-1388.

Steen, H. K. (2001). *The Conservation Diaries of Gifford Pinchot*, Forest History Society, ISBN 9780890300596 Durham, USA.

USGS (1934). Résumé of geology and occurrence of petroleum in the United States, In: *Hearings before a Subcommittee of the Committee on Interstate and Foreign Commerce*, USGS (Ed.), 884-907, Government Printing Office, ISBN: 1153526220, Washington, USA, Part 1.

Villarello, J.D. (1908). Algunas regiones petroliferas de Mexico. Imprenta y Fototipia de la Secretaria de Fomento. Ciudad de Mexico, ISSN: 0188-316X.

White, D. (1919). Mr White sees danger of exhaustion. *The Oil & Gas Journal*, (May 2 1919), pp. 8-9, ISSN 0030-1388.

Williams, N. (1928). Geophysics big factor on Gulf Coast. *The Oil & Gas Journal*, (November 8 1928), pp. 35, 88, ISSN 0030-1388.

Williamson H. F. et al. (1968). *The American Petroleum Industry. The age of energy 1899-1959*, Northwestern University Press, SBN-13: 978-0313227899 , Evanston, USA.

8

Modern Transitions in Saving Energy and the Environment

Shahriar Khan

Independent University, Bashundhara R/A, Dhaka,
Bangladesh

1. Introduction

Although there have been numerous programs to reduce fossil fuel consumption, there is still tremendous potential for further reduction of energy consumption. Reduced consumption today will both lower prices in future, and increase availability for future generations. Renewable energies can hardly compare to fossil fuel in convenience and energy intensity. The developed world has moved only slightly from incandescent to fluorescent and LED lamps. Larger automobiles, owing to their greater safety in collisions, and the status they convey, have contributed to a "size race" among consumers for larger automobiles. Urban sprawl, as practiced in much of the world, has contributed to spread-out cities, and a car-dependent culture. Owing to rising fuel prices, urban sprawl is largely unsustainable, as could be evidenced by the recent crash in the suburban housing market in the developed world. Back-up generators and UPS by individual households and consumers may worsen problems, and even cause a vicious cycle of power shortages. Over the last few decades, forest and vegetation densities have increased, but marine life has decreased, for reasons which are unclear at this time.

2. Nature of the decline of fossil fuel

While fossil fuel is known to be in decline, the nature of the decline and future trends in prices is unclear. Since 1900, the world population has more than quadrupled, and primary energy consumption has increased by a factor of 22.5 (BP Statistical Review 2011). The consumption of fossil fuel has been increasing at the rate of about 1.6 % annually for many years, and oil companies expect consumption at this rate for another two decades (BP Energy Outlook 2030, 2011).

From these estimates of known and future reserves, it is only a question of time, before oil and gas reserves become largely depleted. At the present rates of consumption, it is generally acknowledged that fossil fuel, especially oil and gas, will decline greatly in a few decades. Oil and gas are being produced from increasingly deep reserves, leading to higher production costs. These greater expenses have been at the cost of disposable incomes, and quality of life.

2.1 Expected time to depletion

The size of fossil fuel reserves and their times to depletion are fundamental issues with conflicting answers given by experts. According to one estimate, the fossil fuel reserve depletion times for oil, coal and gas are approximately 35, 107 and 37 years, respectively (Shafiee, 2009). Accordingly, coal is expected to be available up to 2112, and will be the only fossil fuel remaining after 2042.

The consumption of fossil fuel in the next few decades is complicated by uncertainties about known and future reserves, and our consumption rates. Speculation over the next few decades is further complicated by the large numbers of variables and unknowns, some of which are identified below:

a. uncertainty about known and yet-unknown reserves
b. uncertainty about the rise of fuel prices, over the next few decades, in the face of diminishing reserves
c. uncertainty as to the trends of the diminishing consumption of oil and gas, after it has reached well into its decline.
d. uncertainty as to how renewable energy such as solar and bio-fuels would replace fossil fuel in the next few decades.
e. importance of oil to the chemical industry

In spite of these uncertainties and unknowns about the future of fossil fuel, there is consensus as to the importance of conserving fuel today.

2.2 Coal

The issue of coal is somewhat different from oil and gas, because it is expected that coal will last for perhaps another 100 years, well beyond the expected life of oil and gas. The huge reserves of coal make coal the likely major replacement of oil and gas. As coal is particularly well adapted to electricity production at power stations, it is expected that coal-powered plants will be available well after oil and gas have gone into sharp decline. On the other hand, coal cannot be used for most of the transportation industry, such as for powering automobiles and airplanes (Table 1).

	Oil	Gas	Coal	Hydroelectric	Nuclear
Electricity Generation	Well suited	Well suited	Well suited	Well suited	Well suited
Automobiles	Well suited	well suited	Unsuitable	Almost impossible	Almost impossible
Trucks and Lorries	Well suited	Suitable	Barely suitable	Almost impossible	Almost impossible
Trains and Locomotives	Suitable but not for underground	Barely suitable	Barely suitable	Almost impossible	Almost impossible
Ships	Well suited	Suitable	Barely suitable	Almost impossible	Suitable only for large ships
Aeroplanes	Well suited	Barely suitable	Almost impossible	Almost impossible	Almost impossible

Table 1. Conventional Power Sources and their Applicability

2.3 Continuous decline, rather than total depletion

One popular perception about oil and gas (and coal) is that they will be suddenly depleted (Figure 1).

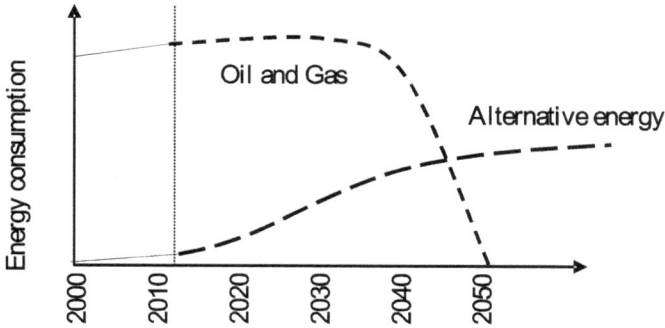

Fig. 1. Popular perception about Oil and Gas, and Alternative Energy. Fallacies are that there will be (a) a mostly complete depletion of oil and gas (b) Alternative energy production will be comparable to former levels of oil and gas.

In reality, the effect of diminishing supply and rising prices will soon cause consumption to decrease. Contrary to most existing speculations, the issue of depletion is complicated by the fact that, today it is commercially viable to extract at best about two-thirds of fuel in a reserve. There are large residual reserves of oil under the ground, which are not commercially feasible to extract today, but will become very attractive several decades later.

The reality is that rather than being completely depleted, fossil fuel will go into a long decline for many decades, during which prices will continue to rise. During this time, both fossil fuel and renewable energy will exist side by side (Figure 2). However, renewable energy will never come close to the energy production formerly by fossil fuel.

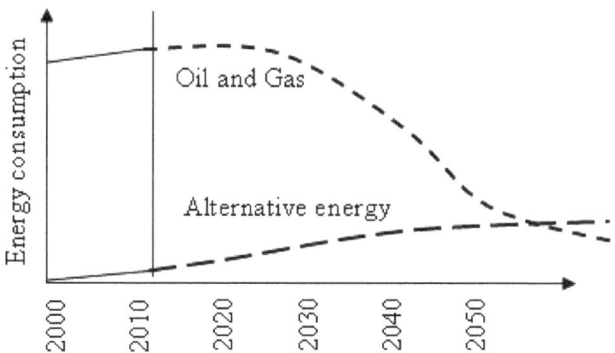

Fig. 2. A more realistic expectation of the decline where Oil and gas, rather than being completely depleted, will continue to decline, with rising prices, for a long time. Alternative energy will rise to levels much lower than former levels of oil and gas.

Petroleum is also one of the main sources of raw materials for the chemical industry today. An example is the bitumen used in paving roads, which is produced from petroleum. There are very few substitutes for petroleum-based bitumen, comparable to it in price and availability. In future, with the shortage and depletion of oil, there will also be a great shortage of other petroleum-based raw materials. The abandoned reservoirs of today will look very promising when there are real shortages of oil in future.

2.4 Nuclear power

After the three major nuclear reactor accidents (Three Mile Island, Chernobyll, and Fukushima), caution with nuclear power is at a high. However, with the impending decline of oil and gas, nuclear power will be an inexpensive option, which people and the government are more likely to accept.

3. Renewable energy

The focus today appears to have shifted from conserving fossil fuel to renewable energy. More literature today is dedicated to renewable energy, than to conserving the remaining fossil fuel reserves. This focus on renewables may create false expectations among consumers about the true capabilities of renewables. There is generally little awareness among consumers about the the limitations of alternative energy. The switch to alternative and renewable energy involves large capital investment, especially for solar and wind power. In spite of the research, renewable energy accounts for as little as 1.8% of global energy consumption today, up from 0.6% in 2000.

3.1 Solar energy

In spite of efforts at solar power for at least two decades, implementation has been difficult because of low power intensities, large capital costs, and difficulities incorporating with existing technology. Silicon panels are much the same as the silicon chips used for microprocessors, and have similar requirements and constraints for manufacture. Silicon panels are expensive, and the area needed for a household (excluding air-conditioning) is barely met by panels all over its roof. Consumers are mostly unaware that solar panels can at best about convert about 25 % of the solar energy falling on it. Usage at night requires expensive and bulky batteries, which must be replaced every few years. Dependence on batteries can be reduced or avoided, by selling solar energy to the grid, as being by practiced by some household users in Europe. The utilities are cautious about accepting solar energy from others, as it introduces noise into the power grid.

3.2 Wind energy

Limitations of wind energy include that it can only be implemented in areas of high wind, requires large investment and maintenance, and has relatively low energy densities compared to fossil fuel. Renewable energy used in power generation has grown this year by 15.5%, driven by continued robust growth in wind energy (+22.7%). The increase in wind energy in turn was driven by China and the US, which together account for nearly 70% of global growth. Opponents argue that wind turbines clutter scenic countrysides.

3.3 Biofuels

In areas of North America, biofuels account for up to 10% of automobile fuel, for the purpose of cleaner emissions. Compared to fossil fuel oil, production rates for biofuels are low and costs are high. Large areas of ancient rainforests have been cleared in Brazil for biofuels. Also, biofuels divert land which could otherwise have been used for food crop production. Globally biofuel production grew at 13.8 %, driven mostly by the US and Brazil. Biofuel may be the only substitute for oil and gas for transportation applications such as automobiles, airplanes, and shipping (Table 2).

	Solar panels	**Wind energy**	**Biodiesel.**	**Battery power**
Automobile	Barely suited, with batteries	Unsuitable	Suitable	Suitable
Trucks and lorries	Almost impossible	Almost impossible	Suitable	Barely suitable
Locomotives	Almost impossible	Almost impossible	Suitable	Highly unsuitable
Ships	Almost impossible	Almost impossible	Suitable	Highly suitable
Airplanes	Almost impossible	Almost impossible	Suitable	Almost impossible

Table 2. Renewable sources and their applicability for transportation

4. The transportation sector

The present prosperity of the world is much dependent on the rapid transportation of people across large distances. The transportation sector involves the automobile, trucking, locomotive (train), shipping and aviation industries. The vast majority of this transportation industry runs on oil. The exceptions are mass transit, such as city trains running on electricity, and large defense ships running on nuclear power. There is an ongoing transition from oil to natural gas power and electric power for automobiles.

4.1 Automobiles

As oil becomes more scarce, gas powered, hybrid, and fully electric vehicles are expected to gain popularity. Fully electric vehicles, which have been confined to special applications, are now moving on to everyday use. It is expected that conventional automobiles will be largely replaced by hybrid and fully electric vehicles. These may be rechargeable from distant power stations (figure 3). Owing to the inherent difficulties with electric vehicles, compared to oil-powered vehicles, the number of automobiles is expected to decline over future decades.

4.2 Locomotives and trains

Trains mostly run on diesel, with the exception of electric trains in city areas for mass transit. As oil and gas continue to be depleted, electric trains will become better options. However, the capital investment is prohibitive for copper and aluminum conductors over long distances, or cross country. Under these circumstances, the economical option may be a return to coal-powered trains, built much like coal-powered power stations.

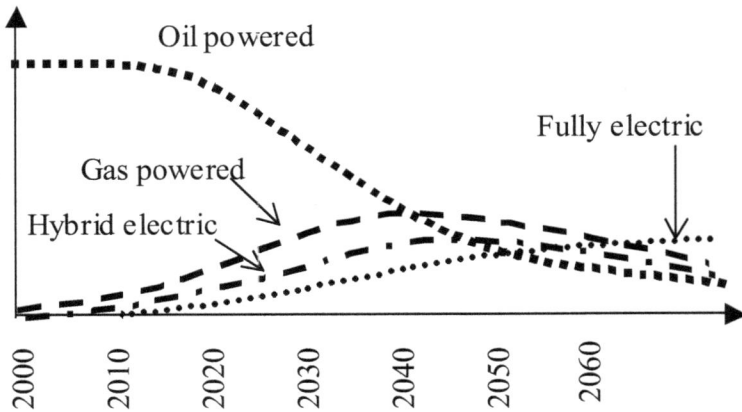

Fig. 3. A speculation into the number of automobiles powered by (a) oil (b) hybrid electric (c) gas (d) fully electric.

4.3 Shipping vessels

Shipping vessels range from recreational boats to the largest oil tankers and cruise ships. The vast majority of these are run by liquid fossil fuel (diesel etc.), the exception being nuclear powered defense ships and submarines. Compressed gas and its accompanying gas cylinders, while largely feasible for marine power, would present engineering and safety issues for vessels and ships.

4.4 Airplanes

As liquid (aviation) fuel continues its path to depletion, and rises in cost, we look to sources besides fossil fuel for the powered flight of airplanes. A gas powered airplane, would have the engineering problem of having a large number of gas cylinders to store the equivalent of the large quantities of aviation fuel formerly stored in the wings. The inherent risks of having gas cylinders all over the body and wings would be a major engineering problem, and a prohibitive risk for airplanes in the air.

A coal powered plane would require the likes of a coal-powered plant right inside an airplane. This would be prohibitive in various ways, such as in weight of the plant, electric motors, propeller-driven, and not jet-powered flight.

Solar panels on a plane may produce only about the order of a hundredth of the energy needed for flight, and would be propellor-driven, and have electric motors, clearly too heavy for flight. Also, night time flight would require batteries, which is another prohibitive addition to weight. For demonstration purposes, a solar-powered planes has circumnavigated the globe. (solarimpulse.com). This plane had the wingspan of an Airbus A340, the weight of a family car, and the power of a scooter.

A nuclear powered airplane is almost an impossibility, considering the great weight and space needed for a nuclear reactor. For powered flight, the only real alternative to liquid fossil fuel is biodiesel (Table 2).

5. Saving energy programs

The awareness for energy conservation has driven various programs and initiatives for long. A primary focus of this chapter is to increase fossil fuel conservation. Saving energy would tend to suppress the rise in prices, and cause the fuel to last longer (Figure 4).

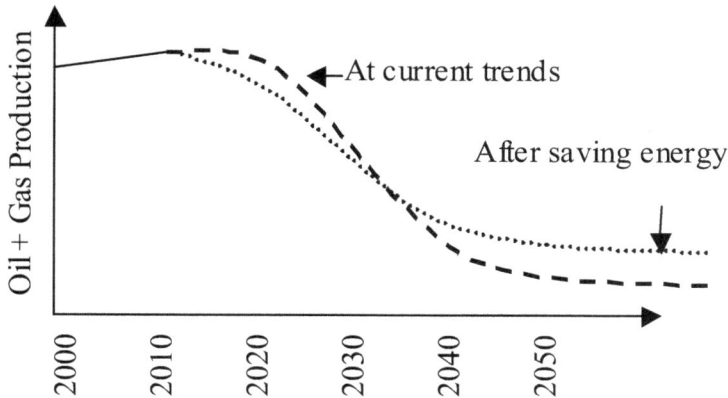

Fig. 4. Oil and Gas production at current trends, and after saving energy programs

5.1 North america

North America, besides being the driving force of the world economy, is also a proportionately large consumer of fossil fuel. A number of initiatives have been taken in North America for energy conservation (aceee.org). The Department of Energy has an official website on saving energy, energysavers.gov, which speaks of saving energy at home, at the office, in the vehicle, workplace, etc. It ranges from simple issues such as the advantages of CFL (compact fluorescent lamps), and LEDs to details such as how one would establish a hydroelectric plant in one's own property. The Environmental Protection Agency has sponsored at least two programs that promote the adoption of energy-efficient technologies through voluntary agreements with private sector firms (Howartha, 2000). Companies which advocate the saving of energy through their websites include Tampa Electric (tampaelectric.com) and Sempra Energy Utility (sdge.com/homerebates). In Canada, companies have similar partnerships with the government for reducing energy consumption. In spite of all these attempts, there is plenty of scope for further reducing energy consumption.

5.2 Europe

Saving energy is pursued at the government level in Europe and Asia. Outside of the USA, the government is more likely to be involved in saving energy programs (Martinot, 1998). Numerous energy conservation programs and studies have been conducted in Europe (Kaygusuz, 1999), of which the attempts in Hungary are among those well documented (Aunan, 1998, 2009). As in North America, there is still plenty of scope for reducing energy conservation.

5.3 Asia

China has rapidly increased its consumption to become the second largest largest fuel consumer in the world. Quite appropriately, there have been studies of saving energy in China also (YunXia, 2008). Developing and Third world countries, such as Bangladesh, present some different factors on the issue of saving energy (S. Khan, 2011, 2008).

5.4 Back-up generation and power supplies

In developing countries, the discrepancies between demand and supply may cause rolling blackouts, otherwise known as load shedding. This inability to meet load requirements induces consumers such as households and offices to install back-up gnerators and UPS (Uninterruptible Power Supplies). Back-up generators, create further problems, such as those described below.

1. Back-up generators require capital investment by consumers, likely to be much greater than the extra investment required in power stations.
2. The efficiencies of such back-up generators are much lower than that of power plants.
3. The smaller genertors normally operate at less than full load, causing the efficiencies to be even lower.
4. Back-up generators require diesel whereas power plants have the potential to operate on the more available coal.
5. Back up generators bring exhaust and pollution to the premises of the consumers, rather than having them at the distant power station.
6. Back-up generators create greater dependence on diesel or gas, which may create further fuel shortages for power station, creating even more black-outs. This may induce consumers to invest more on back-up generators, creating a vicious cycle shown below
7. Back-up UPS are at best about 50 % efficient, and therefore end up consuming twice as much electricity from the power stations.
8. Back-up UPS may contribute to a similar vicious cycle as back-up generators, inducing even greater rolling black-outs (figure 6)

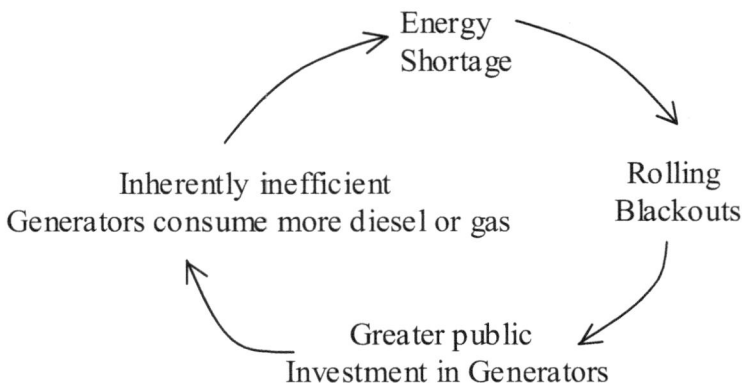

Fig. 5. The vicious cycle, where rolling blackouts induce more installation of generators, which consume more diesel or gas, otherwise usable in more efficient power stations.

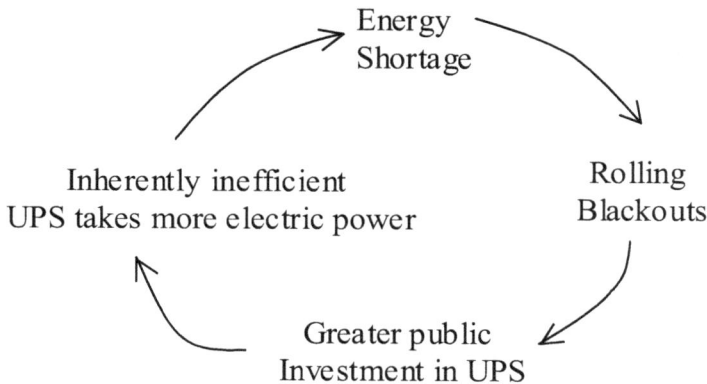

Fig. 6. The vicious cycle, where fuel shortage cause more rolling blackouts, causing installation of more UPS, which absorbs more electric power, causing more shortages.

6. Consumer behavior

In spite of all the energy-saving programs, there is great scope of reducing energy consumption by individuals. Since about 1981, surveys have indicated that a significant portion of North Americans believe the energy problem is real and serious. Since 1977, state agencies and universities in Virginia have been funded to provide energy conservation information to the general public (Geller, 1981). Large portions of the public also support relatively strong conservation policies. Two major reasons cited for conserving energy have been to save money, and to solve the energy problem (Marvin, 1981).

Energy may be saved by improving consciousness about the energy consumption of appliances and equipment. Along these lines, there have been attempts at "smart metering" which gives gives real time energy consumption information (Ehrhardt-Martinez, 2010). In Bangladesh, pre-paid metering has been implemented, which stops energy, when the pre-paid amount runs out.

According to a preliminary survey in developed and underdeveloped countries (S. Khan, 2011), offices and public buildings were more likely to have lights and air-conditioners on needlessly. Shopping complexes were found to use lights much brighter than required. The preliminary surveys indicated that air conditioning was used throughout the year, even during months or days, when plain ventilation was sufficient for health and comfort. Cooling fans were not installed or used, when in conjunction with outside ventilation, they were clearly sufficient to counter the warm weather.

The essence of saving energy programs is changing consumer behavior. Awareness must be created that saving energy will prolong availability of fossil fuel, and help keep prices low.

6.1 Telecommunication

While the land phone has had almost complete penetration in North America for the last few decades, there has been a dramatic rise in cell phones in more recent times. The rise in telecommunication has been particularly dramatic in developing countries, where penetration has increased from a small minority of the population a decade ago, to most of the population today.

Personal and business communication over the phone and internet have gained popularity over the last few years, allowing people to be in touch across cities or continents. Better phone communication and video-conferencing (Skype.com etc.) reduce fossil-fuel required for personal travel. In this sense, telecommunication has greatly helped fossil-fuel conservation.

The improvements in telecommunication and the internet are a mixed blessing to energy conservation. The ease of internet communication induces personal and casual communication over the internet for prolonged periods. This induces consumers to keep their computers when not needed otherwise. The widespread popularity of websites like facebook.com induce people to keep their computers on for long periods. While laptops and mobile devices consume less power, desktops and their CRT monitors may consume several hundred watts of power. This energy has become a little-recognized cause for increases in power consumption today. The increasing popularity of energy-efficient hand held devices may favorably improve energy conservation over the next few years.

6.2 Air coolers and acclimatization

A major portion, or most of the energy consumed by households and offices is spent on air coolers. In view of this, there have been public service messages informing consumers of the great savings in reducing air-cooler thermostat temperatures by a few degrees. However, there have been very few messages, asking people to replace air-cooling with overhead fans and plain ventilation when possible. It should be recognized that air-coolers normally operate in closed rooms, and air-coolers must work extra to bring down closed-room temperatures to the outside ambient temperatures.

Owing to their great power consumption, air-coolers are unsustainable over the next few decades. It should be noted that humans are very capable of acclimatization, and can quickly adjust to high ambient temperatures when needed. Residents of colder Europena countries can quickly acclimatize to the higher temperature of the tropics, such as in Asia and Africa.

The notion that computer-related equipment last longer in cooler temperatures is now being changed. In a related research conducted in 1994 in Japan, it was seen that computers fared well in higher ambient temperatures of 28 C, as compared to the previous 23-24 C. Mobile phone Base Transceiver Stations (BTS) in Bangladesh are being installed without any air-coolers, as compelled by energy shortages in the country.

6.3 Consumerism

The lifestyle in much of the developed world consists of rapid consumption and replacement of consumer goods. Such consumer items include automobiles, appliances, electronics, computers, garments, etc. Spending by consumers generates revenue, jobs, taxes for the government, and stimulates the economy in general. Media, such as newspapers, and TV consist of much advertisements, which promote a lifestyle with rapid replacement of consumer items. The problem arises as the production of consumer goods heavily consumes fossil fuel.

While proponents of such a lifestyle argue that such consumption is a choice made by consumers, it is clear that such practices are unsustainable beyond the next few decades. In the past, increases in fossil fuel prices have contributed to price increases in consumer goods. Depletion of fossil fuel will tend to continue the upward trend in prices of consumer goods, forcing consumers to ultimately move away from such a lifestyle. A program to save energy should attempt to discourage unsustainable lifestyles promoting consumerism.

7. Barriers to energy conservation

One of the greatest barriers to energy conservation today is that it energy consumption generates revenue, wealth, and taxes. While individuals have an inherent and inbuilt concern for their own future, and the future of their children, companies are generally driven by the bottom line and have little inherent concern for the welfare of humanity. Even though the top management may have concern for the future, when these same officers think collectively as a company, the concern for the future takes second priority over maximizing the bottom line, or profits. Commercial organizations are not very well-built to recognize the potential fuel shortages and depletion many decades down the line.

Growth in fossil fuel consumption may be viewed positively by the oil-related industry. When oil and gas shortages cause prices to rise, the higher prices contribute further to the revenue. This may again be viewed positively by commercial forces, which generally direct themselves to maximizing immediate profits.

Companies related to fossil fuel production and distribution benefit from increased extraction and consumption through increased revenue, and consequently increased profits. Increased fossil fuel production generates greater wealth, and benefits the large numbers of companies related to the fuel industry.

It should be noted that much data on remaining reserves and future expected consumption is provided by oil companies. Their data should be considered as optimistic keeping in mind the financial forces which drive their growth.

The revenue from increased fossil fuel production in turn generates more taxes for the government. The government, which are entrusted by the people, to protect the assets of the country, may not be inherently sufficiently far-sighted to prioritze fossil fuel conservation. Governments have mandates for only a few years, leading them to be more concerned about what happens during their tenure, rather than what is good in the long term. To governments, increased oil revenue may be a quick solution for improving the economy and increasing employment, consequently improving popularity among voters.

7.1 Individuals

Individuals generally have the greatest concern for saving fossil fuel, as they may be concerned about what happens in their own lifetime, the lifetime of their children, or their children's children. There is general consensus, that consumers of today should have reduced consumption so that there is more fossil fuel left in the coming decades, and for future generations. The interest of individuals to conserve energy cannot be left entirely to companies or even governments for implementation. Efforts to save energy must include participation by conscious citizens, and academia.

8. Larger automobiles

There is much potential of reducing energy consumption in automobiles, by moving to smaller and more efficient automobiles. A recent StatsCan report indicated that Canadians spend 18 % of disposable income on transportation. The fundamental issue is that automobiles do not need to be as large with low gas mileage, as they are today. The present popularity of larger cars and SUVs are unsustainable, over the next few decades. Studies have shown that the popularity of larger automobiles and SUVs is largely due their perceived and real safety in case of a collision. This popularity is in addition to other factors, such as comfort and space, and the status they convey about the owners.

During an impact, the impulse on passengers of a heavier automobile is less than the impact on a lighter car. We consider an automobile of mass M_1 travelling at velocity V_I hitting a stationary automobile of mass M_2. We assume they move away with a common velocity V_F. From the law of conservation of momentum, their final velocity will be

$$V_F = \frac{M_1 V_I}{M_1 + M_2}$$

The impulse on passengers in the first car is

$$M_1 [V_1 - \frac{M_1 V_I}{M_1 + M_2}]$$

The impulse on passengers in the second car is

$$M_2 \frac{M_1 V_I}{M_1 + M_2}$$

Analysis of the above shows the smaller impact forces on the passengers of the larger car, and consequently greater safety.

This safety during collision, and the status conveyed by larger automobiles, has at least contributed to a "size race" among consumers in pursuit of larger cars. There is much scope of reducing this "size race" among consumers, greatly favoring fossil fuel conservation.

8.1 Sports utility vehicles

The Sports Utility Vehicle (SUV) has been popular for at least 2 decades. Today, the largest SUV weighs as much as two mid-size SUVs and gets just 13 mpg on the highway. While there have great research and incentives on decreasing the size of cars, the SUV had initially exploited a legal loophole by being built on the chasis of trucks, bypassing the fuel efficiency required of cars. Consumers, eager for increased safety, had contributed to the great popularity of SUVs, owing to their increased safety arising from greater weight.

Numerous federally sponsored studies (speakout.com), show that car occupants are more likely to be killed when struck with by a SUV than by another car. Many SUV drivers have said they buy the big vehicles because they make them "feel safe."

These SUVs and large cars alongside more efficient cars today, are examples that there is great scope of improving energy conservation by implementing policies to discourage this "size race" by consumers in their pursuit of safety in collisions.

9. Urban sprawl

Another area with great scope for improvement of fuel consumption is what is popularly known as Urban Sprawl. Urban sprawl is the outwards spreading and development of a city and its suburbs (Wikipedia, 2011). It is characterized by low-density and auto-dependent development on rural land. It also has high segregation of uses (e.g. stores and residential), and various design features that encourage car dependency. Residents of sprawling neighbourhoods tend to consume more fossil fuel per person.

Urban sprawl is controversial, with supporters claiming that consumers prefer lower density neighbourhoods, with a suburban lifestyle with two or more cars. An opposing viewpoint is that urban sprawl forces residents to drive to conduct daily activities. Whatever the preferences of residents, it is clear that urban sprawl, and its car-dependent culture are very much unsustainable.

9.1 Employment sprawl

Employment sprawl is closely associated with urban sprawl and car-dependent communities. This is where jobs are located in areas of urban sprawl. This leaves employees little choice but to participate in the urban sprawl, by relocating to close to their suburban offices. Employment sprawl is prevalent both in cities with and without mass transit such as subways. Companies have set up employment centers well outside the subway system of Toronto, Canada. Employment sprawl causes the interesting phenomena, where residents living near city centers, must reverse-commute outside the city by cars. The mismatch of residences with employment areas is known as spatial mismatch.

9.2 Housing

Accompanying urban sprawl is the use of large housing that usually exceeds the physical needs of residents. Large housing costs more in heating and cooling costs. According to StatsCan, housing costs a further 30 per cent of disposable income of Canadians (Times Colonist, Dec. 18, 2011)

Compared to North America, urban and employment sprawls are less prevalent in Europe and Asia. The International Association of Public Transport policy indicates that in European and Asian counties 5-8 % of GDP is spent on transportation, compared to 13 % in North America.

The causes of urban sprawl are acknowledged to be mainly zoning laws. Sprawl generates much revenue for real estate, automobile and oil companies. A small number of policy-makers have spoken out about urban sprawl, but at this time, they are a minority (NZ Herald, Dec. 17, 2011).

Simply reducing urban sprawl by allowing employment and residences to be located close to city centers would greatly reduce fuel consumption by reducing dependence on cars, and encouraging mass transit like buses and subways. This would be more sustainable in the long run.

9.3 Recent decline in housing market in North America

In the ongoing decline in the housing market in North America, prices in distant suburbs have gone down by 66 %, whereas prices in the urban areas have gone down by about 20 %. While many explanations have been given for this crash in the housing market, it could also be evidence of the non-sustainability of urban sprawl, in the face of rising fuel costs. Faced with fuel costs at $ 3 - 4 per gallon, commuting over long distances is becoming an increasingly expensive option. With high fuel costs, housing in distant suburbs, and an automobile-dependent lifestyle has lost much of its former appeal.

9.4 Role of trees to reduce cooling costs

The presence of trees and vegetation around and on houses and buildings is another simple proven method of keeping energy costs low (Akbari, 1992, 1997, 2001; Raeissi, 1999). Trees cool buildings both by shading the walls and roof from the sun, and by cooling the ambient temperatures by their evaporation processes. A rooftop garden on a five-story commercial building gave savings of 0.6–14.5% in the annual energy consumption (Wonga, 2003). Increasing of soil thickness and its moisture content would further reduce the building energy consumption substantially

10. Increasing density of forests

Increased fossil fuel consumption causes global warming. (Figure 7), and increased greenhouse gases, especially CO_2 (Figure 8). These may be used to analyze the changes in vegetation and forestry around the world, and the changes in marine life.

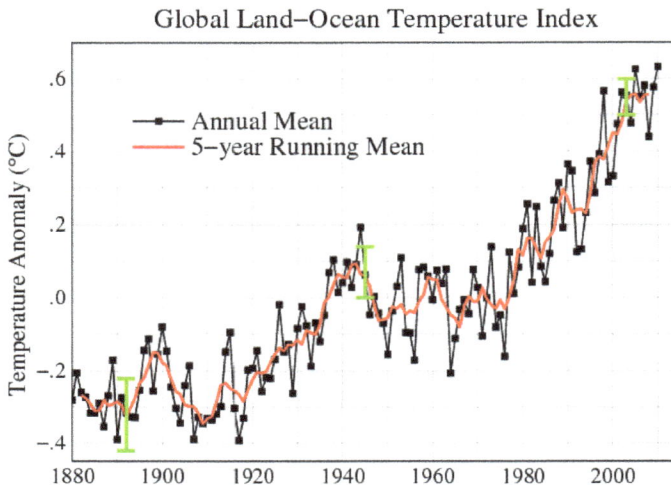

Fig. 7. Global mean land-ocean temperature index, 1880 to present, with the base period 1951-1980. The green bars show uncertainty estimates. [Courtesy: Wikipedia.com]

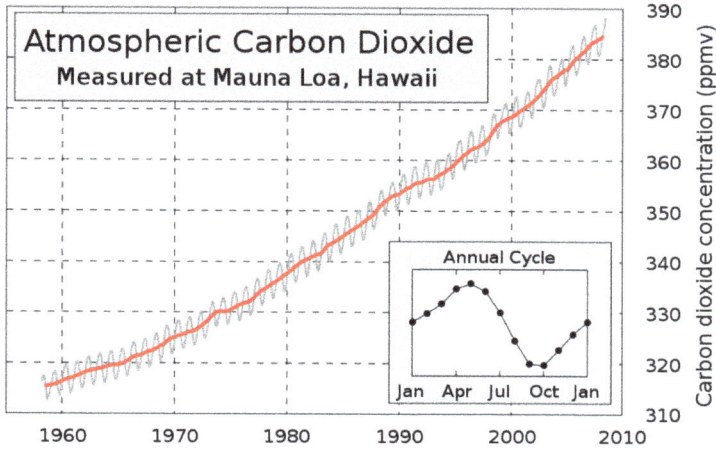

Fig. 8. Carbon dioxide concentrations as directly measured at Mauna Loa, Hawaii. (courtesy Wikipedia).

Deforestation over the decades and centuries has brought down naturally forested areas to less than 10 % of their prehistoric levels. Paradoxically, it has been observed that forest densities have significantly increased over the decades (Figure 9). Forests in many regions of the world have actuallly become of higher density, according to researchers in Rockefeller's Program for the Human Environment.

The reason cited in the above study is better management of the forests. One contradiction of the above explanation is that increased density of forests have also been observed in places where there has been no maintenance of forests.

Fig. 9. These photos from the same spot in Finland, taken in 1893 (left) and in 1997 (right) show that while the forest area is the same, the trees have become larger (Courtesy: newswire.rockefeller.edu).

10.1 Malaysia, 1975 - 2002

The author personally noticed significantly increased density of rainforests in Malaysia during the period 1975 - 2002. In the period of 1975-78 in the area of Tanah Rata, Cameron Highlands, it was noticed it was mostly possible to walk upright around in the rainforests without being significantly obstructed by the undergrowth. Paths in the forests frequented by people were relatively clear, allowing people to walk upright with little obstruction.

In 2002, in the same Tanah Rata region, it was found that the same natural rainforests had become significantly more dense. It had become much more difficult to walk around in the rainforests, because of obstruction from the undergrowth and vines. It was observed that passages in the forests had become more of tunnels allowing crouching and crawling through the undergrowth, crouching and crawling, rather than upright walking, through the undergrowth.

This expected increasing density of forests over the last few decades has also been observed by the author in the Sedona area of Arizona in 2012. This area has also been the backdrop of movie Westerns from close to 50 years ago, which form a basis of comparison for the vegetation densities we see today.

10.2 Global warming as cause of increased density of forests

One reason often cited for the increased forest densities is global warming. Plants are known to thrive with a number of factors, such as increased rainfall, humidity, higher temperatures and increased carbon dioxide. Freezing temperatures are known to kill off the plants, allowing mainly coniferous plants to thrive.

Gobal warming should cause shifts of vegetation and forestation patterns away from the equator to the poles as shown below (Figure 10). This changing patterns of vegetation and forests has generally been observed.

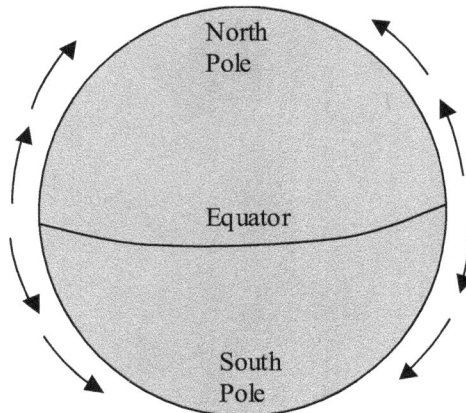

Fig. 10. Expected migration of vegetation and marine life, if global warming were the cause. Vegetation densities have increased as expected, whereas marine life has declined dramatically.

10.3 Increased carbon dioxide as cause of increase of forestation

An additional reason for increased density of forestation is proposed here to be increased carbon dioxide (CO_2) in the atmosphere. Besides increased temperatures, vegetation and forests are very dependent on CO_2 for photosynthesis, and creation of living tissue. The fossil fuel burnt artificially releases increased carbon dioxide into the atmosphere over the decades. This increasing carbon dioxide would cause a shift in the dynamic equilibrium of forest density, favoring the growth of forests. With other factors remaining unchanged, the increase in carbon dioxide levels would cause more dense vegetation and forests. It is known among aquarium enthusiasts that pumping CO_2 from a cylinder into an aquarium increases the density of aquatic plants. An increase in CO_2 into the atmosphere would be expected to cause a similar increase in forests and vegetation.

10.4 Wildlife

The global decline in wildlife over the last few decades is largely attributed to loss of habitat such as rainforests, and the encroachment of man. The decrease in wildlife has also been attributed to the pollutants and contaminants introduced into the atmosphere. **The large decline in the numbers predatory birds such as eagles and hawks was first indicated in the study some decades ago, where DDT introduced into the food chain was found to have been the cause.** Owing to a lack of other explanations, the huge decline of birds today could be attributed to their great sensitivity to toxins in the environment.

Also sensitive to environmental toxins are amphibians such as frogs. About a decade ago, mutations in frogs causing extra limbs were found to be have increased alarmingly. Pollutants and contaminants were cited as the reason for these mutations.

Animals such as bison, elephants, etc. deal with depleting grasses/vegetation and annual weather patterns by migrating over large geographic distances. The artificial fences criss-crossing our continents could be another reason for the large decline in wildlife populations.

11. Declining marine life

Having looked at the changes in global forests, we look at the present widespread decline in marine life, fisheries, coral ecosystems, etc. It is widely acknowledged that marine life has been decreasing at an alarming rate over the last few decades (Reynolds, 2005; Marinebio.com, 2011).

The most common explanation for the decline of marine life is global warming (SFGate.com, Dec. 2011). Global warming as the cause of declining marine life raises questions which cannot be easily answered. It implies that fish which used to live at the equator would shift to higher latitudes (Figure 10). Fish formerly of higher latitudes would now shift to the arctic circle. Instead, marine life, including plankton, is found to have decreased dramatically worldwide, strongly indicating a cause other than global warming.

The issue arises as to why trees and vegetation have benefited, at least marginally, whereas marine life has declined alarmingly over the decades.

11.1 Overfishing

Some decades ago, the decline in marine life was attributed to overfishing. International agreements were put into place to limit fishing in the oceans. Particularly well known are the bans on whaling, which were adhered to by most countries. In spite of these voluntary limits on fishing, marine life rather than rebounding, has continued it's steady decline. With the restrictions on fishing and whaling, plankton at the bottom of the food chain should actually have increased. The question arises as to why there has not been a corresponding increase in the plankton in the oceans.

Ocean plankton are at the bottom of the food chain, and are the ultimate source of food for most marine life. It is estimate that phytoplankton is responsible for about half of Earth's photosynthesis, a process that removes carbon dioxide from the atmosphere and converts it into organic carbon and oxygen that feeds nearly every ocean ecosystem.

Paradoxically, the plankton densities over the world have continued their steady decline. Clearly, there is some major factor besides global warming and overfishing which are causing the alarming decline in marine life. The increase in CO_2 has clearly not been sufficient to visibly benefit the CO2 dependent marine plankton.

Nine years of NASA satellite data published in the journal Nature show that the growth rate and abundance of phytoplankton around the world decreases in warm ocean years and increases in cooler ocean years. This has been used to support that global warming is the cause of the decline of the plankton. However, this explanation is insufficient, and the variation of plankton could have some other cause besides annual temperature cycles. Contrary to their explanation, global warming is clearly not the main factor leading to the decline of plankton.

11.2 Contaminants and pollutants in the sea

The evidence is clear that it is not global warming, but some other underlying reason behind the declining marine life. A possible cause, and that explanation offered here, is the pollutants and toxins being discharged into the oceans.

We now look at the effect of the garbage, sewage, and industrial wastes being dumped into the oceans in much of the world. The dumping of such waste into oceans was legal until the early 1970's when it became regulated and restricted. However, dumping still occurs illegally everywhere. The peak of sewage dumping was 18 million tons in 1980, a number that has decreased to 12 million tons in the 1990s.

Rivers, canals, and harbors are dredged to remove silt and sand buildup or to establish new waterways. About 10% of all dredged material is polluted with heavy metals such as cadmium, mercury, and chromium, and pesticides which ultimately find their way into the sea.

Today it is acknowledged that accumulation of waste in the ocean is detrimental to marine health. Ocean dumping can destroy entire habitats and ecosystems when excess sediment builds up and toxins are released. Although ocean dumping in critical habitats and at critical times is regulated, toxins are still spread by ocean currents.

One of the best explanations for the declining marine life could be this increasing contamination of the oceans, caused by disposal of these wastes into the sea. It is known that the fishes are very sensitive to contamination in the water. Those in charge of aquariums, especially marine aquariums, are aware that fishes may be harmed or killed by even quantities of toxins. As any hobbyist with a fish tank may know, dropping a small contaminant such as a cigarette butt, could easily kill a tankful of fish. The exception to this could be fish which have coexisted with humans for a long time, (such as gold fish) and those that have learnt to survive well in contaminated waters (farmed fish).

Large numbers of fishes are known to wash up on shores, often for no apparent reason; a phenomena for which contamination is a likely explanation. A satisfactory explanation has not been found for why whales and other fish beaching, and ultimately killing themselves. This act of self-destruction is clearly highly destructive to the continuity of their genes, from an evolutionary point of view. One possible explanation for this counter-evolutionary behaviour of fishes beaching themselves could be to get away from chemically contaminated waters. The chemicals in the water may also confuse the fish into getting away from the water into the beach.

12. Rising ocean levels

A major concern associated with global warming, the melting of polar ice, and rising ocean levels, is flooding of low lying areas of the world. It is feared that within decades, ocean levels may rise over inches or feet, inundating low lying areas, especially river deltas such as Bangladesh.

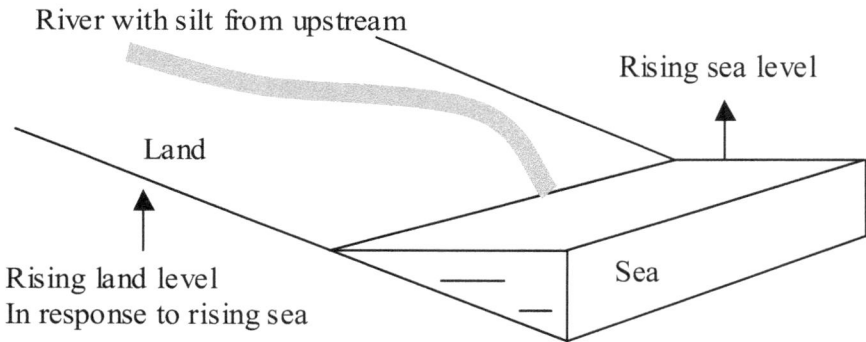

Fig. 11. Rising sea levels causing a shift in the equilibrium of the land river and sea in a river delta. The land level rises as sea levels rise.

It has been seen that river deltas (such as the Ganges delta of Bangladesh) are dynamic systems, where the land levels are the equilibrium point of the opposing forces of silt deposition and erosion (S. Khan, 1991) . This equilibrium concept of land level is also applicable to other deltas, such as for the Mississippi (in Louisiana), the Nile (in Egypt). It has been seen that for the last few decades, coastal land is being added at the rate of several feet every year. Considering that global warming rates have been mostly constant over decades, and is expected to be constant for the next few decades, it is likely that the trend of coastal land addition will continue at its present rates. It is unlikely that the coastal land addition will reverse over the next few decades causing coastal land loss or inundation. A rise of sea level will shift the land-river-sea equilibrium, resulting in faster deposition of silt in and around the rivers. This will cause a rise of the land elevations of the river delta, that will at least match the rise of sea levels. This implies that the coastal land addition of river deltas will continue at their present rates over the next few years. Modern concerns of coastal land inundation is therefore not applicable to river deltas, as a rise of sea will cause land levels to continue to rise, and coastal land will continue to be added at their present rates.

13. Conclusion

Lack of fossil fuel conservation affects us both through increasing energy prices, and a faster depletion of the fossil fuel reserves meant for future generations. Although popular perception expects a sudden depletion of oil, reduced oil consumption at high prices is likely to continue for many decades, or even a century. Over the next few decades, coal is expected to replace gas and oil in producing electricity. The transportation industry, especially automobiles, which are heavily dependent on liquid fuel today are unsustainable, and are expected to decline in numbers. Renewable energy and biofuels can not compare with the ease and convenience of fossil fuel, which is a great incentive to fossil fuel consumption. Fossil fuel oil is the only feasible source of raw materials for the chemical industry. The status conveyed by larger automobiles, and their safety in collisions with smaller cars, has contributed to a "size race" for bigger cars by consumers. Modern consumer lifestyles with air-coolers, large automobiles, and generous consumption of consumer equipment needlessly consume fossil fuel, and are unsustainable in the long run. The increased carbon dioxide in the atmosphere from fossil fuel, may actually have contributed to more dense forestry and vegetation in the world. On the other hand, pollutants and toxins (not global warming or CO_2) are likely to be the cause for the alarming decline in marine life today. The alarm regarding global warming causing sea water encroaching into river deltas is mostly unfounded, as there will be a shift in the dynamic equilibrium of the river, land, and sea. This will cause faster silt deposition, causing a rise in land levels, and continued addition of coastal land.

14. References

American Council for an Energy-Efficient Economy, 529 14th Street, Suite 600, Washington, D.C. 20045, aceee.org

B. Howartha, Brent M. Haddadb, Bruce Paton The economics of energy efficiency: insights from voluntary participation programs, Energy Policy, Volume 28, Issues 6-7, June 2000, pp. 477-486

BP Energy Outlook 2030, London 2011, www.bp.com, Accessed October 18, 2011.

BP Statistical Review of World Energy, June 2011, www.bp.com, Accessed October 18, 2011.

Corinna Fischer, Feedback on household electricity consumption: a tool for saving energy? Energy Efficiency, Volume 1, Number 1, pp. 79-104.

E. Marvin, Consumers' Attitudes Toward Energy Conservation, Journal of Social Issues, Volume 37, Issue 2, pages 108–131, Spring 1981

E. Scott Geller, Evaluating Energy Conservation Programs: Is Verbal Report Enough? Journal of Consumer Research , Vol. 8, No. 3, Dec., 1981.

Eric Martinot , Nils Borg, Energy-efficient lighting programs: Experience and lessons from eight countries, Energy Policy, Volume 26, Issue 14, December 1998, Pages 1071-1081

H Akbari , M Pomerantz, H Taha (2001). Cool surfaces and shade trees to reduce energy use and improve air quality in urban areas, Solar Energy, Volume 70, Issue 3, pp. 295-310

Hashem Akbari, Dan M. Kurna, Sarah E. Bretza, James W. Hanforda (1997), Peak power and cooling energy savings of shade trees, Energy and Buildings, Volume 25, Issue 2, 1997, Pages 139-148

Hashem Akbari, Haider Taha, The impact of trees and white surfaces on residential heating and cooling energy use in four Canadian cities, Energy, Volume 17, Issue 2, February 1992, Pages 141-149

John D Reynolds, Nicholas K Dulvy, Nicholas B Goodwin, Jeffrey A Hutchings, "Biology of extinction risk in marine fishes," Proc. Royal Soc. Biological Sciences, 22 November 2005 vol. 272 no. 1579, pp. 2337-2344

John M. Pandolfi1, Roger H. Bradbury, Enric Sala, et. al. "Global Trajectories of the Long-Term Decline of Coral Reef Ecosystems, Science 15 August 2003, Vol. 301 no. 5635 pp. 955-958.

K. Aunan, , G. Pátzay, H. Asbjørn Aaheima, H. Martin Seipa, Health and environmental benefits from air pollution reductions in Hungary, Science of The Total Environment, Volume 212, Issues 2-3, 8 April 1998, Pages 245-268

K. Aunan, G. Pátzay, et. al, Health and Environmental Benefits from the Implementation of an Energy Saving Program in Hungary, 26-Feb-2009, Publisher: Center for International Climate and Environmental Research - Oslo (CICERO)

Kamil Kaygusuz, Energy Situation, Future Developments, Energy Saving, and Energy Efficiency in Turkey, Energy Sources, Volume 21, Issue 5, 1999 pages 405-416

Karen Ehrhardt-Martinez, John A. "Skip" Laitner, Kat A. Donnelly, Chapter 10 - Beyond the Meter: Enabling Better Home Energy Management, Energy, Sustainability and the Environment, 2011, Pages 273-303

Karen Ehrhardt-Martinez, Kat A. Donnelly, John A. "Skip" Laitner Advanced Metering Initiatives and Residential Feedback Programs: A Meta-Review for Household Electricity-Saving Opportunities, June 2010, Report Number E105

Linda Steg, Charles Vlek, Talib Rothengatter A review of intervention studies aimed at household energy conservationWokje Abrahamse, Journal of Environmental Psychology Volume 25, Issue 3, September 2005, Pages 273-291

Marinebio.org,, Threatened and Endangered Species, Accessed Nov. 10, 2011.

Michael J. Behrenfeld1, Climate-driven trends in contemporary ocean productivity, Nature 444, Letters, pp. 752-755, 7 December 2006.

Minoru Yarnamoto and Takashi Abe, The New Energy-Saving Way Achieved by Changing Computer Culture, IEEE Transactions on Power Systems, Vol. 9, No. 3, August 1994 Are SUVs a menace to other cars on the road? http://speakout.com/activism/issue_briefs/1257b-1.html, Accessed Dec. 20, 2011. http://www.timescolonist.com/business/Sprawl+plunges+Canadians+into+deb t/5878873/story.html

N.H Wonga, D.K.W Cheonga, H Yana, J Soha, C.L Ongb, A Siab, The effects of rooftop garden on energy consumption of a commercial building in Singapore, Energy and Buildings, Volume 35, Issue 4, May 2003, pp. 353-364.

NZ Herald, Urban sprawl not the answer for Auckland, says councillor, Dec 17, 2011, www.nzherald.co.nz/nz/news/article.cfm?c_id=1&objectid=10773736

Ocean warming's effect on phytoplankton / NASA satellite data show how global climate change hurts marine food chain, SFGate.com, Dec. 7, 2011.

Richard B. Howarth, Brent M. Haddad, Bruce Paton, The economics of energy efficiency: insights from voluntary participation programs, Energy Policy, Volume 28, Issues 6-7, June 2000, Pages 477-486

S. Raeissi, M. Taheri, Energy saving by proper tree plantation, Building and Environment, Volume 34, Issue 5, 1 September 1999, pp. 565-570

Shahriar Khan, "Planting Trees for Mitigating Flooding, Erosion, and Cyclonic Damage," Proceedings of Natural Disasters in Bangladesh, 1991, Carleton University, Ottawa.

Shahriar Khan, An energy saving program for bangladesh, for reducing load shedding, and for continuity of power for IT sector, Proceedings of the 11th International Conference on Computers and Information Technology, 2008, Khulna, Bangladesh.

Shahriar Khan, An Energy Saving Program for Bangladesh, Journal of Petroleum Technology and Alternative Fuels, Vol. 2, No. 6, pp. 86 - 94, June 2011.

Shahriar Shafiee, Erkan Topal, (2009). "When will fossil fuel reserves be diminished?" Energy Policy, Volume 37, Issue 1, Pages 181-189, January 2009.

Stephen J DeCanio, The efficiency paradox: bureaucratic and organizational barriers to profitable energy-saving investments, Energy Policy, Volume 26, Issue 5, April 1998, Pages 441-454

The Great Reversal, an increase in forest density worldwide, is under way, Posted June 6, 2011, The Rockefeller University, newswire.rockefeller.edu.

Wikipedia.com, Urban Sprawl, accessed Dec. 20, 2011. Sprawl plunges Canadians into Debt, The Times Colonist, Dec. 18, 2011

Wokje Abrahamse, Linda Steg, Charles Vlek, Talib Rothengatter , "A review of intervention studies aimed at household energy conservation," Journal of Environmental Psychology, Volume 25, Issue 3, September 2005, Pages 273-291.

www.marinebio.org, Ocean Pollution, Accessed Oct. 9, 2011.

Yates, Suzanne M.; Aronson, Elliot, A social psychological perspective on energy conservation in residential buildings. American Psychologist, Vol 38(4), Apr 1983, 435-444

YunXia Wangb, Tao Zhao, Analysis of interactions among the barriers to energy saving in China, GuoHong Wanga, Energy Policy, Volume 36, Issue 6, June 2008, pp. 1879-1889

9

Global Trends of Fossil Fuel Reserves and Climate Change in the 21st Century

Bharat Raj Singh[1] and Onkar Singh[2]
[1]SMS Institute of Technology, Lucknow
[2]Harcourt Butler Technological Institute, Kanpur
India

1. Introduction

Today's energy markets are dominated by a substantial increase in energy demand due to the strong economic growth in the developing countries especially in China and India. At the same time it is also observed that the capacity to deliver fossil energy may be limited due to limited production capacity and lack of infrastructure such as pipeline, refining and terminal capacities (CERA, A global sense of energy insecurity). A number of nations are concerned with their security of supply with respect to delivery of power, oil and gas, and we see a development toward more nationalization of energy production and distribution in several nations. Huge investments in production capacity and infrastructure are needed in many countries to secure necessary access to energy (*IEA, World Energy Outlook, 2004, p 32*).

Emissions of carbon dioxide due to our use of fossil energy will change the climate and the temperature is estimated to increase by 2 to 6º Celsius within year 2100, which is a tremendous increase from our current average temperature of 1.7º Celsius (IPCC). This will probably cause huge changes to our society, both positive and negative, but the total impact on our society is currently very uncertain.

The global population is expected to increase by 30% the next 25 years, where 80-90% of the increase is expected to be in developing countries (*IEA, World Energy Outlook 2004, p 43-46*). To be able to establish a sustainable global development, with growth in population and living standard, it will probably be necessary to develop renewable and cleaner energy sources, improved energy efficiency and mechanisms that make it attractive to utilize new technology.

The 30 year update claims that the global system is currently in an un-sustainable situation, and that there are limits to growth on our planet – on resources, food, environment, and also in the population the earth can supply over time. If we do not act soon to establish a sustainable world, we will probably face enormous challenges in providing goods, energy and food to the population and we will probably experience recession, hunger, conflicts, reduced living conditions and maybe a significant reduction in population.

This study describes some of the background for the scenario analysis such as: potential impacts of changes in the energy resource situation, both fossil and renewable, impact on global climate, important geo-political issues and major global trends which can have an impact on the energy as well as climate.

2. Global trends of fossil fuel reserves

United States, Russia and China are leading producers and consumers of World Energy. These three countries together produced 31% and consumed 41% world total energy as per International Energy Agency (IEA) 1999. United Sates consumed three times the energy than China, the second largest consumer of World. Fossil fuels will remain the most important energy source, at least until 2030, and the use of oil, gas and coal is expected to grow in volume (IEA, 2009) over this period. Coal is not scarce, but is problematic for pollution and climate change reasons. The production costs of oil continue to rise with the expansion of the share of deepwater exploration in the supply (IEA, 2008). Although coal and gas are abundantly available, environmental and logistical reasons prevent a substantial shift away from oil to these energy sources.

Fossil fuel reserves are concentrated in a small number of countries. 80 % of the coal reserves are located in just six countries; the European Union (EU) has 4 % of the global stock. The EU share of the world's gas reserves decreased from 4.6 % in 1980 to 1.3 % in 2009. These reserves are expected to be exhausted before 2030. More than half of the global stock is found in only three countries: Iran, Qatar and Russia (24 % in 2009), which is a major gas supplier for the EU. Ten countries (of which eight are OPEC members) have 80 % of the world's oil reserves. Some of these countries may exercise their power to restrict supply or influence the price (NIC, 2008). EU dependence on imported fossil fuels is slowly rising and presently amounts to about 55 %. Some EU countries (for instance Estonia, Italy, France and Sweden) have sizeable oil shale stocks. Reduced foreign supply may encourage them to exploit these sources. The Arctic region is expected to contain a substantial amount of oil, probably up to 90 billion barrels (EU: about 12 billion barrels).

2.1 Will fossil fuel reserves be effectively depleted by 2050?

Crude oil, coal and gas are the main resources for world energy supply. The size of fossil fuel reserves and the dilemma that when non-renewable energy will be diminished, is a fundamental and doubtful question that needs to be answered. Here a new formula for calculating when fossil fuel reserves are likely to be depleted is presented along with an econometrics model to demonstrate the relationship between fossil fuel reserves and some main variables (*Shahriar Shafiee et.al. 2009*). The new formula is modified from the Klass model and thus assumes a continuous compound rate and computes fossil fuel reserve depletion times for oil, coal and gas of approximately 35, 107 and 37 years, respectively. This means that coal reserves are available up to 2112, and will be the only fossil fuel remaining after 2042.

In the Econometrics model, the main exogenous variables affecting oil, coal and gas reserve trends are their consumption and respective prices between 1980 and 2006. The models for oil and gas reserves unexpectedly show a positive and significant relationship with consumption, while presenting a negative and significant relationship with price. The econometrics model for coal reserves, however, expectedly illustrates a negative and significant relationship with consumption and a positive and significant relationship with price. Consequently, huge reserves of coal and low-level coal prices in comparison to oil and gas make coal one of the main energy substitutions for oil and gas in the future, under the assumption of coal as a clean energy source.

Fossil fuels play a crucial role in the world energy market *(Goldemberg, 2006)*. The world's energy market worth around 1.5 trillion dollars is still dominated by fossil fuels. The World Energy Outlook (WEO) 2007 claims that energy generated from fossil fuels will remain the major source and is still expected to meet about 84% of energy demand in 2030. There is worldwide research into other reliable energy resources to replace fossil fuel, as they diminish; this is mainly being driven due to the uncertainty surrounding the future supply of fossil fuels. It is expected, however, that the global energy market will continue to depend on fossil fuels for at least the next few decades.

World oil resources are judged to be sufficient to meet the projected growth in demand until 2030, with output becoming more concentrated in Organization of Petroleum Exporting Countries on the assumption that the necessary investment is forthcoming IEA, 2007. According to WEO 2007 oil and gas supplies are estimated to escalate from 36 million barrels per day in 2006 to 46 million barrels per day in 2015, reaching 61 million barrels per day by 2030. In addition, oil and gas reserves are forecast at about 1300 billion barrels and 6100 trillion cubic feet in 2006, respectively (BP, 2007). The World Energy Council (WEC) in 2007 estimated recoverable coal reserves of around 850 billion tonne in 2006.

Table 1 shows the distribution of remaining reserves of fossil fuels. Firstly, as seen in Table 1, coal constitutes approximately 65% of the fossil fuel reserves in the world, with the remaining 35% being oil and gas. Secondly, while the size and location of reserves of oil and gas are limited in the Middle East, coal remains abundant and broadly distributed around the world. Economically recoverable reserves of coal are available in more than 70 countries worldwide and in each major world region. In other words, coal reserves are not limited to mainly one location, such as oil and gas in the Middle East. These two geological reasons support the fact that coal reserves have potential to be the dominant fossil fuel in the future.

Region	Fossil fuel reserve (giga tonnes of oil equivalent)				Fossil fuel reserve (%)			
	Oil	Coal	Gas	Sum	Oil	Coal	Gas	Sum
North America	8	170	7	185	0.86	18.20	0.75	19.81
South America	15	13	6	34	1.61	1.39	0.64	3.64
Europe	2	40	5	47	0.21	4.28	0.54	5.03
Africa	16	34	13	63	1.71	3.64	1.39	6.75
Russia	18	152	52	222	1.93	16.27	5.57	23.77
Middle East	101	0	66	167	10.81	0.00	7.07	17.88
India	1	62	1	64	0.11	6.64	0.11	6.85
China	2	76	2	80	0.21	8.14	0.21	8.57
Australia and East Asia	2	60	10	72	0.21	6.42	1.07	7.71
Total	165	607	162	934	17.67	64.99	17.34	100.00

[*Source*: WCI (2007) and BP (2006)]

Table 1. Location of the world's main fossil fuel reserves in 2006

Fossil fuel reserve trends tend to mainly depend on two important parameters: consumption and price. The Energy Information Administration (EIA) has projected that energy consumption will increase at an average rate of 1.1% per annum, from 500 quadrillion Btu in 2006 to 701.6 quadrillion Btu in 2030. Currently, the growth in world energy consumption is approximately 2% per annum. "In terms of global consumption, crude oil remains the most important primary fuel accounting for 36.4% of the world's primary energy consumption (without biomass)". The International Energy Agency (IEA) claims oil demand as the single largest consumable fossil fuel in the global energy market will fall from 35% to 32% by 2030. Coal is the second largest consumable fossil fuel relative to the three main fossil fuels; in part largely due to consumption over the past couple of years. According to WEO 2007, "coal is seen to have the biggest increase in demand in absolute terms, jumping by 73% between 2005 and 2030". "Coal accounted for about 28% of global primary energy consumption in 2005; surpassed only by crude oil" (BGR, 2007). Reserves of gas in comparison to oil and coal will moderately increase for the next two decades, from 21% to 22%. Although other energy resources are expanding in the world, the rate of fossil fuel consumption for energy will also continue to increase through to 2030.

The next important issue after global consumption of fossil fuels is fossil fuel price movement. Proven fossil fuel reserves will fluctuate according to economic conditions, especially fossil fuel prices. In other words, proven reserves will shrink when prices are too low for fossil fuels to be recovered economically and expand when prices deem fossil fuels economically recoverable. In addition, the trend of fossil fuel prices significantly affects fossil fuel consumption. On the other hand, fossil fuel price fluctuations affect other variables such as international inflation, global GDP growth, etc. Consequently, the size of fossil fuel reserves depends on their prices.

The oil price is currently very high at around $140 per barrel in nominal terms. This is much higher than after several other oil price crises, such as the Iran/Iraq war, Gulf war and 9/11 as per WTRG, 2008. According to OPEC (2007), OPEC benchmark crude price is assumed to remain in the $50 to $60 per barrel range in nominal terms for much of the projected period and rising further in the longer term with inflation. Therefore, the oil price at the moment is much higher than the OPEC prediction. Moreover, WEC (2007) forecast the oil price based on the assumption that the average crude oil price will fall back from recent highs of over $75 per barrel to around $60 (in year 2006 dollars) by 2015 and then recover slowly, reaching $62 (or $108 in nominal terms) by 2030. Coal prices have had less fluctuation in comparison to oil in the last 50 years. The coal market depicts relatively constant coal prices in historical data. WEC (2007) assumes that this trend will remain flat until the middle of the next decade, then increase very slowly, reaching just over $60 per tonne by 2030. Gas prices have generally followed the increase in oil prices since 2003, typically with a 1 year lag. Annual Energy Outlook 2007 predicted that the average transmission and distribution margin for delivered gas is projected to change from $2.38 per thousand cubic feet in 2006 to between $2.07 and $2.44 per thousand cubic feet in 2030 (2005 dollars). As a result, forecasting fossil fuel prices are uncertain and unpredictable.

2.2 Fossil fuel reserve versus consumption

The trends of fossil fuel reserves versus consumption are discussed. As can be seen from Fig.1, the trend of oil and gas reserves with their consumption increased. This means that reserve and consumption for oil and gas over the last 26 years have an unusual positive correlation.

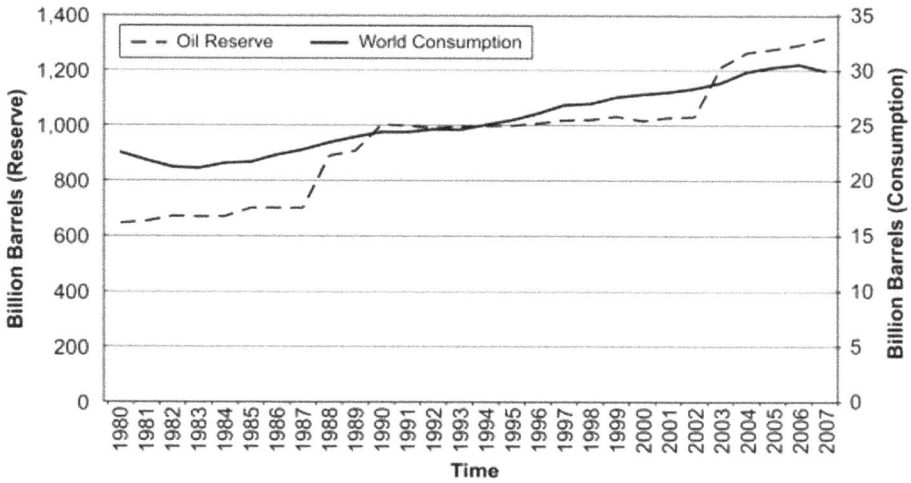

Fig. 1. Trends of world crude oil proven reserves and oil consumption from 1980 to 2007. [*Source*: EIA and BP]

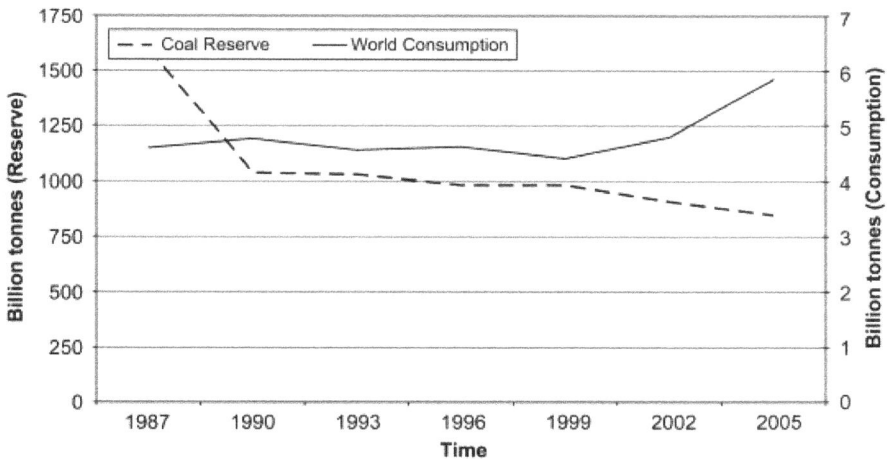

Fig. 2. Trends of world coal proven reserves and coal consumption from 1987 to 2005. [*Source*: EIA and BP]

Fig. 2 shows reserve versus consumption of coal. This graph shows a negative correlation between coal reserve and consumption. In spite of the fact that the data for coal were less available and more volatile in comparison to oil and gas, the relation between coal reserve and coal consumption is still negative and significant. According to *Shihab-Eldin (2004)*, the increase in fossil fuel resources is due to the availability of improved data, as well as technological improvements. Consequently, the reserves of oil and gas have not shown any decreasing trend during the last couple of decades and predictions that they were about to run out are not substantiated *for the last 26 years (1980 to 2007) as seen from Fig. 1 and Fig. 3.*

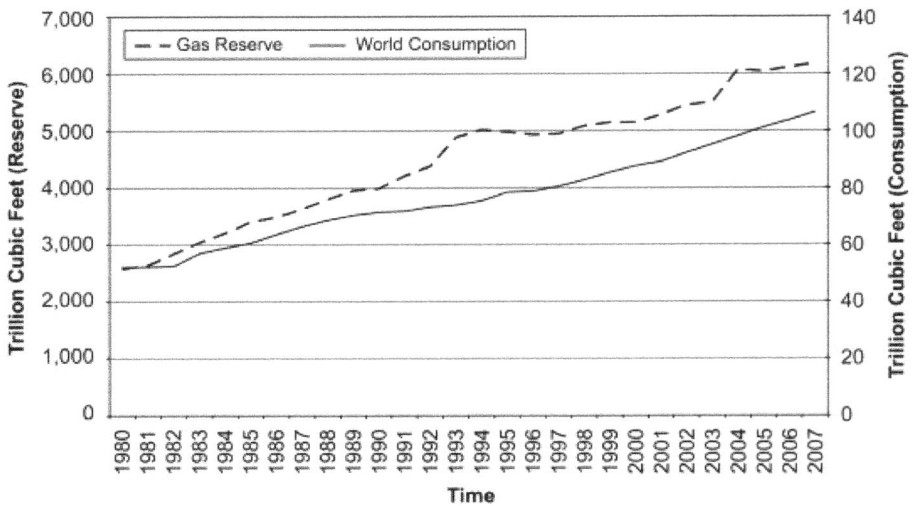

Fig. 3. Trends of world natural gas proven reserves and gas consumption from 1980 to 2007. [*Source*: EIA and BP]

2.3 Fossil fuel reserves versus depletion time

Most researchers estimate reserve depletion time by assuming constant production rates. For example, WEO 2006 estimated a ratio for oil of between 39 and 43 years, 164 year for coal and 64 years for gas. *Lior (2008)* assumed constant fossil fuel production rates and then estimated the ratio of production to reserves to be approximately 40, 60 and 150 for oil, gas and coal, respectively. As can be seen, none of the research modified the rate of production or consumption of fossil fuel to calculate the ratio of consumption to reserves. Consequently, the new model added this assumption and adjusted new formula to calculate fossil fuel reserve depletion time.

Model	Ratio of consumption to reserves			Klass model			New model		
	Oil	Coal	Gas	Oil	Coal	Gas	Oil	Coal	Gas
Year	40	200	70	34	106	36	35	107	37

[*Source*: EIA and BP, and computed]

Table 2. Fossil fuel reserves depletion times

Table 2 illustrates the time that fossil fuels will be depleted using the Klass model and new model. As can be seen in this table, the Klass model for oil, coal and gas depletion times is calculated to be about 34, 106 and 36 years, respectively, compared to 35, 107 and 37 in the new model. Ultimately, the reserve of coal using either approach still has a longer availability than oil and gas. This means that the coal reserves will be available until at least 2112 at this rate, and it will be the single fossil fuel in the world as indicated in Table 2. Fossil fuel reserves depletion times after 2042.

The second method tries to calculate the time that fossil fuels will be depleted by computing ratio of consumption to reserves. Thus, the average ratios of world consumption to reserves for oil, coal and gas can be computed from Fig. 1, Fig. 2 and Fig. 3.The graphs Fig.4, shows the trend of ratio of world consumption to reserves for oil, coal and gas from 1980 to 2006.

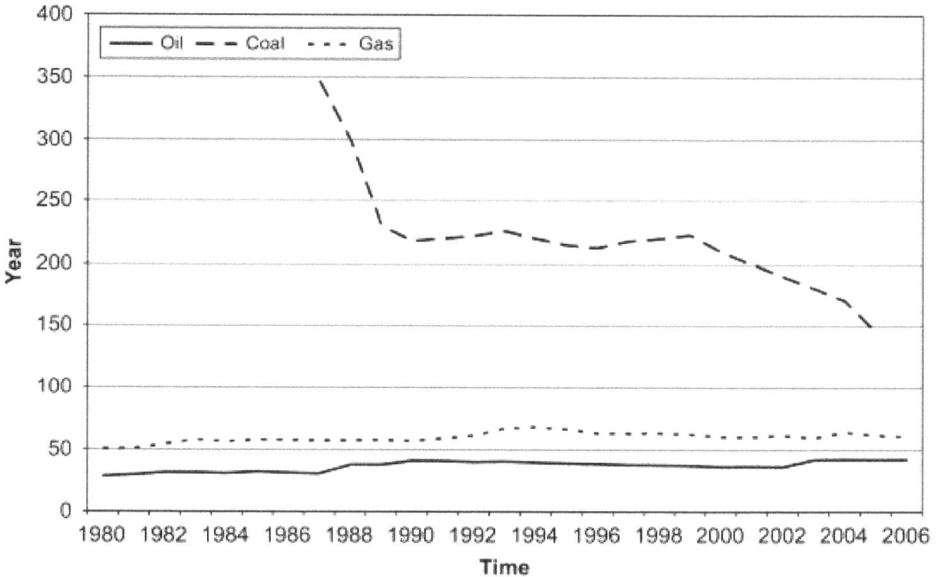

Fig. 4. The ratios of world consumption to reserves for oil, coal and gas from 1980 to 2006.

As can be seen in this figure these ratios for oil and gas were constant, around 40 and 60 years, respectively. This means that during the last 26 years, the reserves of oil and gas have not shown any decreasing trend during the last couple of decades and predictions that they were about to run out are not substantiated for the last 26 years (1980 to 2007) as seen from Fig. 1 and Fig. 3. This means that if the world continues to consume oil, coal and gas at 2006 rates, their reserves will last a further 40, 200 and 70 years, respectively.

3. Nature of global warming and climate change

Global warming and climate change refer to an increase in average global temperatures. Natural events and human activities are believed to be contributing to an increase in average global temperatures. This is caused primarily by increases in "greenhouse" gases such as Carbon Dioxide (CO_2).A warming planet thus leads to a change in climate which can affect weather in various ways, as shown below.

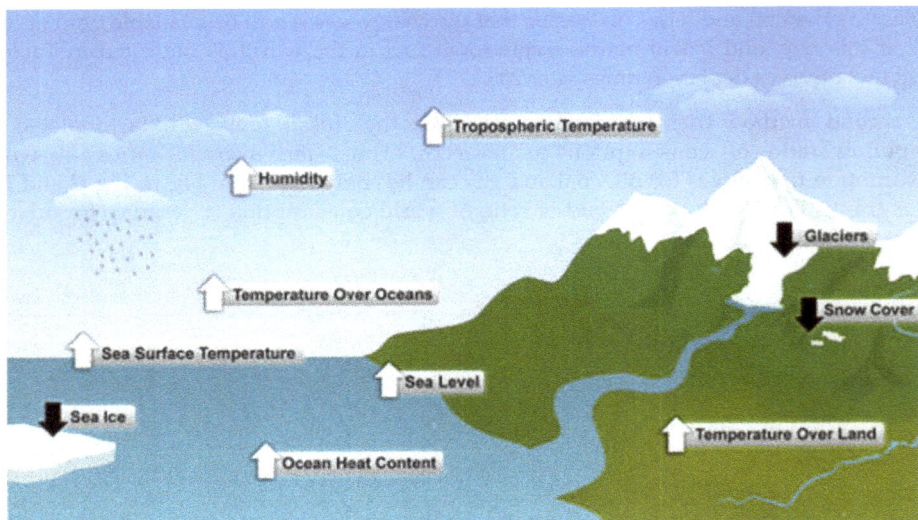

Fig. 5. Ten indicators for a warming world, Past Decade Warmest on Record According to Scientists in 48 Countries, NOAA, July 28, 2010

For decades, scientists and environmentalists have warned that the way we are using Earth's resources is not sustainable. Alternative technologies have been called for repeatedly, seemingly upon deaf ears (or, cynically, upon those who don't want to make substantial changes as it challenge their bottom line and takes away from their current profits).

Global warming in today's scenario is threat to the survival of mankind. In 1956, an US based Chief consultant and oil geologist *Marion King Hubert, (1956)* predicted that if oil is consumed with high rate, US oil production may peak in 1970 and thereafter it will decline. He also described that other countries may attain peak oil day within 20-30 years and many more may suffer with oil crises within 40 years, when oil wells are going to dry. He illustrated the projection with a bell shaped *Hubert Curve* based on the availability and its consumptions of the fossil fuel. Large fields are discovered first, small ones later. After exploration and initial growth in output, production plateaus and eventually declines to zero.

In India, vehicular pollution is estimated to have increased eight times over the last two decades. This source alone is estimated to contribute about 70 per cent to the total air pollution. With 243.3 million tons of carbon released from the consumption and combustion of fossil fuels in 1999, India is ranked fifth in the world behind the U.S., China, Russia and Japan. India's contribution to world carbon emissions is expected to increase in the coming years due to the rapid pace of urbanization, shift from non-commercial to commercial fuels, increased vehicular usage and continued use of older and more inefficient coal-fired and fuel power-plants (*Singh, BR, et al., 2010*).

Thus, peak oil year may be the turning point for mankind which in turn led to the end of 100 year of easy growth and may end up a better world, if self-sufficiently and sustainability of energy is not maintained on priority. Although the worldwide efforts are being made to explore non-conventional energy resources such as: solar energy, wind energy, bio-mass and bio-gas, hydrogen, bio-diesel which may help for the sustainable fossil fuel reserves and reduce the tail pipe emission and other pollutants like: CO_2, NO_X etc., but special emphasis should also be laid upon to the storage of energy such as compressed air stored from solar, wind and or other resources like: climatic / disaster energy to be tapped down to maintain energy sustainability of 21st century which may also lead to environmentally and ecologically better future.

3.1 Effect of global warming

The various effects of climate change pose risks that increase with global warming (i.e., increases in the Earth's global mean temperature). The IPCC (*2001d and 2007d*) has organized many of these risks into five "reasons for concern:

- Threats to endangered species and unique systems,
- Damages from extreme climate events,
- Effects that fall most heavily on developing countries and
- The poor within countries, global aggregate impacts (i.e., various measurements of total social, economic and ecological impacts), and large-scale high-impact events.

The effects, or impacts, of climate change may be physical, ecological, social or economic. Evidence of observed climate change includes the instrumental temperature record, rising sea levels, and decreased snow cover in the Northern Hemisphere. According to the Intergovernmental Panel on Climate Change (*IPCC, 2007a:10*), "[most] of the observed increase in global average temperatures since the mid-20th century is *very likely* due to the observed increase in [human greenhouse gas] concentrations". It is predicted that future climate changes will include further global warming (i.e., an upward trend in global mean temperature), sea level rise, and a probable increase in the frequency of some extreme weather events. United Nations Framework Convention on Climate Change has agreed to implement policies designed to reduce their emissions of greenhouse gases.

3.2 Effect of climate change

The phrase climate change is used to describe a change in the climate, measured in terms of its statistical properties, e.g., the global mean surface temperature. In this context, climate is taken to mean the average weather. Climate can change over period of time ranging from months to thousands or millions of years. The classical time period is 30 years, as defined by the World Meteorological Organization. The climate change may be due to natural causes, e.g., changes in the sun's output, or due to human activities, e.g., changing the composition of the atmosphere. Any human-induced changes in climate will occur against the background of natural climatic variations.

The most general definition of *climate change* is a change in the statistical properties of the climate system when considered over long periods of time, regardless of cause, *whereas* Global warming" refers to the change in the Earth's global average surface temperature. Measurements show a global temperature increase of 1.4 °F (0.78 °C) between the years 1900 and 2005. Global warming is closely associated with a broad spectrum of other climate changes, such as:

- Increases in the frequency of intense rainfall,
- Decreases in snow cover and sea ice,
- More frequent and intense heat waves,
- Rising sea levels, and
- Widespread ocean acidification.

3.2.1 Impacts of climate change

According to different levels of future global warming, impacts of climate has been used in the IPCC's Assessment Reports on climate change (*Schneider DH, et al., 2007*). The instrumental temperature record shows global warming of around 0.6 °C over the entire 20th century (*IPCC 2007d.1*). The future level of global warming is uncertain, but a wide range of estimates (projections) have been made (*Fisher, BS et al., 2007*). The IPCC's "SRES" scenarios have been frequently used to make projections of future climate change (*Karl, 2009*). Climate models using the six SRES "marker" scenarios suggest future warming of 1.1 to 6.4 °C by the end of the 21st century (above average global temperatures over the 1980 to 1999 time period) (*IPCC 2007d.3*). The projected rate of warming under these scenarios would very likely be without precedent during at least the last 10,000 years (*IPCC 2001-SPM*). The most recent warm period comparable to these projections was the mid-Pliocene, around 3 million years ago (*Stern N., 2008*). At that time, models suggest that mean global temperatures were about 2–3 °C warmer than pre-industrial temperatures (*Jansen E., et al., 2007*).

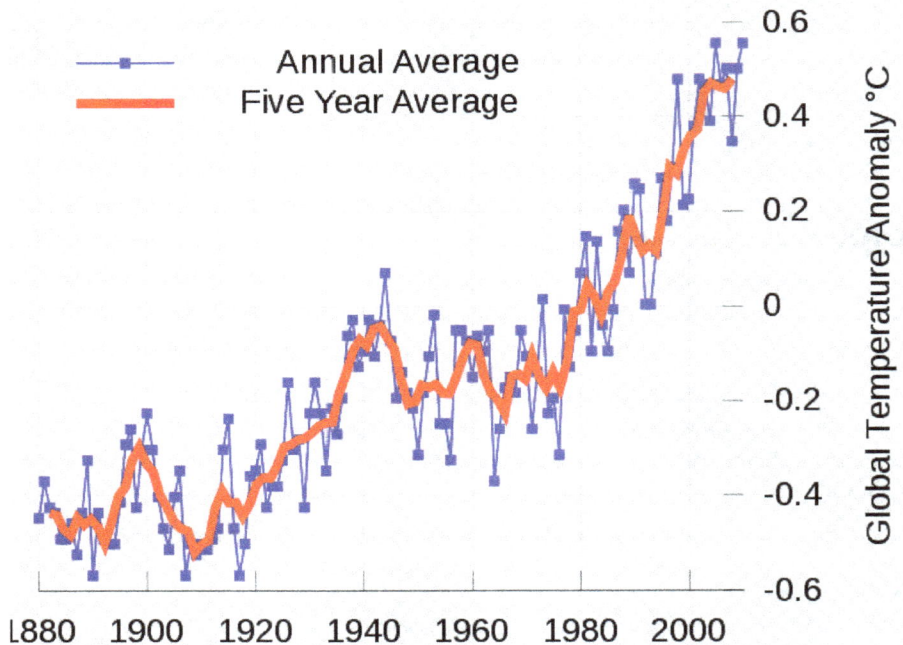

Fig. 6a. Global mean surface temperature difference from the average for 1880–2009

The most recent report IPCC projected that during the 21st century the global surface temperature is likely to rise a further1.1 to 2.9 °C (2 to 5.2 °F) for the lowest emissions scenario used in the report and 2.4 to 6.4 °C (4.3 to 11.5 °F) for the highest

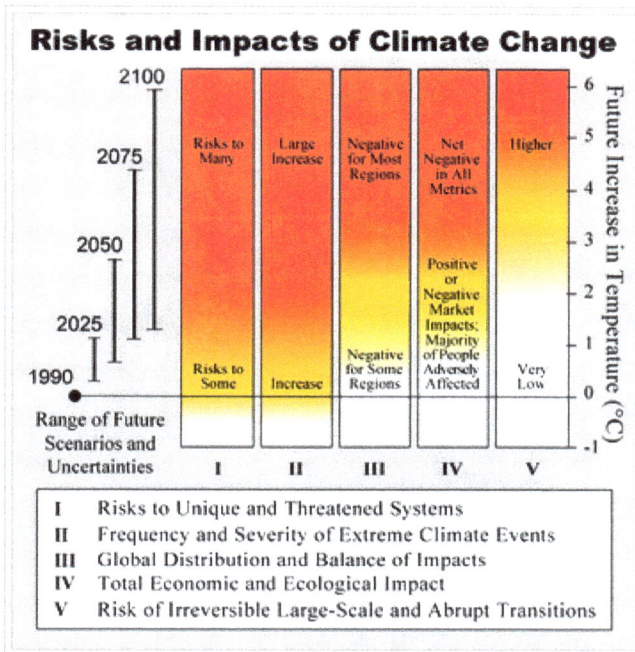

Fig. 6b. Projected Global Temperature Rise 1.1 to 6.4 °C during 21st century

3.2.2 Physical impacts of climate change

Working Group I's contribution to the IPCC Fourth Assessment Report, published in 2007, concluded that warming of the climate system was "unequivocal" (*Solomon S, 2007a*). This was based on the consistency of evidence across a range of observed changes, including increases in global average air and ocean temperatures, widespread melting of snow and ice, and rising global average sea level(*Solomon S, 2007b*).

Human activities have contributed to a number of the observed changes in climate (*Hegerl GC, et. al., 2007*). This contribution has principally been through the burning of fossil fuels, which has led to an increase in the concentration of GHGs in the atmosphere. This increase in GHG concentrations has caused a radiative forcing of the climate in the direction of warming. Human-induced forcing of the climate has likely to contributed to a number of observed changes, including sea level rise, changes in climate extremes (such as warm and cold days), declines in Arctic sea ice extent, and to glacier retreat.

Human-induced warming could potentially lead to some impacts that are abrupt or irreversible. The probability of warming having unforeseen consequences increases with the rate, magnitude, and duration of climate change.

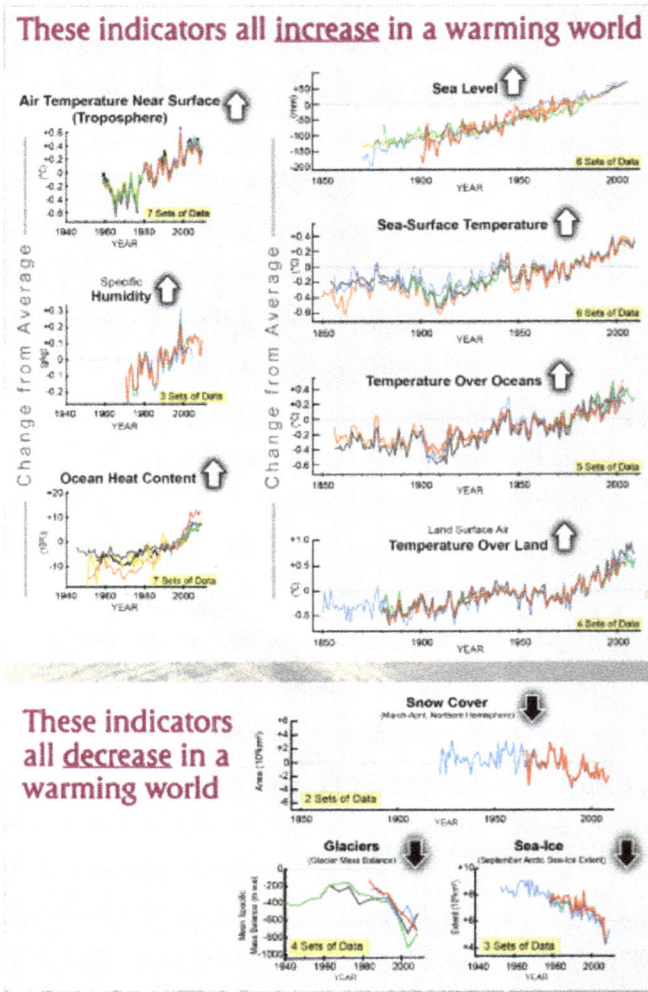

Fig. 7. Key climate indicators that show global warming

3.2.3 Effects on weather

Observations show that there have been changes in weather (*Le Treut H, et. al., 2007*). As climate changes, the probabilities of certain types of weather events are affected. Changes have been observed in the amount, intensity, frequency, and type of precipitation. Widespread increases in heavy precipitation have occurred, even in places where total rain amounts have decreased. IPCC (2007d) concluded that human influences had, more likely than not (greater than 50% probability, based on expert judgement), contributed to an increase in the frequency of heavy precipitation events. Projections of future changes in precipitation show overall increase in the global average, but with substantial shifts in where and how precipitation falls. Climate models tend to project increasing precipitation at high latitudes and in the tropics (e.g., the south-east monsoon region and over the tropical Pacific) and decreasing precipitation in the sub-tropics (e.g., over much of North Africa and the northern Sahara).

Evidence suggests that, since the 1970s, there have been substantial increases in the intensity and duration of tropical storms and hurricanes. Models project a general tendency for more intense but fewer storms outside the tropics.

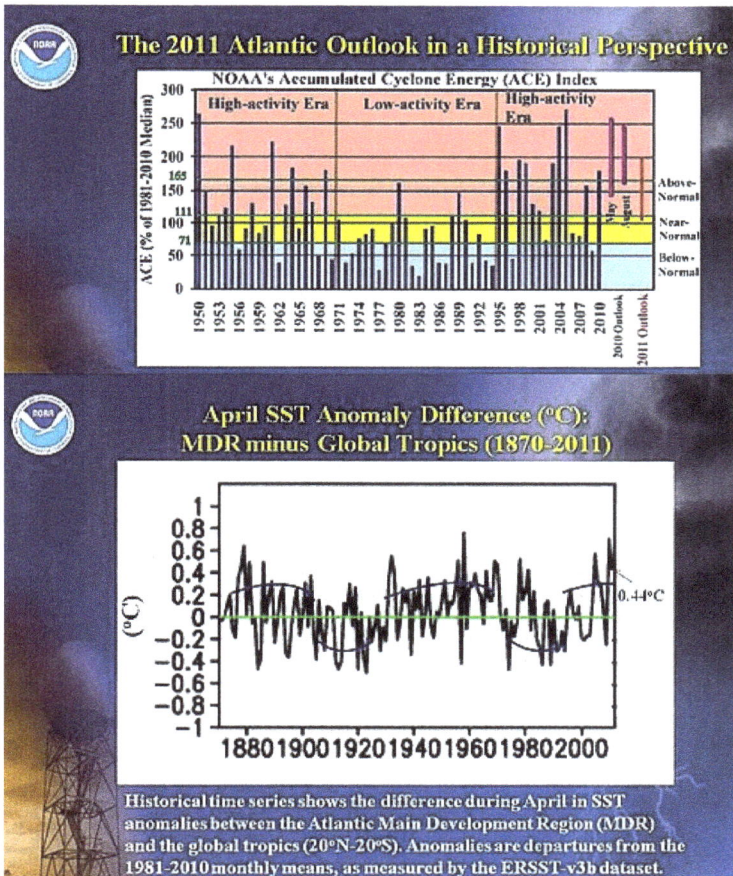

Fig. 8. Accumulated cyclone energy in the Atlantic Ocean and the sea surface temperature difference which influences such, measured by the U.S. NOAA.

3.2.4 Extreme weather, tropical cyclone, and list of atlantic hurricane records

Since the late 20th century, changes have been observed in the trends of some extreme weather and climate events, e.g., heat waves. Human activities have, with varying degrees of confidence, contributed to some of these observed trends. Projections for the 21st century suggest continuing changes in trends for some extreme events. Solomon *et al.* (2007), for example, projected the following likely (greater than 66% probability, based on expert judgment) changes:

- Increase in the areas affected by drought;
- Increased tropical cyclone activity and
- Increased incidence of extreme high sea level (excluding tsunamis).

Projected changes in extreme events will have predominantly adverse impacts on ecosystems and human society.

3.2.5 Glacier retreat since 1850 and disappearance

IPCC (2007a:5) found that, on average, mountain glaciers and snow cover had decreased in both the northern and southern hemispheres.

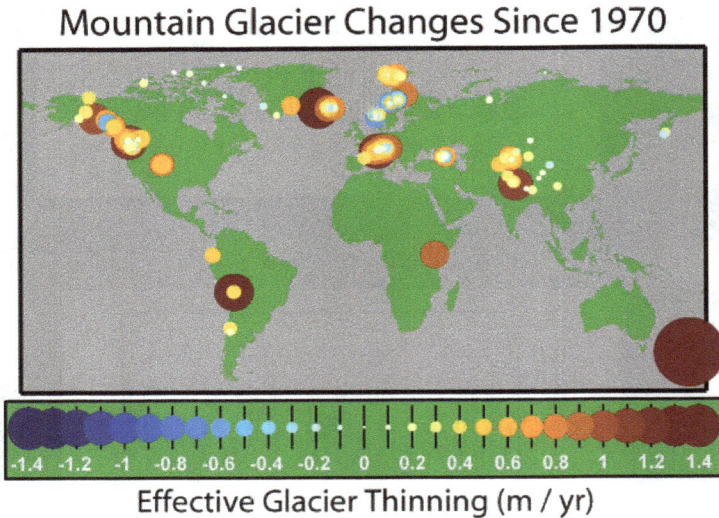

Fig. 9. A map of the change in thickness of mountain glaciers since 1970 (Thinning in orange and red, thickening in blue).

This widespread decrease in glaciers and ice caps had contributed to observed sea level rise. With very high or high confidence, *IPCC (2007d: 11)* made a number of projections relating to future changes in glaciers:

- Mountainous areas in Europe will face glacier retreat
- In Polar Regions, there will be reductions in glacier extent and the thickness of glaciers.
- More than one-sixth of the world's populations are supplied by melt-water from major mountain ranges. Changes in glaciers and snow cover are expected to reduce water availability for these populations.
- In Latin America, changes in precipitation patterns and the disappearance of glaciers will significantly affect water availability for human consumption, agriculture, and energy production.

3.2.6 Role of the oceans in global warming

The oceans serve as a sink for carbon dioxide, taking up much that would otherwise remain in the atmosphere, but increased levels of CO_2 have led to ocean acidification. Furthermore, as the temperature of the oceans increases, their absorptivity for excess CO_2 decreases. The oceans have also acted as a sink in absorbing extra heat from the atmosphere. This extra heat has been added to the climate system due to the build-up of GHGs. More than 90 percent of warming that occurred over 1960–2009 is estimated to have gone into the oceans.

Global warming is projected to have a number of effects on the oceans. Ongoing effects include rising sea levels due to thermal expansion and melting of glaciers and ice sheets, and warming of the ocean surface, leading to increased temperature stratification. Other possible effects include large-scale changes in ocean circulation.

a. *Number of tropical storms and hurricanes per season*: This bar chart shows the number of named storms and hurricanes per year from 1893-2010.

•   Costliest tropical cyclone: Hurricane Katrina - 2005 - $81.2 billion in damages.

•   Smallest tropical cyclone on record: Marco - 2008 - gale force winds extended 10 mi (20 km) from storm center (previous record: Cyclone Tracy 1974 - 30 mi (50 km))

b. *Ocean Acidification*: About one-third of the carbon dioxide emitted by human activity has already been taken up by the oceans. As carbon dioxide dissolves in sea water, carbonic acid is formed, which has the effect of acidifying the ocean, measured as a change in pH. The uptake of human carbon emissions since the year 1750 has led to an average decrease in pH of 0.1 units (*IPCC 2007d. "3.3.4 Ocean acidification*). Projections using the SRES emissions scenarios suggest a further reduction in average global surface ocean pH of between 0.14 and 0.35 units over the 21st century.

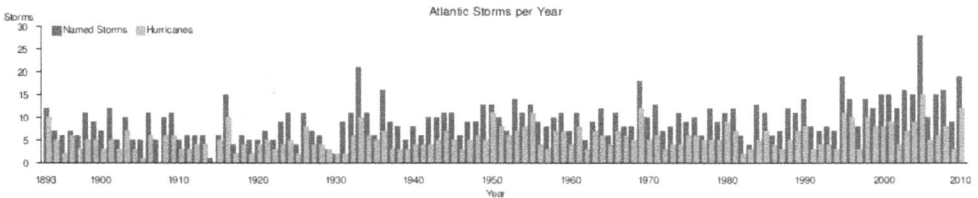

Fig. 10. Worldwide cyclone records set by Atlantic storms

The effects of ocean acidification on the marine biosphere have yet to be documented. Laboratory experiments suggest beneficial effects for a few species, with potentially highly detrimental effects for a substantial number of species. With medium confidence, Fischlin *et al.* (2007) projected that future ocean acidification and climate change would impair a wide range of plank tonic and shallow benthic marine organisms that use aragonite to make their shells or skeletons, such as corals and marine snails (pteropods), with significant impacts particularly in the Southern Ocean.

c. *Oxygen Depletion*: The amount of oxygen dissolved in the oceans may decline, with adverse consequences for ocean life (*Crowley TJ, 1988; Shaffer G, et al., 2009*).

d. *Sea level rise*: Sea level has been rising 0.2 cm/yr, based on measurements of sea level rise from 23 long tide gauge records in geologically stable environments. Sea level was projected to rise by 18 to 59 cm (7.1 to 23.2 in) for the time period 2090–99. This projection is with the increase in level relative to average global sea level over the 1980–99 periods (*Bindoff, NL, et al., 2007*).

The IPCC (2007d, p. 5) reported that between 1961 and 2003, global average sea level rose at an average rate of 1.8 mm per year (mm/yr), with an uncertainty range of 1.3–2.3 mm/yr. Between 1993 and 2003, the rate increased above the previous period to 3.1 mm/yr (uncertainty range of 2.4–3.8 mm/yr).

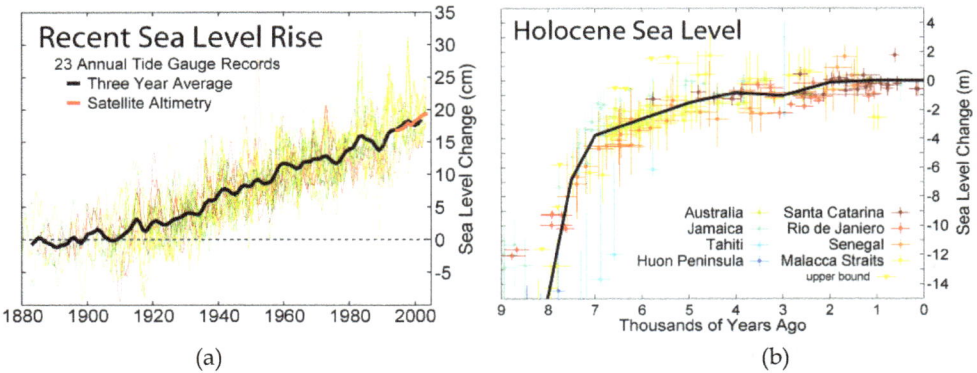

Fig. 11. (a) Current sea level rise and (b) Sea level rise during the Holocene.

A range of projections suggested possible sea level rise by the end of the 21st century of between 0.56 and 2 m. These projections are based on the same measurement range, with the increase in sea level by 2090-99 measured against average global sea level over the 1980–99 time periods.

e. *Ocean temperature rise*: From 1961 to 2003, the global ocean temperature has risen by 0.10 °C from the surface to a depth of 700 m. There is variability both year-to-year and over longer time scales, with global ocean heat content observations showing high rates of warming for 1991–2003, but some cooling from 2003 to 2007 (*Bindoff, NL, 2011*). The temperature of the Antarctic Southern Ocean rose by 0.17 °C (0.31 °F) between the 1950s and the 1980s, nearly twice the rate for the world's oceans as a whole (*Gille, Sarah T, 2002*). As well as having effects on ecosystems (e.g. by melting sea ice, affecting algae that grow on its underside), warming reduces the ocean's ability to absorb CO_2.

f. *Regional effects of global warming*: Some are the results of a generalised global change, such as rising temperature, resulting in local effects, such as melting ice. In other cases, a change may be related to a change in a particular ocean current or weather system. In such cases, the regional effect may be disproportionate and will not necessarily follow the global trend.

There are three major ways in which global warming will make changes to regional climate: melting or forming ice, changing the hydrological cycle (of evaporation and precipitation) and changing currents in the oceans and air flows in the atmosphere. The coast can also be considered a region, and will suffer severe impacts from sea level rise.

1999-2008 Mean Temperatures

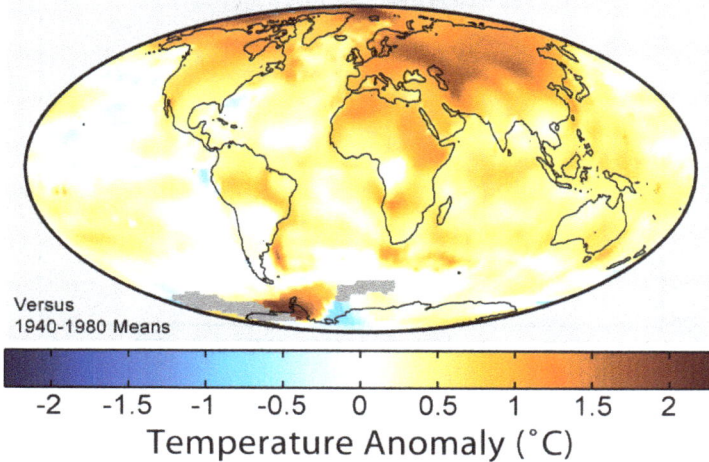

Fig. 12. Mean surface temperature change for 1999–2008 relative to the average temperatures from 1940 to 1980

- *Social systems*: The impacts of climate change can be thought of in terms of sensitivity and vulnerability. "Sensitivity" is the degree to which a particular system or sector might be affected, positively or negatively, by climate change and/or climate variability. "Vulnerability" is the degree to which a particular system or sector might be adversely affected by climate change.

  The sensitivity of human society to climate change varies. Sectors sensitive to climate change include water resources, coastal zones, human settlements, and human health. Industries sensitive to climate change include agriculture, fisheries, forestry, energy, construction, insurance, financial services, tourism, and recreation.

- *Food supply*: Climate change will impact agriculture and food production around the world due to: the effects of elevated CO_2 in the atmosphere, higher temperatures, altered precipitation and transpiration regimes, increased frequency of extreme events, and modified weed, pest, and pathogen pressure (Easterling *et al.*, 2007). In general, low-latitude areas are at most risk of having decreased crop yields (Schneider *et al.*, 2007).

  So far, the effects of regional climate change on agriculture have been relatively limited. Changes in crop phenology provide important evidence of the response to recent regional climate change. Phenology is the study of natural phenomena that recur periodically, and how these phenomena relate to climate and seasonal changes. A significant advance in phenology has been observed for agriculture and forestry in large parts of the Northern Hemisphere.

- *Health*: Human beings are exposed to climate change through changing weather patterns (temperature, precipitation, sea-level rise and more frequent extreme events) and indirectly through changes in water, air and food quality and changes in ecosystems, agriculture, industry and settlements and the economy (Confalonieri *et al.*, 2007a). According to a literature assessment, the effects of climate change to date have been small, but are projected to progressively increase in all countries and regions.

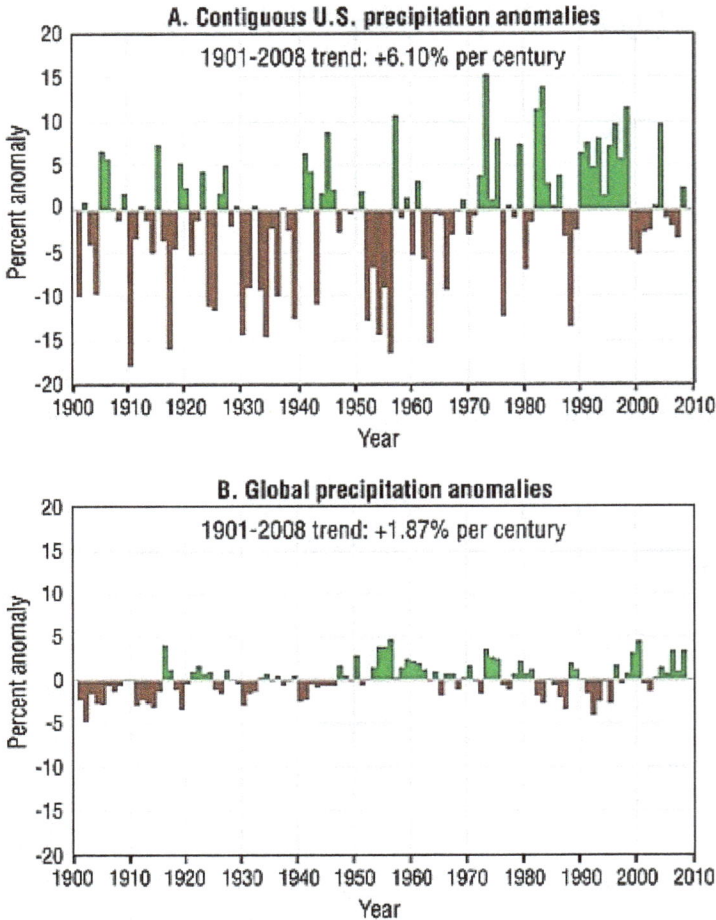

Fig. 13. Precipitation during the 20th century and up through 2008 during global warming, the NOAA estimating an observed trend over that period of 1.87% global precipitation increase per century.

A study by the World Health Organization (*WHO, 2009*) estimated the effect of climate change on human health. Not all of the effects of climate change were included in their estimates, for example, the effects of more frequent and extreme storms were excluded. Climate change was estimated to have been responsible for 3% of diarrhoea, 3% of malaria, and 3.8% of dengue fever deaths worldwide in 2004. Total attributable mortality was about 0.2% of deaths in 2004; of these, 85% were child deaths.

- *Projections*: With high confidence, IPCC (2007d:48) projected that climate change would bring some benefits in temperate areas, such as fewer deaths from cold exposure, and some mixed effects such as changes in range and transmission potential of malaria in Africa. Benefits were projected to be outweighed by negative health effects of rising temperatures, especially in developing countries.

- *Extreme events*: With high confidence, Confalonieri *et al.* (2007b) projected that climate change would increase the number of people suffering from death, disease and injury from heatwaves, floods, storms, fires and droughts.
- *Floods and weather disasters*: Floods are low-probability, high-impact events that can overwhelm physical infrastructure and human communities (Confalonieri *et al.*, 2007c). Major storm and flood disasters have occurred in the last two decades. The impacts of weather disasters are considerable and unequally distributed. For example, natural disasters have been shown to result in increased domestic violence against - and post-traumatic stress disorders in – women. In terms of deaths and populations affected, floods and tropical cyclones have the greatest impact in South Asia and Latin America. Vulnerability to weather disasters depends on the attributes of the person at risk, including where they live and their age, as well as other social and environmental factors. High-density populations in low-lying coastal regions experience a high health burden from weather disasters.
- *Heatwaves*: Hot days, hot nights and heatwaves have become more frequent (*Confalonieri et al., 2007d*). Heatwaves are associated with marked short-term increases in mortality. For example, in August 2003, a heat wave in Europe resulted in excess mortality in the range of 35,000 total deaths.
  Heat-related morbidity and mortality is projected to increase (Confalonieri *et al.*, 2007e). The health burden could be relatively small for moderate heatwaves in temperate regions, because deaths occur primarily in susceptible persons.
- *Drought*: The effects of drought on health include deaths, malnutrition, infectious diseases and respiratory diseases. Countries within the "Meningitis Belt" in semi-arid sub-Saharan Africa experience the highest endemicity and epidemic frequency of meningococcal meningitis in Africa, although other areas in the Rift Valley, the Great Lakes, and southern Africa are also affected (*Confalonieri et al., 2007f*). The spatial distribution, intensity, and seasonality of meningococcal (epidemic) meningitis appear to be strongly linked to climate and environmental factors, particularly drought. The cause of this link is not fully understood.
- *Fires*: In some regions, changes in temperature and precipitation are projected to increase the frequency and severity of fire events (*Confalonieri et al., 2007g*). Forest and bush fires cause burns, damage from smoke inhalation and other injuries.
- *Infectious disease vectors*: With high confidence, *Confalonieri et al. (2007h)* projected that climate change would continue to change the range of some infectious disease vectors such as: Dengue, Malaria, Diarrhoeal diseases etc. Vector-borne diseases, (VBD) are infections transmitted by the bite of infected arthropod species, such as mosquitoes, ticks, triatomine bugs, sandflies, and blackflies. There is some evidence of climate-change-related shifts in the distribution of tick vectors of disease, of some (non-malarial) mosquito vectors in Europe and North America. Climate change has also been implicated in changes in the breeding and migration dates of several bird species. Several species of wild bird can act as carriers of human pathogens as well as of vectors of infectious agents.
- *Ground-level ozone*: With high confidence, *Confalonieri et al. (2007i)* projected that climate change would increase cardio-respiratory morbidity and mortality associated with ground-level ozone. Ground-level ozone is both naturally occurring and is the primary constituent of urban smog (*Confalonieri et al., 2007j*). Ozone in smog is formed through

chemical reactions involving nitrogen oxides and other compounds. The reaction is a photochemical reaction, meaning that it involves electromagnetic radiation, and occurs in the presence of bright sunshine and high temperatures. Exposure to elevated concentrations of ozone is associated with increased hospital admissions for pneumonia, chronic obstructive pulmonary disease, asthma, allergic rhinitis and other respiratory diseases, and with premature mortality.

Background levels of ground-level ozone have risen since pre-industrial times because of increasing emissions of methane, carbon monoxide and nitrogen oxides (*Confalonieri et al., 2007k*). This trend is expected to continue into the mid-21st century.

- *Cold-waves*: Cold-waves continue to be a problem in northern latitudes, where very low temperatures can be reached in a few hours and extend over long periods (*Confalonieri et al., 2007l*). Reductions in cold-deaths due to climate change are projected to be greater than increases in heat-related deaths in the UK.

3.2.7 Water resources crisis

A number of climate-related trends have been observed that affect water resources. These include changes in precipitation, the crysosphere and surface waters (e.g., changes in river flows). Observed and projected impacts of climate change on freshwater systems and their management are mainly due to changes in temperature, sea level and precipitation variability. Sea level rise will extend areas of salinization of groundwater and estuaries, resulting in a decrease in freshwater availability for humans and ecosystems in coastal areas. In an assessment of the scientific literature, *Kundzewicz et al. (2007)* concluded, with high confidence, that:

- The negative impacts of climate change on freshwater systems outweigh the benefits. All of the regions assessed in the IPCC Fourth Assessment Report (Africa, Asia, Australia and New Zealand, Europe, Latin America, North America, Polar regions (Arctic and Antarctic), and small islands) showed an overall net negative impact of climate change on water resources and freshwater ecosystems.
- Semi-arid and arid areas are particularly exposed to the impacts of climate change on freshwater. It has been assessed that many of these areas, e.g., the Mediterranean basin, Western United States, Southern Africa, and north-eastern Brazil, would suffer a decrease in water resources due to climate change.

3.2.8 Environmental migration and conflict

General circulation models project that the future climate change will bring wetter coasts, drier mid-continent areas, and further sea level rise (*Scott, MJ, et.al., 1996*). Such changes could result in the gravest effects of climate change through sudden human migration. Millions might be displaced by shoreline erosions, river and coastal flooding, or severe drought.

Migration related to climate change is likely to be predominantly from rural areas in developing countries to towns and cities. In the short term climate stress is likely to add incrementally to existing migration patterns rather than generating entirely new flows of people. It is quite likely that environmental degradation, loss of access to resources (e.g., water resources) (*Desanker, P., et al., 2001*) and resulting human migration could become a source of political and even military conflict. Factors other than climate change may, however, be more important in affecting conflict. For example, Wilbanks *et al.* (2007) suggested that major environmentally influenced conflicts in Africa were more to do with the relative abundance of resources, e.g., oil and diamonds, than with resource scarcity. Scott *et al.* (2001) placed only low confidence in predictions of increased conflict due to climate change.

3.2.9 Aggregate economic impacts of climate change

Aggregating impacts adds up the total impact of climate change across sectors and/or regions. Examples of aggregate measures include economic cost (e.g., changes in gross domestic product (GDP) and the social cost of carbon), changes in ecosystems (e.g., changes over land area from one type of vegetation to another), human health impacts, and the number of people affected by climate change (*Smith, J.B., et al., 2001*).

Economic impacts are expected to vary regionally. For a medium increase in global mean temperature (2-3 °C of warming, relative to the average temperature between 1990-2000), market sectors in low-latitude and less-developed areas might experience net costs due to climate change. On the other hand, market sectors in high-latitude and developed regions might experience net benefits for this level of warming. A global mean temperature increase above about 2-3 °C (relative to 1990-2000) would very likely result in market sectors across all regions experiencing either declines in net benefits or rises in net costs.

Aggregate impacts have also been quantified in non-economic terms. For example, climate change over the 21st century is likely to adversely affect hundreds of millions of people through increased coastal flooding, reductions in water supplies, increased malnutrition and increased health impacts.

3.2.10 Climate change and ecosystems

Beyond the year 2050, climate change may be the major driver for biodiversity loss globally. It was projected by Fischlin *et al.* (2007a) that approximately 20 to 30% of plant and animal species assessed so far would likely be at increasingly high risk of extinction should global mean temperatures exceed a warming of 2 to 3 °C above pre-industrial temperature levels (Fischlin *et al.*, 2007b). The uncertainties in this estimate, however, are large: for a rise of about 2 °C the percentage may be as low as 10%, or for about 3 °C, as high as 40%, and depending on biota(Parry 2007a) (all living organisms of an area, the flora and fauna considered as a unit) the range is between 1% and 80%. As global average temperature exceeds 4 °C above pre-industrial levels, model projections suggested that there could be significant extinctions (40-70% of species that were assessed) around the globe.

3.2.11 Biogeochemical cycles

Climate change may have an effect on the carbon cycle in an interactive "feedback" process. Using the A2 SRES emissions scenario, Schneider *et al.* (2007:789) found that this effect led to additional warming by 2100, relative to the 1990–2000 period, of 0.1–1.5 °C. This estimate was made with high confidence. The climate projections made in the IPCC Fourth Assessment Report of 1.1–6.4 °C account for this feedback effect. On the other hand, with medium confidence, Schneider *et al.* (2007) commented that additional releases of GHGs were possible from permafrost, peat lands, wetlands, and large stores of marine hydrates at high latitudes.

3.2.12 Greenland and West Antarctic ice sheets

With medium confidence, the IPCC (2007b) concluded that with a global average temperature increase of 1-4 °C (relative to temperatures over the years 1990-2000), at least a partial deglaciation of the Greenland ice sheet, and possibly the West Antarctic ice sheets would occur *(Parry 2007b)*. The estimated timescale for partial deglaciation was centuries to millennia, and would contribute 4 to 6 metres (13 to 20 ft) or more to sea level rise over this period.

4. Conclusion

From the present study it is observed that substantial increase in energy demand in the developing countries, due to the strong economic growth is dominating energy markets today and subsequently increasing global population deplete fossil fuel reserves is making the sustainability vulnerable. Other major findings of the study are as under:

- The global population is expected to increase by 30% the next 25 years, where 80-90% of the increase is expected to be in developing countries.
- Limits to Growth – the 30 year update claims that the global system is currently in an un-sustainable situation, and that there are limits to growth on our planet – on resources, food, environment, and also in the population the earth can supply over time.
- Fossil fuel reserve depletion times for oil, coal and gas of approximately would be 35, 107 and 37 years, respectively. This means that coal reserves are available up to 2112, and will be the only fossil fuel remaining after 2042.
    - The oil price is currently very high at around $140 per barrel in nominal terms. This is much higher than after several other oil price crises, such as the Iran/Iraq war, Gulf war and 9/11 as per WTRG, 2008. According to OPEC (2007), OPEC benchmark crude price is assumed to remain in the $50 to $60 per barrel range in nominal terms for much of the projected period and rising further in the longer term with inflation.
    - Coal prices have had less fluctuation in comparison to oil in the last 50 years. The coal market depicts relatively constant coal prices in historical data. This trend will remain flat until the middle of the next decade, and then increase very slowly, reaching just over $60 per tonne by 2030.
    - Gas prices have generally followed the increase in oil prices since 2003, typically with a 1 year lag. The average transmission and distribution margin for delivered gas is projected to change from $2.38 per thousand cubic feet in 2006 to between $2.07 and $2.44 per thousand cubic feet in 2030.
- Due to fast energy requirement and subsequent high consumption of fossil fuel, greenhouse" gases such as Carbon Dioxide (CO_2) are tremendously increasing and causing Global warming is threat to the survival of mankind in today's scenario.
- The effects, or impacts, of climate change may be physical, ecological, social or economic and closely associated with Global warming. Climate change includes the instrumental temperature record, rising sea levels, and decreased snow cover in the Northern Hemisphere and sea ice, more frequent and intense heat waves, rising sea levels, and widespread ocean acidification, frequency of intense rainfall, and widespread ocean acidification.

- World Health Organization (WHO, 2009) estimated to have been responsible for 3% of diarrhea, 3% of malaria, and 3.8% of dengue fever deaths worldwide in 2004. Total attributable mortality was about 0.2% of deaths in 2004; of these, 85% were child deaths.
- Approximately 20 to 30% of plant and animal species would likely be at increasingly high risk of extinction when global mean temperatures exceed a warming of 2 to 3 °C above pre-industrial temperature levels. As global average temperature exceeds 4 °C above pre-industrial levels, there could be significant extinctions (40-70% of species that were assessed) around the globe.

Beyond the year 2050, climate change may be the major driver for biodiversity loss globally. The climate problem affects everyone, and everyone has a stake in deciding what should be done. It is for you to decide what actions you should take as an individual (in your home, your car, and so forth). Equally important, as a citizen you must decide which policies to support or oppose and make every one aware about disastrous consequences of the situation.

5. Acknowledgments

Authors indebted to extend their thanks to the Management of School of Management Sciences, Technical Campus, Lucknow and Harcourt Butler Technological Institute, Kanpur for providing the support of Library.

6. References

BGR, 2007 BGR, 2007, Reserves, resources and availability of energy resources 2005, In: Federal Institute for Geosciences and Natural Resources, H. (Ed.), Bundesanstalt für Geowissenschaften und Rohstoffe.

Bindoff, N.L., *et al.* "FAQ 5.1 Is Sea Level Rising?" In Solomon 2007, Chapter 5: Observations: Oceanic Climate Change and Sea Level. CH: IPCC Retrieved: 2011-06-17.

Bindoff, NL, J Willebrand, V Artale, A, Cazenave, J Gregory, S Gulev, K Hanawa, C Le Quéré, S Levitus, Y Nojiri, CK Shum, LD Talley and A Unnikrishnan. "5 Observations: Oceanic Climate Change and Sea Level". In Solomon 2007 (PDF), CH: IPCC.

BP, 2006 BP, BP Statistical Review of World Energy 2006, British Petroleum (2006).

BP, 2007a BP, BP Annual Review 2007, British Petroleum (2007).

BP, 2007b BP, BP Statistical Review of World Energy 2007, British Petroleum (2007).

Chang, Hasok (2004), *Inventing Temperature: Measurement and Scientific Progress*. Oxford: Oxford University Press, ISBN 978-0-19-517127-3.

Confalonieri *et al.*, 2007a,"Executive summary", In Parry 2007, Chapter 8: Human health, Retrieved 2011-07-13.

Confalonieri *et al.*, 2007b, "8.2.3 Drought, nutrition and food security", In Parry 2007, Chapter 8: Human health, Retrieved: 2011-07-13.

Confalonieri *et al.*, 2007c, "8.2.2 Wind, storms and floods", In Parry 2007, Chapter 8: Human health, Retrieved: 2011-07-13.

Confalonieri *et al.*, 2007d, "8.2.1.1 Heatwaves", In Parry 2007, Chapter 8: Human health, Retrieved: 2011-07-13.

Confalonieri *et al.*, 2007e, "8.4.1.3 Heat- and cold-related mortality", In Parry 2007, Chapter 8: Human health, Retrieved: 2011-07-13.

Confalonieri *et al.*, *2007f*, "8.2.3.1 Drought and infectious disease", In Parry 2007, Chapter 8: Human health, Retrieved: 2011-07-13.

Confalonieri *et al.*, *2007g*, "8.2.6.3 Air pollutants from forest fires", In Parry 2007, Chapter 8: Human health, Retrieved 2011-07-13.

Confalonieri *et al.*, *2007h*, "8.2.8 Vector-borne, rodent-borne and other infectious diseases", In Parry 2007, Chapter 8: Human health, Retrieved 2011-07-13.

Confalonieri *et al.*, *2007i*, "8.2.5 Water and disease", In Parry 2007, Chapter 8: Human health, Retrieved 2011-07-13.

Confalonieri *et al.*, *2007j*, "8.2.6 Air quality and disease", In Parry 2007, Chapter 8: Human health, Retrieved 2011-07-13.

Confalonieri *et al.*, *2007k*, "8.4.1.4 Urban air quality", In Parry 2007, Chapter 8: Human health, Retrieved 2011-07-13.

Confalonieri *et al.*, *2007l*, "8.2.1.2 Cold-waves", In Parry 2007, Chapter 8: Human health, Retrieved 2011-07-13.

Crowley, T. J.; North, G. R. (May 1988), "Abrupt Climate Change and Extinction Events in Earth History", *Science* 240 (4855): 996–1002, Bib code: 1988Sci...240...996C, doi:10.1126/science.240.4855.996, PMID 17731712.

Desanker, P., *et al.*, "Executive summary", In McCarthy 2001, Chapter 10: Africa, Retrieved 2011-06-20.

Easterling, WE, *et al.*, "5.4.1 Primary effects and interactions", In Parry 2007, Chapter 5: Food, Fibre, and Forest Products, pp. 282, Retrieved 2011-06-25.

Fischlin, A, *et al.*, "Executive summary", In Parry 2007 (PDF), Chapter 4: Ecosystems, their properties, goods, and services, pp. 213, Retrieved 2011-06-11.

Fischlin, A., *et al.*, 2007a, "Executive summary", In Parry 2007 (PDF), Chapter 4: Ecosystems, their properties, goods and services. pp. 213–214, Retrieved: 2011-09-07.

Fischlin, A., *et al.*, 2007b, "4.4.11 Global synthesis including impacts on biodiversity", In Parry 2007, Chapter 4: Ecosystems, their properties, goods and services, Retrieved: 2011-10-01.

Fisher, BS, *et al* (2007). "3.1 Emissions scenarios", In B Metz, *et al. Issues related to mitigation in the long term context*. Climate Change 2007: Mitigation, Contribution of Working Group III to the Fourth Assessment Report of the Intergovernmental Panel on Climate Change, Cambridge, UK & New York, NY, USA; CH: Cambridge University Press; IPCC, Retrieved: 2011-05-04.

Gille, Sarah T (February 15, 2002). "Warming of the Southern Ocean Since the 1950s", *Science* 295 (5558): 1275–7, Bibcode 2002Sci...295.1275G, doi:10.1126/science.1065863, PMID 11847337.

Goldemberg, 2006 J. Goldemberg, The promise of clean energy. *Energy Policy*, 34 (2006), pp. 2185–2190.

Hegerl, GC, *et al.* "Executive Summary", In Solomon 2007, Chapter 9: Understanding and Attributing Climate Change. CH: IPCC.

Hubbert M.K., 1956, Nuclear energy and the fossil fuels; American Petroleum Institute, Drilling and Production Practice, Proc. Spring Meeting, San Antonio, Texas. 7-25

IEA, 2006 IEA, World Energy Outlook 2006, Organisation for Economic Co-operation and Development, International Energy Agency, Paris and Washington, DC (2006).

IEA, 2007a, IEA, Coal Information 2007, Organisation for Economic Co-operation and Development, International Energy Agency, Paris and Washington, DC (2007).

IEA, 2007b, IEA, World Energy Outlook 2007, China and India, Organisation for Economic Co-operation and Development, International Energy Agency, Paris and Washington, DC (2007).

Intergovernmental Panel on Climate Change (2007d), "Climate Change 2007: Synthesis Report, Contribution of Working Groups I, II and III to the Fourth Assessment Report of the Intergovernmental Panel on Climate Change (Core Writing Team *et al.* (eds.))". IPCC, Geneva, Switzerland, Retrieved: 2009-05-20.

IPCC (2001b), "Figure SPM-2", In McCarthy 2001, Summary for Policymakers, Retrieved: 2011-05-18.

IPCC 2001d, "3.16", Question 3, Retrieved: 2011-08-05.

IPCC 2007d, "1. Observed changes in climate and their effects", Summary for Policymakers, CH: IPCC, Retrieved: 2011-06-17.

IPCC 2007d, "3. Projected climate change and its impacts", Summary for Policymakers. CH: IPCC.

IPCC 2007d, "3.3.4 Ocean acidification", Synthesis Report, Retrieved: 2011-06-11.

IPCC 2007d, "5.2 Key vulnerabilities, impacts and risks-long-term perspectives", Synthesis report, Retrieved: 2011-08-05.

IPCC, 2001, SPM Question 3

Jansen, E, *et al.*, "6.3.2 What Does the Record of the Mid-Pliocene Show?", In Solomon 2007, Chapter 6: Palaeoclimate. CH: IPCC Retrieved: 2011-05-04.

Karl 2009, ed., "Global Climate Change".

Klass, 1998 D.L. Klass, Biomass for Renewable Energy, Fuels, and Chemicals, Academic Press, San Diego (1998).

Klass, 2003 D.L. Klass, A critical assessment of renewable energy usage in the USA. *Energy Policy*, 31 (2003), pp. 353–367.

Kundzewicz, Z.W., *et al.*, "Executive Summary", In Parry 2007, Chapter 3: Fresh Water Resources and their Management, pp. 175, Retrieved: 2011-08-14.

Le Treut, H, *et al.* "FAQ 1.2 What is the Relationship between Climate Change and Weather?", In Solomon 2007, Historical Overview of Climate Change. CH: IPCC.

Lior, 2008, N. Lior, Energy resources and use the present situation and possible paths to the future, *Energy*, 33 (2008), pp. 842–857.

Parry 2007a, ed. "Definition of "biota"". Appendix I: Glossary Retrieved: 2011-10-01.

Parry 2007b, ed., "Magnitudes of impact", Summary for Policymakers, CH: Intergovernmental Panel on Climate Change, pp. 17, Retrieved: 2011-05-08.

Schneider, SH, *et al.*, "19.3.1 Introduction to Table 19.1", In Parry 2007, Chapter 19: Assessing Key Vulnerabilities and the Risk from Climate Change. CH: IPCC, Retrieved: 2011-05-04.

Schneider, SH, *et al.*, "19.3.2.1 Agriculture", In Parry 2007, Chapter 19: Assessing Key Vulnerabilities and the Risk from Climate Change, pp. 790, Retrieved: 2011-06-25.

Scott, M., *et al.*, "7.2.2.3.1 Migration", In McCarthy 2001, Chapter 7: Human Settlements, Energy, and Industry, Retrieved: 2011-08-29.

Scott, M.J., *et al.*, "12.3.1 Population Migration", In Watson 1996, Chapter 12: Human Settlements in a Changing Climate: Impacts and Adaptation.

Shaffer, G.; Olsen, S. M.; Pedersen, J. O. P. (2009), "Long-term ocean oxygen depletion in response to carbon dioxide emissions from fossil fuels", *Nature Geoscience* 2 (2): 105–109, Bibcode 2009NatGe...2..105S, doi:10.1038/ngeo420.

Shahriar Shafiee, Erkan Topal, When will fossil fuel reserves be diminished? Energy Policy, Volume 37, Issue 1, January 2009, Pages 181-189.

Singh, B.R., et al., A Study on Sustainable Energy Sources and its Conversion Systems Towards Development of an Efficient Zero Pollution Novel Air Turbine to Use as Prime-Mover to the Light Vehicle, ASME Conf. Proc., 2008, Volume 8, Paper no. IMECE2008-66803 pp. 371-378, doi: 10.1115/IMECE2008-66803.

Solomon 2007a, ed. "Direct Observations of Recent Climate Change", Summary for Policymakers, CH: Intergovernmental Panel on Climate Change.

Solomon, S, et al. "Table TS.4", In Solomon 2007, Technical Summary, p. 52.

Solomon, S, et al. 2007b "TS.3.4 Consistency Among Observations", In Solomon 2007, Technical Summary. CH: IPCC.

Stern, N (May 2008). "The Economics of Climate Change" (PDF). American Economic Review: Papers & Proceedings (UK: LSE) 98 (2): 6, doi:10.1257/aer.98.2.1, Retrieved: 2011-05-04.

WHO (2009), "2.6 Environmental risks", 2 Results, Global health risks: mortality and burden of disease attributable to selected major risks. Produced by the Department of Health Statistics and Informatics in the Information, Evidence and Research Cluster of the World Health Organization (WHO). World Health Organization, 20 Avenue Appia, 1211 Geneva 27, Switzerland, WHO Press, ISBN: 978 92 4 156387 1, Retrieved: 2011-07-14.

Wilbanks, T., et al., "7.4.1 General effects: Box 7.2, Environmental migration", In Parry 2007, Chapter 7: Industry, Settlement and Society, Retrieved 2011-08-29.

10

Presence of Polycyclic Aromatic Hydrocarbons (PAHs) in Semi-Rural Environment in Mexico City

Salvador Vega[1], Rutilio Ortiz[1], Rey Gutiérrez[1],
Richard Gibson[2] and Beatriz Schettino[1]
[1]*Laboratorio de Análisis Instrumental, Departamento de Producción Agrícola y Animal*
Universidad Autónoma Metropolitana Unidad Xochimilco, Colonia, Coyoacán
[2]*Institute of Agri-Food and Land Use*
School of Biological Sciences Queen's University Belfast
[1]*México*
[2]*Ireland*

1. Introduction

The quality of the environment in big cities depends on its population and their domestic, transport, and industrial activities. In some places agricultural land use coexists with urban areas and as a result of this urbanization and the presence of infrastructure for services like water, electricity, drainage, and the use of fossil fuels etc, contamination problems in the atmosphere, soil and water (Wilcke, 2000), that lately lead some ills on organisms such as respiratory malaises, liver-lung-skin cancer, irritation on eyes and others discomforts. The growth of urban environments presents a major challenge. However, Mexico City as center of economic growth, education, technological advancement, and culture, large city also offer opportunities to manage the growing population in a sustainable way.

These concentrations of people and activity are exerting increasing stress on the natural environment, with impacts at urban, regional and global levels. In the last few decades, air pollution has become one of the most important problems of megacities. The nitrogen and sulphur compounds are main air pollutants, photochemical smog-induced primarily from traffic, but also from industrial activities, power generation, and solvents-has become the main source of concem for air quality. Air pollution has serious impact on public health, causes urban and regional haze, and has the potential to contribute significantly to climate change (Molina & Molina, 2004).

Mexico City and metropolitan area (MCMA), often simply called Mexico City, consists of 16 delegations of the Federal District and 37 contiguous municipalities from the State of Mexico and one municipality from the State of Hidalgo, some with populations over 1 million, that make up the total population of above 20 million for this megacity (Escobedo et al., 2000).

Polycyclic aromatic hydrocarbons (PAHs) are compounds with two or more aromatic rings (benzene) produced by both natural and anthropogenic pathways although anthropogenic activities generally release much greater amounts to the environment (Eom et al., 2007). They originate from combustion, coke production, oil derivates and high temperature industrial processes. PAHs are considered as persistent organic pollutants (POPs) according to the Stockholm Convention. In many studies of contamination, they have been found in air, water, food and soil. There is evidence that some PAHs are carcinogenic, mutagenic and toxic. Monitoring of the PAHs in the environment is important in the evaluation of risk to the health of organisms.

With this problematic situation, the food production may be contaminated with different classes of organic and inorganic residues and contaminants (García-Alonso et al., 2003). For Mexico City case, the presence of contaminants in rural environment highlights persistent organic pollutants (POPs), for example polycyclic aromatic hydrocarbons (PAHs). The sources of these compounds are variable, for example vegetation and fossil fuel combustion, heating (Finizio et al., 1998). PAHs with high persistent in the environment are benze(a)pyrene, anthracene, crysene and others with molecular high weight (> 4 rings aromatics). The occurrence of PAHs is widespread in environmental compartments as air, water, soil and food.

Soils are large reservoirs of hazardous contaminants derived from anthropogenic activities. Some studies of wet and dry atmospheric deposition of PAHs have found values of >10 mg/kg, mainly in urban soils and tropical areas, for example in Brazil (Krauss et al., 2000; Wilcke, 2000). Soils are contaminated with PAHs mainly from atmospheric deposition from stationary sources (gas burning, industrial and municipal organic residues incineration, forest fires) and mobile sources (mainly from fuel fossil combustion for terrestrial transportation) (Mastral & Callen, 2000).

The presence of PAHs in soils has been found to be increasing in industrial and urban developments over the last few decades. Some studies have indicated that vehicle exhausts are major sources of PAHs in soils along with increased use of wastewater for irrigation of crops. Soil contamination by PAHs is considered to be a good indicator of the level of environmental pollution by human activities (Chung et al., 2008). PAHs from soil and water are possibly dangerous to human health because plant root uptake can result in bioconcentration (Samsoe et al., 2002).

The quality of air, water and soil are important for the production of vegetables and animals, and of course for humans as well. The occurrence of contaminants in the environment above certain levels may entail multiple negative consequences in the ecosystems as well as for the human food chain (Liu & Korenaga, 2001).

Our objectives were to investigate the occurrence of PAHs on semi rural terrains within Mexico City and identify the sources of these organic contaminants in crops, water for irrigation and soil in two areas (Tlahuac and Milpa Alta), which are considered important as aquifer recharge zones.

2. Material and methods

The Metropolitan Zone of Mexico Valley (MZMV), comprises Mexico City which is considered a Megacity (Molina & Molina, 2004) located in a basin on the central Mexican plateau with a population around 20 million, 4 million vehicles, and 35,000 industries

(Figure 1). It is situated at a tropical latitude, has an urban area of about 3500 km², is at 2240 m altitude, and is surrounded by high mountains on three sides, all of which contribute to poor air quality (Fast et al., 2007).

We collected a composite sample (2 kg) of apple for each location from an area of approximately 1500 m² during both dry and wet months and steam cactus stem in Milpa Alta in 2008-2009. For irrigation water we took 1 L during 2008 from each location with glass previously cleaned with solvents. Finally, we collected a composite sample (1 kg) of soil from each location during both dry and wet months in 2009. The samples were conserved according to standard methods of conservation.

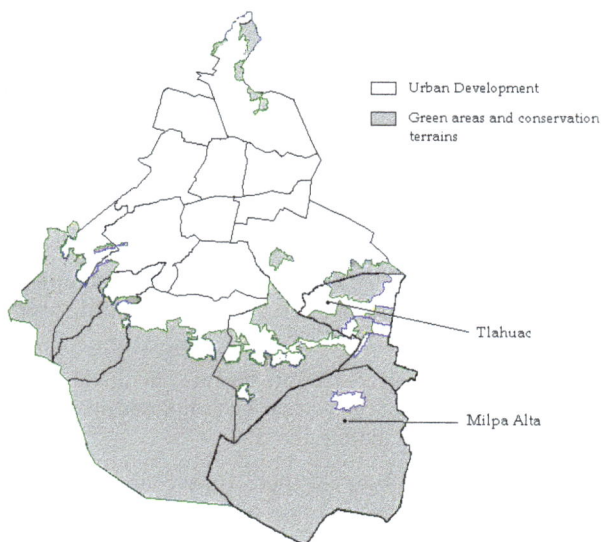

Fig. 1. Distribution of urban and green-conservation terrains in Mexico City (From GDF, 2003).

Soxhlet extraction was used for fruits and soil to extract 10 g of sample mixed with anhydrous sodium sulfate, using a solvent mix of hexane-dichloromethane (1:1) according to the method of Samsoe et al. (2002). For water samples liquid-liquid extraction was used with a mix of hexane-acetone. Chromatographic columns were prepared with chromatographic absorbents to obtain PAH extracts. The organic extract was concentrated in a rotary evaporator to 1 mL and transferred to a vial for gas chromatographic analysis according to EPA method 8100. The concentrations and profiles of PAH compounds were analyzed using a Perkin Elmer AutoSystem gas chromatograph with capillary column HP-5. The oven temperature was initially set at 90°C and the final temperature was 300°C. Detector and injector temperature were 320°C. The carrier gas was high purity helium (99.99%). A sample of 1 µL was injected in splitless mode.

Identification of PAH compounds was based on matching their retention time with a mixture of PAH standards (Chem Service). The 16 PAH compounds were naphthalene (Nap), acenaphthylene (Acy), acenaphthene (Ace), fluorene (Flu), phenanthrene (Phe), anthracene (Ant), fluoranthene (Fla), pyrene (Pyr), benzo(a)fluorene (BaF), benzo(a) anthracene (BaA), chrysene (Cry), benzo(b)fluoranthene (BbF), benzo(k) fluoranthene (BkF), indeno(1,2,3-cd)pyrene (Ind), dibenzo(ah+ac)anthracene (DaA), and benzo(ghi)perylene (Bghi). In general the detection limits were 0.01 mg/kg.

Quality control was carried out by analysis of fortified blanks and samples together with the performance of the GC. Recoveries were 80-95% except for naphthalene for which a value of 50% was obtained. Quantification of individual PAHs was made by an external standard method.

3. Results and discussion

3.1 Water

The occurrence of low weight molecular compounds in water is due to wet and dry deposition of particles from the atmospheric that contain adsorbed PAHs such as naphthalene and phenanthrene. The probable source of these compounds is organic matter combustion to low temperatures (Nagy et al., 2007); relatively high concentrations in comparison to high molecular weight compounds can be explained by the relative solubility of the PAHs. This pattern of concentrations has been seen before (Bishnoi et al., 2005; Ma et al., 2005; Chung et al., 2008)

The mean value of total PAHs in Tlahuac and Milpa Alta water samples for 2008 were both 0.04 µg/L. This did not exceed the permissible value of 1.0 µg/L proposed by European Union for irrigation of crops.

In 2009, the values were lower than detection limits. The PAHs were concentrated in suspended particles or in sediment particles that accumulated in the reservoirs with high content of fine sediments and organic matter.

Within results in water samples, we observed that values changed according to the wet and dry season. In the dry season, concentration and number of PAHs were higher than in the wet season, probably due to a dissolution effect where concentration reduced. There is a great interaction of contaminants between atmospheric, water superficial bodies and soils. The scarcity of water in some months of year for crops is limiting in these rural terrains

The use of wastewater for this zone does not present a problem of contamination for crops and soils, as there is no association between the PAHs and type of water. In some areas of China there a direct association between residual water and contamination and degradation of arable soils and an associated drop in quality of crops over the medium to long term (Cai et al., 2007).

In the figure 2 we appreciated the distribution of PAHs according number of aromatic rings, where in Tlahuac and Milpa has similar percentage of 3, 4 and 6 aromatic rings, the high values were in 3 (33% in both sites) and 4 (35% approximately in both sites). Similar studies have determined that those compounds are derived of combustion of organic matter (vegetation and fuels), so PAHs occurrence in water in both zones is by combustion mainly via deposition for case of Mexico City.

The occurrence of PAHs in water generally is associated by suspended particles or sediments due the high affinity to organic carbon further low solubility of many aromatic compounds. The main source of contamination of PAHs in water is by deposition of airborne and urban storm water runoff. Many PAHs found in water are derived of combustion process such as anthracene, benzo(a)pyrene and others compounds. Some investigations have suggested a monitoring of wastewater a long term domestic wastewater irrigation to evaluate risks about this resource for agricultural terrains (Chung et al., 2008)

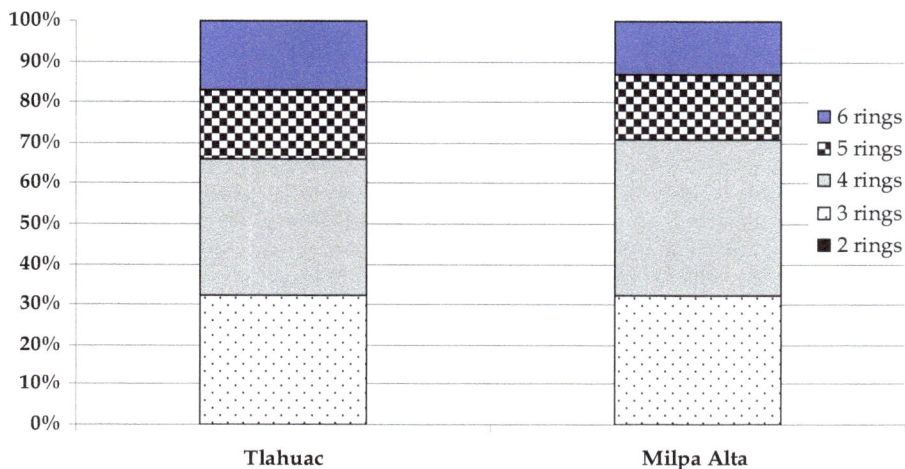

Fig. 2. Distribution of PAHs in water for irrigation in crops semi-rural terrains in Mexico City.

Finally, these areas are considered areas of recharge for aquifers but the PAHs found in the water do not represent a risk for crops (toxicity by translocation) and soils (degradation). Despite the high contamination in Mexico City, wastewater use is adequate for crops although some work has shown that wastewater is a potential source of PAHs contamination (Escobedo et al., 2000; Chung et al., 2008).

3.2 Crops

Our study focus in two crops of economical importance for Tlahuac and Milpa further the government has impulse with economical, technical and material resources for development of farmer. Cactus stem has high acceptation in population because is traditional food from prehispanic to present time while apple crop recently is accepted as attractive and rentable crop. Although there are several agriculture products that grow in these areas and around such as spinach, broccoli, amaranth, olive, ornamental flowers and etc (Grupo Produce, 2006).

Human exposure to PAHs is 88–98% connected with food (meat, seed, vegetable, fruit and others). PAHs can penetrate food indirectly (from air or water) and directly, e.g. during smoking. Once these compounds are released into the atmosphere they can be transported away from their emission sources over long distances and/or deposited to the terrestrial and aquatic environment through dry and wet deposition. A major issue associated with the emission of these compounds is the zone of influence, which determines whether the possible source has predominately local impacts or contributes to regional or global background levels (Rey-Salgueiro et al., 2008).

According to Rey-Salgueiro et al. (2008) the form of vegetable or fruit influenced the PAH concentration over the surface as they found that leaves and quasispherical fruits (grapes and tomatoes) had greater concentrations of PAHs (4 and 5 aromatic rings) than quasi-conical fruits (pepper). Some authors have suggested that lower molecular weight PAHs which dominate in the atmosphere can easily penetrate the cuticle surface of the foliage, while the higher molecular weight PAHs, mainly associated with the atmospheric particulates, are only superficially deposited on plant foliage and are thus more easily washed away by rain.

Kluska (2003) established that the content of PAH in fruits and vegetables depends on pollution in the environment (mainly air pollution) and on area of contact: Apples usually contain 200–500 ng/kg, tomatoes 200 ng/ kg, spinach 6600 ng/kg, and cabbage (savoy) 20,400 ng/kg; this provides a basis for estimating a low potential biological impact associated with the levels found. García-Falcón et al. (2006) screened for the presence of PAHs in soil of rural areas and found thattotal PAHs were always lower than 13,000 ng/kg. As a conclusion, the selected plant foods will probably not cause adverse biological effects to take place. In Tianjin (China), at a PAH contaminated site, total PAHs of rice leaves from various growth stages ranged from 58,900 to 548,000 ng/kg with a mean value of 216,000 ng/kg (Tao et al., 2006).

In Table 1 we compared our results with other vegetables where individual concentrations of PAHs in high molecular weight were highest; we believe that these differences were due at the type of skin (waxes) and morphological structure of crop further there aren't values for apples and cactus stem. Although our concentrations were variables in 2008 and 2009 probable to time and amount of rain, winds and dust storm, mainly.

Vegetation is reportedly an effective media for the entrapment of these and other compounds, mainly through atmospheric deposition. The green parts of vegetables or skin of fruits are provided with an epicuticular wax which acts as a sorbent for lipophylic contaminants (Ratola et al., 2011). It is necessary evaluate crops from semi-rural zones of Mexico City by periodically to assess good quality of products of organic contaminants such as aromatic hydrocarbons. Due some compounds has described as carcinogens at benzo(a)anthracene BaA, chrysene Cry, benzo(b)fluoranthene BbF, benzo(k)fluoranthene BkF, benzo(a)pyrene BaP, indeno(1,2,3-cd)pyrene Ind, dibenzo(ah)anthracene DaA and benzo(ghi)perylene Bghi.

In figure 3 shows dominant individual PAHs in apple were Ant, Fla, Pyr, BaP and BaA mainly in dry season and BaA, BkF, BaP, DaA in wet season. For cactus stem were DaA, Bghi, Pyr, Cry, BaA, Fla, BaP, Ind, Ant and Phe in dry season and BaA, Fla, DaA, Pyr, Bghi and Ant in wet season. In wet season in both crops were high concentrations of PAHs we supposed a diminished by effect of washing but the irregular rains in Mexico City (low content of water and spread out) only causes drop of suspended particles in atmosphere in long time of wet season

It is recognized that the low molecular weight PAHs (two and some three rings) are common in fresh fuels, but also in combustion activities and in some industrial emissions, indicating mostly petrogenic origins, four-ring PAHs (and also some three-ring) are linked with motorized traffic in general and diesel consumption in particular and the high molecular weight PAHs (five and six rings) denote the existence of heavy machine or industrial activities (Yin et al., 2008; Wang et al., 2009)

Compounds	Apple[1] (mg/kg)	Cactus stem[1] (mg/kg)	Apple[2] (mg/kg)	Cactus stem (mg/kg)	Rye rootso[3] (mg/kg)	Potato skin[4] (g/kg)	Carrot skin (g/kg)	Cabbage (g/kg)	Spinach[5] (g/kg)	Fruits[6] (g/kg)	Aerial leafs (g/kg)
Nap (2)*	---	---	0.33	---	---	0.73	0.28	90	41	16.4	589.8
Acy (3)	0.32	0.69	---	---	---	0.42	0.40	63	1.8	2.0	2.3
Ace (3)	1.11	0.58	---	---	---	---	---	---	---	0.4	16.2
Flu (3)	1.11	1.27	---	---	---	---	---	7.0	4.0	4.0	41.5
Phe (3)	1.73	2.77	0.89	0.26	0.6-43.5	3.32	1.34	55	9.2	63.6	100.6
Ant (3)	3.49	2.89	2.04	6.40	---	1.47	1.41	---	22	8.4	85.4
Fla (4)	3.56	3.68	0.01	4.97	---	2.65	10.59	152	746	49.2	66.4
Pyr (4)	3.06	5.04	---	2.89	0.8-272.1	1.63	1.23	56	256	38	8.7
BaA (4)	4.35	4.30	8.46	13.35	---	0.52	0.61	---	13	---	8.0
Cry (4)	1.17	5.23	3.15	2.21	---	---	---	15	45	17.2	12.5
BbF (5)	2.14	2.39	---	0.41	---	0.05	0.66	---	---	1.6	3.1
BkF (5)	4.09	1.99	0.98	2.10	---	0.05	0.62	---	---	1.2	4.8
BaP (5)	3.80	3.79	1.42	0.63	---	0.32	0.91	---	---	2.0	42.4
Ind (6)	0.84	4.95	2.18	2.04	---	0.17	0.40	---	---	---	---
DaA (5)	2.29	10.14	4.28	8.73	---	0.25	0.57	---	---	10	1.1
Bghi (6)	2.78	4.64	0.47	8.71	---	0.42	0.48	---	---	12	0.9

Table 1. Comparison of PAH concentrations in vegetable samples. Note: * Number of aromatic rings; --- No detected; 1. This survey, 2008; 2. This survey, 2009; 3. Zohair et al., 2006; 4. Mo et al., 2008; 5. Li et al., 2008; 6. Tao et al., 2004

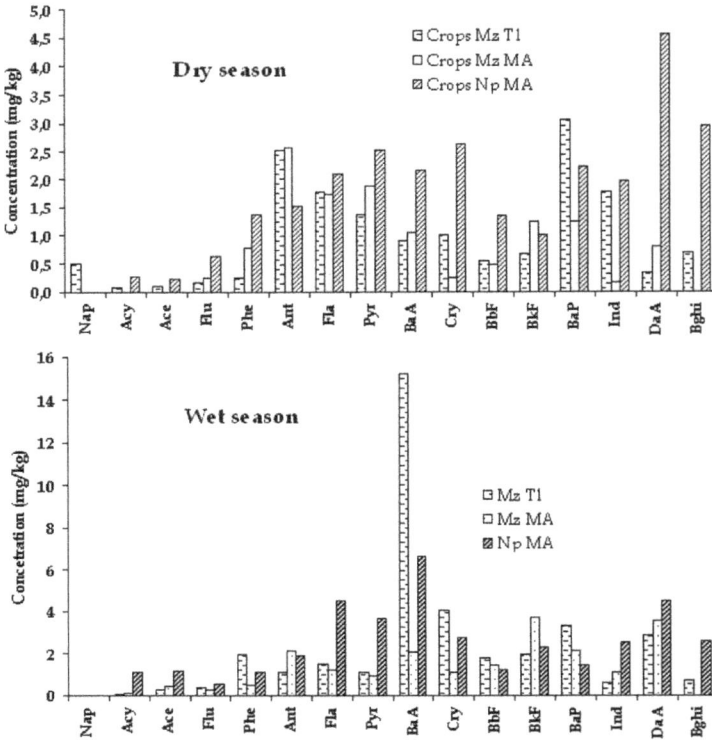

Fig. 3. Values of PAHs for semi-rural terrains from Mexico City (Tlahuac and Milpa Alta).

There routes of PAHs in plants such as take up from soil via the roots or from air via the foliage; uptake rates are dependent on the concentration, solubility, and molecular weight of the PAH and on the plant species (ATSDR, 1995). Some plants have been used to monitor atmospheric deposition of PAHs, example, mosses, lichens and vegetables. In some studies, the atmospheric PAHs such as indeno(1,2,3-c,d)pyrene, fluoranthene, and benzo(a)pyrene are deposited on foliage (leaves and flowers). In general, the atmospheric deposition on leaves often greatly exceeds uptake from soil by roots as a route of PAH accumulation.

In figure 4 we appreciated the average of compounds for 2008-2009 for apple and cactus stem where aromatic compounds were variable in both season. In general dominant compounds for crops were four and five aromatic rings follow by three and six aromatic rings. These occurrence of compounds coincide with before describe by similar studies.

The differences of aromatic compounds are influenced by particularly conditions of each area; in Tlahuac (Tl) coexist mines of construction materials as sand and gravel with diesel machinery and transit of gasoline vehicles, garbage and weed combustion. While Milpa Alta (MA) there gas station, transit of gasoline vehicle, garbage and weed combustion. These sources and particular environmental conditions can explain occurrence of PAHs. For case of apple crop in dry season only Tlahuac has naphthalene and high proportion (~15 %) of compound of six aromatic rings than Milpa Alta (~5 %) while Milpa has high proportion of three, four and five aromatic rings (~30 %, 50 % and 25 % respectively) than Tlahuac; in wet season Tlahuac showed high proportion of four aromatic rings (~60 %) than Milpa Alta (~25 %) and Milpa Alta showed high proportion of three, five and six aromatic rings (~18 %, 52 % and 9 % respectively) than Tlahuac (~ 10 %, 25 % and 5 %). For cactus stem showed differences in four, five and six aromatic rings in dry (~35 %, 30 % and 18 %) and wet season (~45%, 22 % and 10 % respectively).

The differences among crops were due waxes composition of skin where cactus stem has a high concentration of PAHs than in apple. In cactus stem has elevated waxes than apple skin and favour accumulation of particles or contaminants due PAHs are affinity a fat and waxes.

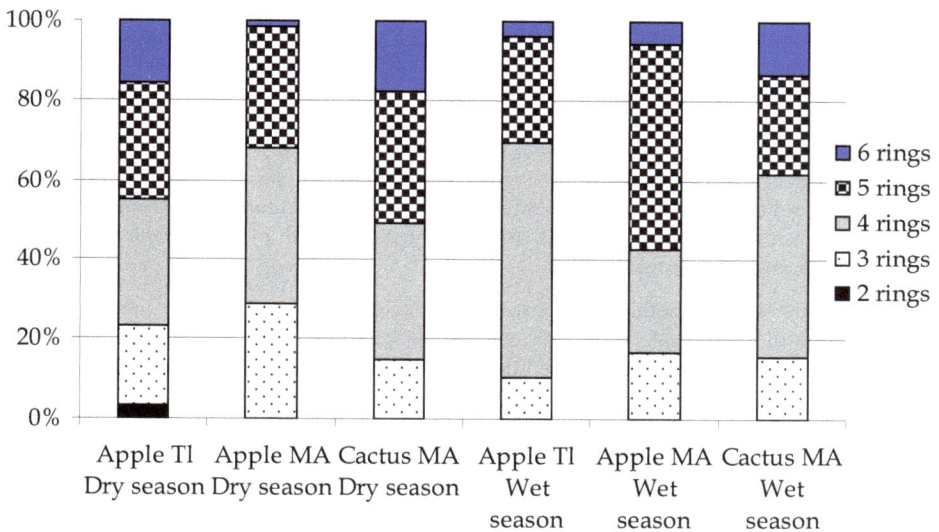

Fig. 4. Distribution of PAHs in crops according number of aromatic rings in Tlahuac (Tl) and Milpa Alta (MA).

Finally vegetables and fruits obtained from a polluted environment may contain higher PAH concentrations than those obtained from nonpolluted environments. According to ATSDR (1995) the PAH content of plants and animals living on the land or in water can be many times higher than the content of PAHs in soil or water.

3.3 Soils

According to the results we found higher values of PAHs in crops than soil. For the case of crops, the individual PAHs were slightly lower in wet than dry season; in the rainy season the irregular intensity and duration of rainfall cleans the atmospheric of suspended particles. In figure 4 the results show greater values in apple from Tlahuac and cactus stem from Milpa Alta in the wet season. In general, the compounds > 3 aromatic rings were present in crops and soil in both season such as BaA, Cry, Fla, Pyr, DaA, BaP and BkF are dominant in apple while cactus stem. According García-Alonso et al. (2003) the presence of Fla and Bghi in the two areas may indicate a common vehicular emission source.

There are many factors influencing the distribution of different PAHs in the environment, such as physicochemical properties of PAHs, physicochemical properties of soil, sources of emission of PAHs, and photochemical degradation of atmospheric PAHs. Two to three ring PAHs are subject to atmospheric transport to remote areas and are considered "multi-hop" chemicals, while higher ring PAHs are associated with particles and undergo "single hop" transport behavior, and higher ring PAHs are prone to rapid deposition and retention close to source regions. So, PAHs may become fractionated from source regions to remote regions during atmospheric transport (Wang et al., 2010).

The concentrations of BaP in the soils is an indication of both pyrogenic and petrogenic sources of PAH pollution on the environment (Essumang et al., 2010) for both seasons. BaP is considered within permissible limits (2 mg/kg) for agricultural and residential land use according to Mexican regulation NOM 138-SEMARNAT/SS-2003. However, the levels of BaP surpassed the limit established by the Danish Environmental Protection Agency (0.1 mg/kg) and Canadian Council of Ministers of the Environment (0.26 mg/kg; Essumang et al., 2010). The BaP concentrations are considered to be a risk to human health due to its potential exposure to carcinogenic PAHs for those living in those areas. According other international regulations our soils has severe problems of contamination, where some prevent measurements have applied such as strict control of emissions of industries, vehicular park and matter combustion.

Within polycyclic aromatic hydrocarbons considered in Mexican regulation, the concentrations of DaA, BaA, BbF, BkF and Ind were lower than the permissible limit for agricultural land use according NOM-138-SEMARNAT/SS-2003 (2 mg/kg). These compounds do not represent a risk for human health in comparison to BaP.

In figure 5 we appreciated the distribution of PAHs in dry and wet season found it a relative high concentration of chrysene in wet season in both sites. In general dominant PAHs for dry season were BaA, Cry, BaP, Fla, Ind and BkF with range of 0.5 to 3.5 mg/kg; Milpa Alta has slightly high BaA, BaP and Fla and Tlahuac were Cry and DaA. In wet season dominant PAHs were Cry and BaP mainly, with range of 1 to 9.5 mg/kg; Tlahuac has high concentration of Cry and BaA and Milpa Alta were Fla, BaP, Bghi, BbF, BkF, Ind and DaA. The variability of individual PAHs is key to recognize the possible source of contamination in the semi-rural area further the environmental condition defines distribution and concentration of contaminants in the soil and other media.

The presence of 4-6 aromatic rings compounds in soils in Mexico City was similar to results from a survey in soil described by Wang et al. (2010) in urban soils from North China. In semirural soil of Mexico the compounds predominant were Cry, BaP and BaA different at found it in North of China. For dry season we found similar percent of four (50 % approximately); five (25 % approximately) and six (10 %) ring aromatic compounds in Tlahuac and Milpa Alta (Figure 6). In wet season, there are significant changes over aromatic compounds, in Tlahuac has 90 % of four ring aromatic compounds while others compounds diminished to 8 % (5 rings compounds) and 2 % (6 ring compounds). So, Milpa Alta shows similar percent of aromatic compounds than dry season.

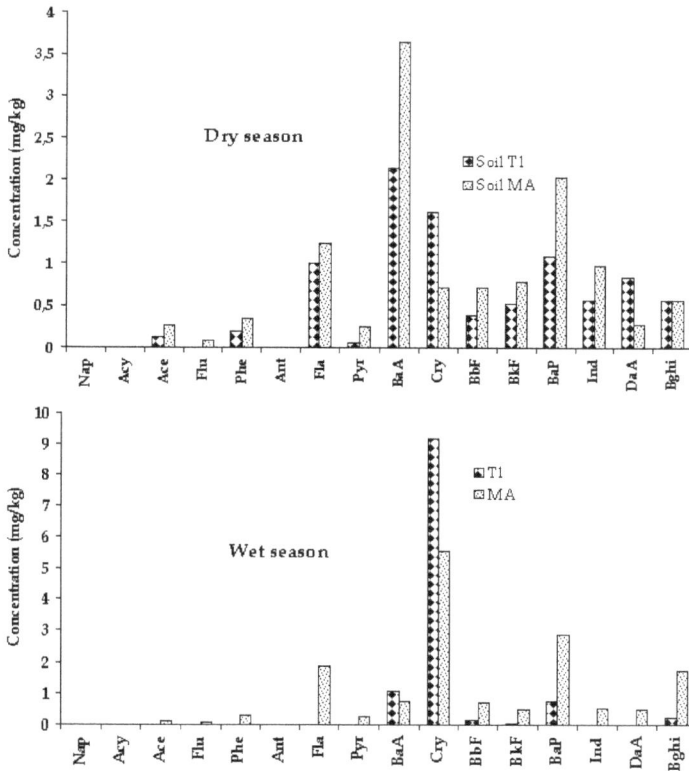

Fig. 5. Values of PAHs in soils from semi-rural terrains from Mexico City.

Nap not found in our analysis we supposed a loss in handling of sample due that is high volatile compound.

The most prominent source of PAHs in the urban environment is the incomplete combustion of biomass (such as vegetation) and fossil fuels (petroleum). Vehicular traffic (mainly diesel-powered) is considered to be the most significant contributor to the atmospheric PAH load within urban areas (Marr et al., 2004). Atmospheric deposition is the most common source of pollution in soil and it is expected that most combustion derived PAHs will be restricted to the top layer of the soil. Urban areas generally have high traffic density which results in heavy contamination of surface soils (Agarwal, 2009). For our study areas there are several asphalt ways that communicate with population areas, it considering as a source of contamination.

According Cram et al (2008), the high concentration of these contaminants in atmospheric is concentrate in South of Mexico City, mainly due to the direction of wind from North to South. As appreciate in figure 1 Tlahuac and Milpa are located in Southeast. These soils are considering as conservation, recharge of aquifer and recreational areas further gives other environmental services as retention, regulation and alleviates the deposition of contaminants derived of fossil fuels combustion that origins heavy metals and ethers, hydrocarbons, etc.

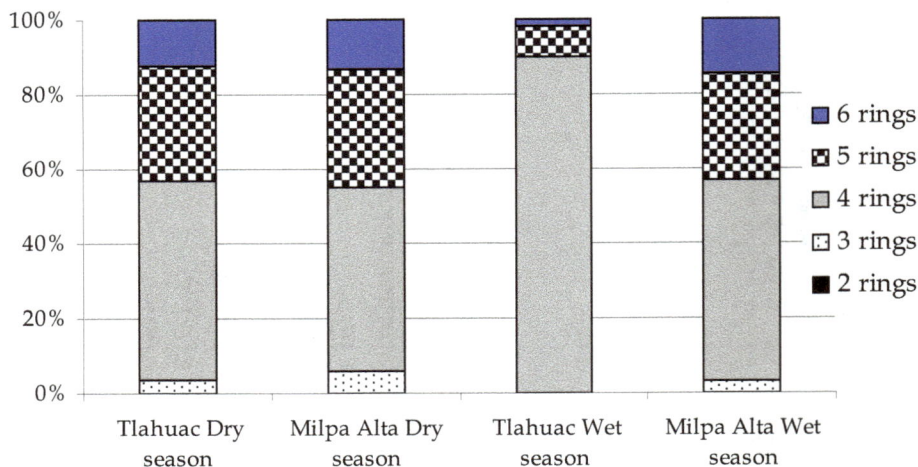

Fig. 6. Distribution of PAHs in soils according to its number of aromatic rings in Tlahuac and Milpa Alta.

Important aspect for these soils in Tlahuac and Milpa are addition of farmyard manure and compost where we appreciated low values of PAHs in comparison crops (Cai et al., 2007). The organic matter plays a role to catch contaminants where can be stabilized within its structure. A long term, organic matter is long reservoirs of several contaminants which must be practice good management of this parameter of soil.

The presence of low molecular weight PAHs (2-3 aromatic rings) is indicative of petroleum (fossil fuel combustion) while high molecular weight PAHs (> 4 aromatic rings) are more likely to be derived from organic material combustion (Ma et al., 2005). There are likely to be multiple contamination sources in vegetables and garbage from these conservation terrains.

The contamination in Mexico City is high due to the environmental conditions, for example high population density and geography (such as altitude and temperature), where atmospheric deposition (diffuse and point sources) leads to high concentrations of PAHs. Similar contamination situations have been reported in Chinese soils (Ping et al., 2007).

In table 2, we appreciated high values of individual PAHs than others areas where compounds with high molecular weight surpass values of others studies of different countries. We supposed that high concentrations are due to pull of contaminants to South of City together certain activities of specific areas as possible explication.

Maliszewska-Kordybach (1996) suggested a soil contamination classification system based on ΣPAH16 as follows: noncontaminated soil (<200 ng/g), weakly contaminated soil (200–600 ng/g), contaminated soil (600–1000 ng/g) and heavily contaminated soil (>1000 ng/g). According to this classification system, all samples in this study were heavily contaminated. This discrepancy among European and Mexican regulations are based in diverse criteria of analytical techniques mainly.

Studies made in past years for Mexico City has showed differences in abundance, type and distribution of PAHs according landscape type and territories of intensive urbanisation/

industrialisation (Marr et al, 2004; Maliszewska-Kordybach et al., 2009). The spatial distribution of soils contaminated with PAHs reflected that rural/forested/ recreation areas decreasing along the South to North of Mexico City. Due in North there high concentration of industries and urban areas while in South of City the soils are considered conservation and recharge of aquifers with low density of population.

Compounds	Soils (µg/kg)						
	Mexico, D.F.[1]		China[2]	India[3], Delhi		Korea[4]	Francia
	Tláhuac	Milpa Alta	Fangcun	Rural zone.	Highway	Rural zone	Park
Nap	---	---		131	---	78.5	
Acy	---	---	50	120	317	41.5	
Ace	60	180	9	198	298	33.7	36.0
Flu	---	90	40	48	152	37.6	9.3
Phe	150	330	281	39	259	141	254
Ant	---	---	34	25	135	33.7	9.8
Fla	500	740	583	101	599	353	8.3
Pyr	30	250	492	45	363	317	581
BaA	1620	2180	232	47	521	284	244
Cry	5400	3150	693	23	332	267	319
BbF	270	720	267	36	540	431	313
BkF	270	630	101	37	661	138	139
BaP	920	2030	136	35	461	294	249
Ind	250	750	47	---	621	248	145
DaA	420	380	42	---	623	120	21
Bghi	380	1150	70	---	1618	221	239
Total	10270	12580	3077	885	7501	2834	3390

Table 2. Comparison of PAH concentrations in rural soils near to urban areas. Note: 1. This study, 2009; 2. Chen et al., 2005; 3. Agarwal, 2009; 4. Nam et al, 2003.

The importance of contamination of soil is due serious risks on population health due:

* Inhalation of smallest particles
* Ingestion of particles or food on contaminate soils
* Direct contact with skin in workers of field or direct markets of commercialization (Sabroso & Pastor, 2004).

Further the contamination of water resource in form superficial or underground that are employ for irrigation for crops or human and animal consumption

Government hopes that in medium time the contaminants reduce significantly in favour the environment and health population of Mexico by programs of improvement of quality air.

3.4 Sources of PAHs in semirural sites

We used the Flu/(Flu+Pyr), Ant/(Ant+Phe), Ind/(Ind+Bghi), BaP/(BaP+Cry), BkF/Bghi and BaP/Bghi ratios and the majority of samples fell into the section identifying pyrogenic sources (fossil fuel, grass and garbage combustion) (Table 3). This is logical considering the presence of heavy machinery (Tlahuac) and traffic jam (Tlahuac and Milpa Alta) and the vegetation combustion (which for example is sometimes used to remove weeds) in the city. Further our results are consistent with reports in Mexico City of the influence of vehicular traffic and industrial activities on atmospheric contamination and contamination of other environmental compartments such as water, soil, crops and organisms (Marr et al., 2004).

Diagnostic ratios	Tlahuac (Tl)		Milpa Alta (MA)		Probable source
	Dry season	Wet season	Dry season	Wet season	
Fla/(Fla+Pyr)	0.94	0.86	---	0.88	Tl: Pyrogenic MA: Pyrogenic
Ant/(Ant+Phe)	---	---	---	---	Tl: Not detected MA: Not detected
Ind/(Ind+Bghi)	0.50	---	0.68	0.23	Tl: Vegetation combustion MA: Vegetation combustion and petrogenic
BaA/(BaA+Cry)	0.57	0.11	0.88	0.03	Tl: Pyrogenic and petrogenic MA: Pyrogenic and petrogenic
BkF/Bghi	0.92	0.09	1.45	0.27	Tl: pyrogenic MA: pyrogenic and petrogenic
BaP/Bghi	1.93	3.55	3.36	1.63	Tl: Traffic and vegetation combustion MA: Traffic and vegetation combustion

Table 3. Diagnostic ratios for identification of contamination source in rural sites from Mexico City.

Further, we employed some statistic tools such as principle components analysis (PCA) and extraction with different factor loadings indicated correlations of each pollutant species with each PC. Each PC was further evaluated and recognized by source markers or profiles as reasonable pollution sources according to Agarwal (2009), Wang et al., (2009) and Zhang et al. (2011).

In this investigation, PCA was performed for PAHs founded in soil samples using Statistica 16.0 software. Within principal components with values greater than 0.7 were retained. Two PCs were finally extracted and explained 58.3% of the total variance for apple case, 93.4% for cactus stem and 71.8% for irrigation water for 2008 (Table 4).

In 2008 for apple, PC 1 explained 31.8% of the total variance and had heavier loadings on Ace, Fla, BaA, Cry and BkF. The presence of Fla, BaA and Cry are typical tracers of traffic emission. Bkf and Bbf are also largely released by both gasoline and diesel engines (Wang et al., 2010). Thus, PC 1 can represent contribution from traffic emission. PC 2 explained 26.4% of the total variance and had heavier loading on Flu, Ind and DaA. As Ind and DaA are considered as predominant emissions of industrial and diesel combustion; PC 2 was deduced to represent industrial combustion (Agarwal, 2009). In cactus stem, PC 1 defined 54.5% and had heavier loadings on Ace, Flu, Phe, Pyr, BbF, BaP, Ind and DaA. The presence of 2 and 6 aromatic rings showed a vegetation and fossil fuel combustion, so PC1 represented a mixed combustion. PC 2 explained 38.8% of total variance and had heavier loading on BaA and Cry, indicators of fuel combustion (Table 4).

For irrigation water, PC 1 explained 52.4% of total variance and had heavier loadings Acy, Ace, Flu, Ant, Cry, BbF, BaP, Ind, DaA and Bghi. These compounds are from vegetation, fuels and industrial combustion. PC1 is classified as mixture combustion. PC 2 explained with 19.3% of total variance and had only Phe, as derived of vegetation combustion. Thus, PC 2 is vegetation combustion.

Compounds	Apple		Cactus stem		Irrigation water	
	Factor 1	Factor 2	Factor 1	Factor 2	Factor 1	Factor 2
Nap	---	---	---	---	---	---
Acy	0,621	-0,596	-0,687	-0,697	**-0,715**	0,446
Ace	**0,773**	-0,483	**-0,701**	-0,697	**-0,795**	0,572
Flu	0,283	**-0,803**	0,829	-0,520	**-0,882**	-0,434
Phe	-0,330	-0,536	0,924	-0,381	-0,309	**-0,810**
Ant	-0,384	-0,382	-0,682	-0,697	**-0,773**	-0,516
Fla	**0,855**	-0,002	0,347	-0,622	-0,593	-0,608
Pyr	0,550	0,143	**0,765**	-0,633	-0,531	0,404
BaA	**0,881**	0,169	0,624	**-0,724**	-0,535	0,299
Cry	**0,808**	-0,104	-0,643	**-0,758**	**-0,715**	0,121
BbF	0,596	0,346	**0,901**	0,371	**-0,891**	-0,327
BkF	**0,722**	0,368	0,687	-0,621	-0,541	0,170
BaP	0,146	-0,471	**0,775**	-0,618	**-0,855**	-0,288
Ind	-0,095	**-0,858**	**-0,734**	-0,673	**-0,718**	0,470
DaA	0,123	**-0,889**	**-0,713**	-0,695	**-0,809**	0,433
Bghi	-0,264	-0,489	**0,875**	-0,473	**-0,909**	-0,094
Explain Variance (%)	31.83	26.47	54.50	38.84	52.42	19.37
Accumulative Variance (%)	31.83	**58.30**	54.50	**93.34**	52.42	**71.79**

Table 4. Principal component analysis on apple, cactus stem and irrigation water in rural areas of Tlahuac y Milpa Alta (2008, first step). Note: --- Not detected

For 2009, second step of sampling we analyze apple and cactus stem as one matrix while Tlahuac and Milpa Alta soil samples with same criteria with intention to corroborate the sources of contamination founded in 2008 sampling in others matrix and to understand the movement of PAHs in soil. In crops we employ three PC that explained 73.4% of the total variance. For soil samples we employed four PC that explained 68.9% of total variance for soils in semirural areas (Table 5).

In crops, PC 1 explained 48.7% of the total variance and had heavier loadings on Acy, Ace, Flu, Ant, Fla, BbF and Bghi. The aromatic compounds are associated with vegetation and fuels combustion. Thus, PC 1 can represent contribution from mixed combustion. PC 2 explained 13.6% of the total variance and had heavier loading on BaA, derived of fuel combustion; PC 2 was deduced to represent fuel combustion. And PC 3 explained 10.9% of the total variance and had heavier loading on Phe and Cry, these derived of fuel combustion. In soils, PC 1 explained 35.8% of total variance and had principal compounds on Ace, Ant, Fla, Pyr, BaA, BbF, BaP and Bghi. These ranges of compounds are representative of vegetation, fossil fuel and industrial combustion (Agarwal, 2009). PC 2 explained 15% with BkF derived of diesel combustion, PC 3 explained with 9.5% on Cry derived fuel combustion and PC 4 explained 8.4% with DaA associated with industrial combustion (Zhang et al., 2011).

Compounds	Apple and Cactus stem			Soils			
	Factor 1	Factor 2	Factor 3	Factor 1	Factor 2	Factor 3	Factor 4
Nap	0,095	-0,519	0,152	0	0	0	0
Acy	**-0,980**	-0,115	0,023	0	0	0	0
Ace	**-0,972**	-0,125	0,024	**0,947**	-0,047	0,203	-0,026
Flu	**-0,915**	-0,154	0,026	0	0,511	-0,028	0,365
Phe	-0,109	-0,134	**-0,884**	0,682	0	0	0
Ant	**-0,824**	0,435	0,233	**0,970**	-0,211	0,008	-0,000
Fla	**-0,966**	-0,070	0,020	**0,912**	0,337	0,045	0,060
Pyr	-0,302	0,651	0,286	**0,849**	0,407	0,033	0,257
BaA	-0,390	**0,729**	-0,177	**0,736**	-0,323	0,363	-0,236
Cry	-0,382	-0,037	**-0,833**	0,123	0,246	**-0,814**	-0,467
BbF	**-0,981**	-0,112	0,022	**0,975**	-0,066	-0,182	-0,069
BkF	**-0,890**	-0,197	0,102	0,675	**-0,696**	-0,046	-0,065
BaP	-0,127	0,381	-0,214	**0,717**	-0,474	-0,395	-0,044
Ind	-0,664	-0,390	0,094	0,584	-0,022	0,327	-0,316
DaA	-0,102	0,580	-0,133	0,080	-0,496	-0,306	**0,715**
Bghi	**-0,903**	0,022	-0,003	**0,763**	0,499	-0,122	0,004
Varianza explicada (%)	48.77	13.61	10.95	35.86	15.07	9.58	8.46
Varianza acumulativa (%)	48.77	62.48	**73.43**	35.86	50.93	60.51	**68.97**

Table 5. Principal component analysis on crops (apple and cactus stem), and irrigation water in rural zones of Tláhuac y Milpa Alta (Step second).

Biomass burning and wildfire are important sources of organic contaminants (PAHs) at a global level. Motor vehicle emission in urban areas where population densities are much higher were found to be high, contribution of PAHs from motor vehicles going to air, dust, water, crops and human exposure; the risk is much higher than in rural areas (Shen et al., 2011). Although recent decades have shown a trend in some big cities of decreasing PAH concentrations due to emission control measures introduced in some countries.

The urban area comprises a wide range of different land uses such as traffic, industry, business, residence, garden and public green space, implying different patterns of human activities and their possible impacts on soil quality. Some work has demonstrated that specific land uses in the urban environment always showed higher PAH concentrations than other land uses. For example, soils collected at the roadside or in busy streets in Shanghai, Dalian and New Orleans all showed much higher levels of PAHs than those collected from parks and residential areas. Haugland et al. (2008) and Jiao et al. (2009) studied PAHs in urban soils from Bergen, Norway and Tianjin, China, respectively, and soils from both cities showed much higher PAHs in the industrial area than other areas.

Although these studies have indicated different levels of PAHs in some land uses of urban areas, research about PAH composition and sources in different land uses of urban environment is scarce; it is thus highly desired to have a better understanding about how different land uses affect PAH distribution in urban soils (Liu et al., 2010).

According to Amador-Muñoz et al (2011) the principal sources were diesel, natural gas and fuel combustion, biogenic emissions and organic matter pyrolysis where PAHs are associated with airborne particles in atmospheric media. Generally, between 80% and almost 100 % of PAHs with 5 rings or more (which are predominately particle-bound in the atmosphere) can be found associated with particles with an aerodynamic diameter of less than 2.5 μm (European Communities, 2001).

The presence of heavy machinery (Tlahuac) and vehicle traffic (Tlahuac and Milpa Alta) with vegetation combustion (which for example is sometimes used to remove weeds) are the sources of PAHs in the city. Further, our results are consistent with reports in Mexico City of the influence of vehicular traffic and industrial activities on atmospheric contamination and contamination of other environmental compartments such as water, soil, crops and organisms.

The presence of low molecular weight PAHs (2-3 aromatic rings) is indicative of petroleum (fossil fuel combustion) while high molecular weight PAHs (> 4 aromatic rings) are more likely to be derived from organic material combustion (Ma et al., 2005). There are likely to be multiple contamination sources in vegetables and garbage from these conservation terrains.

Lastly, recent studies indicate that POPs atmospheric depositions are main source of contamination in big cities derived fossil fuels (diesel, gasoline and natural gas), garbage and vegetation combustion (Rossini et al., 2005). For this reason, the atmospheric compartment must be constantly monitored by supervisory authority and considered by Mexican regulation which, until now, has provided limits for some pollutants such as suspended particles, ozone, carbon monoxide and dioxide, nitrogen oxides and sulphur oxides especially considering the high vehicular units, industrial zones and landfills.

According geography and meteorology conditions play critical roles in the dilution and dispersion of air pollution from source locations, through vertical mixing and horizontal transport in Mexico City (Figure 7). Vertical mixing is facilitated by upward motion of warm air near the surface, to cold air above. In cases where this temperature profile is reversed and warm air lies above colder air at the surface, vertical motion is restricted. This sets up a temperature inversion, which is characterised by stable atmospheric conditions, and results in the accumulation of air pollution (November to March, mainly). Horizontal wind speeds are also reduced, limiting transport of pollutants downwind. The resulting poor air quality and concentrate of organic contaminants may lead to health problems in susceptible populations (Wallace et al., 2010).

Many studies highlight a distinct increase in concentrations of pollution during temperature inversions in Los Angeles, London, Tokyo and others important cities. This seasonal variation also coincides with that of temperature inversions, which are also most frequent in the winter and spring and lead to the accumulation of not just air pollutants, but also allergens and viruses (Wallace et al., 2010).

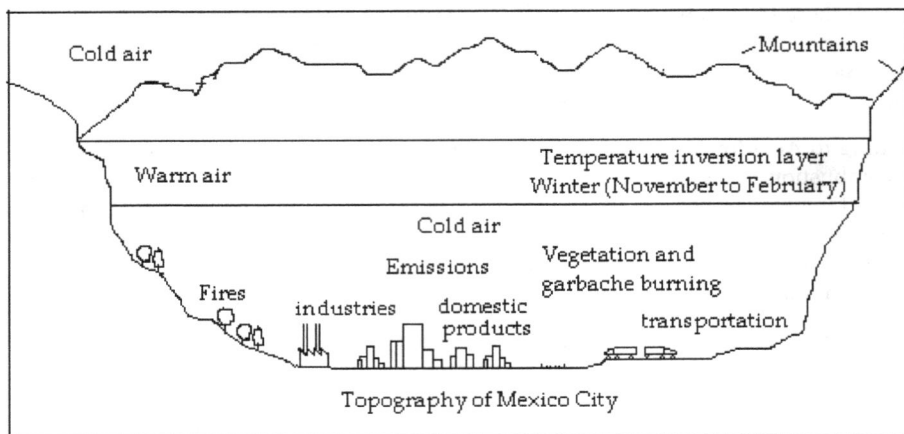

Fig. 7. Scheme contamination for PAHs in Mexico City, while in spring, summer and autumn the behaviour of contaminants change according to environmental conditions.

Thus we found a good description of contamination sources for our studied matrix and the movement of PAHs in the semirural environment for Mexico City. With better data, the Mexican authorities can take more informed decisions in the management of natural resources, legislation and politics, for better control of contaminants and pollution in general.

4. Conclusion

PAH concentrations were variable through the study due to environmental conditions of season, wet (rain) and dry (dust) deposition mainly for crops and soil. The quality of atmospheric conditions defines the contamination in these zones, both for wet and dry deposition. For the case of crops (apple and cactus stem) the values were high over the skins for high and intermediate molecular weights, but values declines with adequate washing or peeling. In soils the values found were within permissible limits of individual PAHs such as benzo(a)pyrene. Organic matter has a high affinity to catch organic contaminants in soils; it is a crucial environmental parameter that regulates the availability of inorganic and organic contaminants in the environment.

This type of study is important to evaluate the degree of contamination in specific environments, considering environmental variables to know the movement the organic contaminants. This will also help guarantee the quality of food produced in semi rural zones nearest to high density population and/or industrial areas such as Mexico City.

Nowadays, the new technologies employ in fuels, gasoline and diesel engines and programs of industrial-vehicular control has improvement the air quality in Mexico City. These actions have the goal to reduce many organic contaminants in favour of human and environment health. In the last years.

5. Acknowledgment

The research was supported by Universidad Autonoma Metropolitana campus Xochimilco.

6. References

Agarwal, T. (2009). Concentration level, pattern and toxic potential of PAHs in traffic soil of Delhi, India. *Journal of Hazardous Materials*, 171: 894–900

Amador-Muñoz, O.; Villalobos-Pietrini, R.; Miranda, J. & Vera-Avila L.E. (2011). Organic compounds of $PM_{2.5}$ in Mexico Valley: Spatial and temporal patterns, behavior and sources. *Science of the Total Environment*, 409: 1453–1465

Agency for Toxic Substances and Disease Registry ATSDR. (1995). *Toxicological profile for Polycyclic aromatic hydrocarbons*. U.S. Department of Health and Human Services, Public Health Service. 458 pp.

Bishnoi, N.R.; Mehta, U.; Sain, U. & Pandit, G.G. (2005). Quantification of polycyclic aromatic hydrocarbons in tea and coffee samples of Mumbai city (India) by high performance liquid chromatography. *Environmental Monitoring and Assessment*, 107: 399–406

Cai, Q.Y.; Mo, C.H.; Li Y.H.; Zeng, Q.Y.; Katsoyiannis, A.; Wu, Q.T.; Férard, J.F. (2007). Occurrence and assessment of polycyclic aromatic hydrocarbons in soils from vegetable fields of the Pearl River Delta, South China. *Chemosphere*, 68: 159-168.

Chung, N.J.; Cho, J.Y.; Park, S.W.; Hwang, S.A. & Park, T.L. (2008). Polycyclic aromatic hydrocarbons in soils and crops after irrigation of wastewater discharged from domestic sewage treatment plants. *Bulletin Environmental Contamination and Toxicology*, 81: 124–127

Cram, S.; Cotler, H.; Morales L.M.; Summer, I. & Carmona, E. (2008). Identificación de los servicios ambientales potenciales de los suelos en el paisaje urbano del Distrito Federal. *Investigaciones Geográficas, Boletín del Instituto de Geografía, UNAM*, 66: 81-104. ISSN 0188-4611.

Eom, I.C.; Rast, C.; Veber, A.M. & Vasseur, P. (2007). Ecotoxicity of a polycyclic aromatic hydrocarbon (PAH)-contaminated soil. *Ecotoxicology Environmental Safety*, 67: 190-205.

EPA (Environmental Protection Agency-US) (1986). *Method 8100 Polynuclear aromatic hydrocarbons*. Revision. September. 10 pp. Available http://www.epa.gov/osw/hazard/testmethods/sw846/pdfs/8100.pdf. Accessed 11 September 2008.

Escobedo, J.F.; Victoria, A.R. & Ramírez A. (2000). *La problemática ambiental en la Ciudad de México generada por las fuentes fijas*. Secretaría del Medio Ambiente. 14 pp. ISSN

Essumang, D.K.; Kowalski, K. & Sogaard, E.G. (2011). Levels, distribution and source characterization of polycyclic aromatic hydrocarbons (PAHs) in topsoils and roadside soils in Esbjerg, Denmark. *Bulletin Environmental Contamination and Toxicology*, 86: 438-443.

European Communities. (2001). *Ambient air pollution by Polycyclic Aromatic Hydrocarbons (PAH)*. Position Paper. Prepared by the Working Group On Polycyclic Aromatic Hydrocarbons. Luxembourg. Pp 49.

Fast, J.D.; de Foy, B.; Rosas, F.A.; Caetano, E.; Carmichael, G.; Emmons, L.; McKenna, D.; Mena, M.; Skarmarock, W.; Tie, X.; Coulter, R.L.; Barnard, J.C.; Wiedinmyer, C. & Madronich, S. (2007). A meteorological overview of the MILAGRO field campaigns. *Atmospheric Chemistry and Physics*, 7: 2233–2257.

Finizio, A.; Di Guardo, A.& Cartmale L. (1998). Hazardous air pollutants (PAHs) and their effects on biodiversity: An overview of the atmospheric pathways of persistent organic pollutants (POPs) and suggestions for future studies. *Environmental Monitoring and Assessment*, 49: 327-336.

García-Alonso, S.; Pérez-Pasto, R.M. & Sevillano-Cataño, M.L. (2003). Occurrence of PCBs and PAHs in an urban soil of Madrid (Spain). *Toxicology Environmental Chemistry*, 85: 193-202.

García-Falcón, M.S.; Soto-González, B. & Simal-Gándara, J. (2006). Evolution of the concentrations of polycyclic aromatic hydrocarbons in burnt woodland soils. *Journal of Environmental Quality*, 40: 759-763.

GDF (Gobierno del Distrito Federal) (2003). *Programa general de ordenamiento ecológico del Distrito Federal*. Available via http://www.sma.df.gob.mx/sma/index.php?opcion=26&id=61. Accessed 11 March 2009

Grupo Produce DF (2006). *Aportes en marcha*. Available vía http://www.grupoproducedf.org.mx/aportes.htm. Accessed 11 March 2008

Haugland, T.; Ottesen, R.T. & Volden, T. (2008). Lead and polycyclic aromatic hydrocarbons (PAHs) in surface soil from day care centers in the city of Bergen, Norway, *Environment Pollution*, 153 266-272.

Jiao, W.T.; Lu, Y.H.; Li, J.; Han, J.Y.; Wang, T.Y.; Luo, W.; Shi, Y.J. & Wang, G. (2009). Identification of sources of elevated concentrations of polycyclic aromatic hydrocarbons in an industrial area in Tianjin, China. *Environmental Monitoring and Assessment*, 10.1007/s10661-008-0606-x.

Krauss, M.; Wilcke, W. & Zech, W. (2000). Polycyclic aromatic hydrocarbons and polychlorinated byphenyls in forest soils: Depth distribution as indicator of different fate. *Environmental Pollution*, 110: 79-88

Kluska, M. (2003). Soil contamination with polycyclic aromatic hydrocarbons in the vicinity of the Ring road in Siedlce City. *Polish Journal of Environmental Studies*, 12: 309-313.

Ma, L.L.; Chu, S.G.; Wang, X.T.; Cheng, H.X.; Liu, X.F. & Xu, X.B. (2005). Polycyclic aromatic hydrocarbons in the surface soils from outskirts of Beijing, China. *Chemosphere*, 58: 1355-1363.

Marr, L. C.; Grogan L.A.; Wohrnschimmel, H.; Molina, L.T.; Molina, M.; Smith, T. J. & Garshick, E. (2004). Vehicle traffic as a source of particulate polycyclic aromatic hydrocarbon exposure in the Mexico City Metropolitan Area. *Environmental Science and Technology*, 38: 2584-2592.

Maliszewska-Kordybach, B. (1996). Polycyclic aromatic hydrocarbons in agricultural soils in Poland: preliminary proposals for criteria to evaluate the level of soil contamination. *Applied Geochemistry*, 11: 121-127.

Maliszewska-Kordybach, B.; Smreczak, B. & Klimkowicz-Pawlas, A. (2009). Concentrations, sources, and spatial distribution of individual polycyclic aromatic hydrocarbons (PAHs) in agricultural soils in the Eastern part of the EU: Poland as a case study. *Science of the Total Environment*, 407: 3746-3753

Mastral, A. & Callen M.S. (2000). A review on polycyclic aromatic hydrocarbon (PAH) emissions from energy generation. *Environment Science and Technology*, 34: 3051-3056.

Mo, C.H.; Cai, Q.Y.; Tang, S.R.; Zeng, Q.Y. & Wu, Q.T. (2008). Polycyclic aromatic hydrocarbons and phthalic acid esters in vegetables from nine farms of the Pearl River Delta, South China. *Archives Environmental Contamination and Toxicology*, 56: 181-189.

Molina, M.J. & Molina, L.T. (2004). Megacities and Atmospheric Pollution (Critical Review). *Journal Air & Waste Management Association*. 54:644-680. ISSN 1047-3289

Nam, J.J.; Song, B.H.; Eom, K.C.; Lee, S.H. & Smith, A. (2003) Distribution of polycyclic aromatic hydrocarbons in agricultural soils in South Korea. *Chemosphere*, 50:1281-1289.

NOM-138-SEMARNAT/SS-2003. (2003). *Límites máximos permisibles de hidrocarburos en suelos y las especificaciones para su caracterización y remediación.* Diario Oficial de la Federación 25/Marzo/2005. 21 pp

Li, Y.T.; Li, F.B.; Chen, J.J.; Yang, G.Y.; Wan, H.F.; Zhang, T.B.; Zeng, X.D. & Liu, J.M. (2008). The concentrations, distribution and sources of PAHs in agricultural soils and vegetables from Shunde, Guangdong, China. *Environmental Monitoring and Assessment*, 139: 61-76.

Liu, X. & Korenaga, T. (2001). Dynamics analysis for the distribution of polycyclic aromatic hydrocarbons in rice. *Journal of Health Science*, 47: 446-451.

Liu, S.; Xia, X.; Yang, L.; Shen, M. & Liu, R. (2010). Polycyclic aromatic hydrocarbons in urban soils of different land uses in Beijing, China: Distribution, sources and their correlation with the city's urbanization history. *Journal of Hazardous Materials*, 177: 1085-1092

Ping, L.F.; Luo, Y.M.; Zhang, H.B.; Li, Q.B. & Wu, L.H. (2007). Distribution of polycyclic hydrocarbons in thirty typical soil profiles in the Yangtze River Delta region, east China. *Environment Pollution*, 147: 358-365.

Ratola, N.; Alves, A.; Lacorte, S. & Barceló, D. (2011). Distribution and sources of PAHs using three pine species along the Ebro River. Environment Monitoring Assessment. DOI 10.1007/s10661-011-2014-x. ISSN: 1573-2959

Rey-Salgueiro, L.; Martínez-Carballo, E.; García-Falcón, M. & Simal-Gándara, J. (2008). Effects of a chemical company fire on the occurrence of polycyclic aromatic hydrocarbons in plant foods. *Food Chemistry*, 108: 347-353.

Rossini, P.; Guerzoni, S.; Matteucci, G.; Gattolin, M.; Ferrari, G. & Raccanelli, S. (2005). Atmospheric fall-out of POPs (PCDD-Fs, PCBs, HCB, PAHs) around the industrial district of Porto Marghera, Italy. *Science of the Total Environment*, 349 : 190- 200.

Sabroso, M.C. & Pastor A. (2004). *Guía sobre suelos contaminados.* CEPYME Aragón-Gobierno de Aragón. Departamento de Economía, Hacienda y Empleo. 109 pp

Samsoe, L.P.; Larsen, E.H.; Larsen, P.B. & Bruun, P. (2002). Uptake of trace elements and PAHs by fruit and vegetables from contaminated soils. *Environment Science and Technology*, 36:3057-3063

Shen, H.; Tao, S.; Wang, R.; Wang, B.; Shen, G.; Li, W.; Su, S.; Huang, Y.; Wang, X.; Liu, W.; Li, B. & Sun, K. (2011). Global time trends in PAH emissions from motor vehicles. *Atmospheric Environment*. 45: 2067-2073

Tao, S.; Cui, Y.H.; Xu, F.L.; Li, B.G.; Cao, J.; Liu, W.X.; Schmitt, G.; Wang, X.J.; Shen, W.R.; Qing, B.P. & Sun, R. (2004) Polycyclic aromatic hydrocarbons (PAHs) in agricultural soil and vegetables from Tianjin. *Science of the Total Environment*, 320: 11-24.

Yin, Ch.Q.; Jiang, X.; Yang, X.L.; Bian, Y.R. & Wang, F. (2008). Polycyclic aromatic hydrocarbons in soils in the vicinity of Nanjing, China. *Chemosphere*, 73: 389-394.

Wallace, J.; Nair, P. & Kanaroglou, P. (2010). Atmospheric remote sensing to detect effects of temperature inversions on sputum cell counts in airway diseases *Environmental Research* 110: 624-632

Wang, K., Shen Y, Zhang S, Ye Y, Shen Q, Hu J,. Wang X (2009). Application of spatial analysis and multivariate analysis techniques in distribution and source study of polycyclic aromatic hydrocarbons in the topsoil of Beijing, China. *Environment Geology*, 56: 1041-1050.

Wang, W.; Simonich, S.L.M.; Xue, M.; Zhao, J.; Zhang, N.; Wang, R.; Cao, J. & Tao, S. (2010). Concentrations, sources and spatial distribution of polycyclic aromatic hydrocarbons in soils from Beijing, Tianjin and surrounding areas, North China. *Environmental Pollution*, 158: 1245–1251

Wilcke, W. (2000). Polycyclic Aromatic Hydrocarbons (PAHs) in soil – a Review. *Journal Plant Nutrition and Soil Science*, 63: 229-248.

Zhang, Y. & Wang, J. (2011). Distribution and source of polycyclic aromatic hydrocarbons (PAHs) in the surface soil along main transportation routes in Jiaxing City, China. *Environment Monitoring Assessment*, 182: 535-543.

Zohair, A.; Salim, A.S.; Soyibo, A.A. & Beck, A.J. (2006). Residues of polycyclic aromatic hydrocarbons (PAHs) polychlorinated biphenyls (PCBs) and Organochlorine pesticides in organically-farmed vegetables. *Chemosphere*, 63: 541-553.

11

Energy and Economy Links – A Review of Indicators and Methods

Mohammad Reza Lotfalipour[1] and Malihe Ashena[2]
[1]Department of Economics, Ferdowsi University of Mashhad
[2]Department of Management and Economics, Tarbiat Mdares University, Tahran
Iran

1. Introduction

Fossil fuels as the most important kind of energy are inevitably linked with the economy and the environment. A stable and continuous supply of fossil fuels and alternative ones is needed while we develop. Economic activity is predominantly related to the energy use, principally fossil fuels, which account for over 60% of global greenhouse gas emissions. This implies an urgent need to decouple economic growth from energy use. Thus this chapter surveys the relation between energy and economy growth.

From development perspective, energy is important for eradicating poverty, improving human welfare and raising living standards (UNDP, et al. 2000). However, in many areas of the world the current patterns of energy supply and use are considered unsustainable, which limits economic development. In other areas, environmental degradation from energy production and consumption inhibits sustainable development. So, energy is critical in the context of sustainable economy and clean environment. It is therefore important for policy makers to understand the implications of different energy programs and alternative policies.

The existing literature on energy and development does show that energy development is an important component of broader development. So we have attempted to pull together some of the ways in which energy might exert a significant influence on the development process. Development involves a number of other steps besides those associated with energy, notably including the evolution of education and labor markets, financial institutions, modernization of agriculture, improving environment and etc. Nevertheless, it is hard to imagine overall economic development succeeding without energy development being one part of the evolution.

Traditional economic theory disregards the importance of energy, because it postulates that the contribution of energy to economic growth is essentially determined by the low share of energy cost in the total cost of capital, labor, and energy. Even if the cost of energy can be neglected, being one of the driving forces in the economic process, the biophysical aspects of the economy should be considered. Therefore, the role of energy in intensifying processes with increasing automation is taken into account, where energy-driven machines replace human labor.

While the mainstream theory of economic growth pays little attention to the role of energy or other natural resources in promoting economic growth, the impact of energy prices on economic activity has attracted significant attention during the last two decades. Resource and ecological economists have criticized this theory in terms of the thermodynamics implications for economic production and the long-term prospects of the economy.

Energy development or increased availability of energy is a part of enhanced economic development. But even with trends toward greater energy efficiency and other dampening factors, total energy use and energy use per capita continue to grow in the advanced industrialized countries and even more rapid growth can be expected in the developing countries as their incomes advance. Generally, economic activities, structural and technological changes are important factors influencing energy use. So some time there is evidence of mutual interactions between economy and energy.

While there is a lack of alternative model of the growth process, extensive empirical work has examined the role of energy in the growth process. The principal finding is that energy used per unit of economic output has declined, but this is to a large extent due to a shift in energy use from direct use of fossil fuels to the use of higher quality fuels.

The role of energy in production is described in the following using neoclassical concept of the production function. This production theory is very general in comparison to the specific models of economic growth including new factors (such as substitution between energy and other inputs, technological change, shifts in the composition of the energy input, and shifts in the composition of economic output) affecting production process (stern, 2004).

Furthermore, energy and economic activity both affect the environment. However, understanding these effects has important implications, while technology has been evolved and improved dramatically. The chapter is structured to cover the key points of energy-economy- environment nexus, too.

The chapter is organized as follows. We begin with a brief review of theory of production and growth that describe the channels through which increased availability of energy might act as a "key" stimulus of economic development.

In the third part conceptual linkage between energy and development and also the role of economic activity in the energy use trend are described. Development of disaggregated energy indicators makes available a powerful set of analytical tools. Those tools reveal the relationship between energy uses and their underlying driving factors.

The next part reviews energy indicators, which are developed to describe the links between energy use and human activity in a disaggregated manner. After reviewing major indicators, we discuss the basic concepts of various indicators and the methodologies used to derive them. Then we review the use of decomposition methods to aid in the analysis of trends in energy use. The key factors in this analysis are (1) substitution between energy and other inputs within an existing technology, (2) technological change, (3) shifts in the composition of the energy input, and (4) shifts in the composition of economic output.

Then energy and economy are related to the environment. Therefore energy indicators can be extended to carbon emissions which play an important role in aiding negotiations over carbon reduction targets and evaluating progress toward meeting abatement goals.

Finally after laying out some conceptual ideas, we examine some empirical evidence.

2. Theory of production and growth

This part reviews the background theory of production and growth. The role of energy in production should be considered in order to understand the role of energy in economic growth. Mainstream economists usually think of capital, labor, and land as the primary factors of production, while goods such fuels and materials are intermediate inputs (which are created during the production period under consideration and are used up entirely in production).

However, the theory of growth and its limitations of consideration of natural resource has been the subject of strong criticism. So the background theory of production and growth can be reviewed from different points of view.

There are essentially three mainstream categories of neo-classical growth models (Stern and Cleveland, 2004). The first one focuses on technological change as the only means by which growth can be achieved (Aghion and Howitt, 1998; Solow, 1956; Stern and Cleveland, 2004). Economic growth beyond equilibrium level (where further returns to capital are no longer possible) is only achievable by increasing returns to existing capital via improvements in technology.

The second category focuses on the consumption of natural capital in determining sustained economic growth. These models assume a priori that it is technically feasible to substitute between physical and natural capital (Stern and Cleveland, 2004). Achieving sustained growth relies on the correct institutional conditions (including property rights, market structure, means of considering future generations) to ensure that any depleted natural capital is substituted for with the corresponding value of man-made capital (ockwell, 2008).

The last category of growth model considers both natural resources and technological change as the determinants of growth. (Stern and Cleveland, 2004).

In all three models of economic growth, the contribution of energy to economic activity is only considered relative to its cost within production process. In economic terms, the models consider energy to be an 'intermediate good' rather than a 'primary input' into production. This implies that decoupling economic growth from energy use is a possible subject, in the case of the latter two models. In the following the mentioned growth models are described briefly.

2.1 The basic growth model

The basic model of economic growth that does not include resources at all is Solow model (1956). This model subsequently was extended with nonrenewable resources, renewable resources, and some waste assimilation services (Kamien and Schwartz, 1982; Toman et al., 1994). These extended models are, however, only applied in the context of debates about environmental sustainability, not in standard macro-economic applications.

Economic growth models examine the evolution of a hypothetical economy over time, as the quantities or the qualities of various inputs into the production process change. In Solow model (1956) using manufactured capital a constant-sized labor force produces output, which is equal to the national income. So, according to neoclassical growth theory, the only cause of continuing economic growth is technological progress. The relationship between productive inputs and output changes, as technological knowledge level rises.

However, the neoclassical the production function can be used to examine the factors that could reduce or strengthen the linkage between energy use and economic activity over time. A general production function can be represented as:

$$Q_i = f(X_i) \tag{1}$$

Where the Q_i is output, and the X_i are capital and labor inputs.

The neo-classical economic worldview sees the economy as a closed system in which goods are produced by capital and labor inputs, and then exchanged between consumers and firms. Economic growth is achieved by increasing inputs of labor or human capital or improvements in technology or quality improvements of capital and labor inputs. More recently, the role of natural capital in economic growth has also been considered.

The simple model does not explain how improvements in technology come about. More recent models explaining technological progress within the growth model as the outcome of decisions taken by firms and individuals.

In endogenous growth models the relationship between capital and output can be written in the following form:

$$Y = A.K \tag{2}$$

Where capital, K, is a composite of manufactured and knowledge-based capital. The key point is that technological knowledge, A, can be regarded of as a form of capital.

2.2 Growth models with natural resources

Most of the natural resources like fossil fuels exist in finite quantities. Finiteness and exhaustibility of fossil fuels make the notion of indefinite economic growth problematic. So far there has been relatively little work including these points in models that also examine the roles of resources in growth (Smulders, 1999).

When there is more than one input there are many alternative paths that economic growth can take. All production involves the transformation or movement of matter in some way and all such transformations require energy. So energy is an essential factor of production (Stern, 1997). Therefore, the role of energy and its availability in the economic production have been emphasized by natural scientists and some ecological economists.

In the neoclassical economics approach, the quantity of energy available to the economy in any period is endogenous (Stern, 1999). Nevertheless, this analytical approach leads to a downplaying of the role of energy as a driver of economic growth and production. Some alternative models of the economy propose that energy is the only primary factor of production. But this means that the available energy in each period is determined exogenously.

The Leontief input-output model can be regarded as an alternative to the neoclassical marginal productivity distribution theory. It represents an economy in which there is a single primary factor of production with prices that are not determined by marginal productivity. This representation of the economy or an ecosystem with energy as the primary factor was proposed by Hannon (1973).

Ecological economists express a more realistic view. They argue that economy should be considered as an open subsystem of the global ecosystem. This accounts for a broader view of inputs of natural capital. These include the absorption of waste from economic activity and the maintenance of the climate that facilitates human life. In the other world, the ecological economists' worldview attempts to account for the laws of thermodynamics.

For ecological economists, energy is a fundamental factor enabling economic production. Some others even argue that energy availability actually drives economic growth, and in turn economic growth resulting in increased energy use (e.g. Cleveland et al., 1984). From this perspective, the possibility of decoupling energy use from economic growth seems more limited.

The relationship between energy and aggregate output can then be affected by some factors such as substitution between energy and other inputs, technological change, shifts in the composition of the energy input and shifts in the composition of output. A common interpretation of standard growth theory is that substitution and technical change can effectively de-couple economic growth from energy and environmental issues.

3. Energy and development: Conceptual linkages

There are significant differences in quality and quantity between energy flows from renewable sources, nonrenewable stocks of fossil fuels and other minerals, and slowly renewable stocks of organic matter (in the form of vegetable and animal biomass), water, etc. in their current and potential contributions to society. Therefore, some inputs to production are non-reproducible, while others can be manufactured at a cost within the economic production system. The primary energy inputs are not given an explicit role in the standard growth theories which focus on labor and capital. However, capital, labor, and in the longer term even natural resources, are reproducible factors of production, while energy (except fuels) is a nonreproducible factor of production (Stern, 1999).

Since the two oil price shocks of the 1970s, there has been extensive debate concerning the trend of energy use and economic activities. It is commonly asserted that there has been decoupling of economic output and resources, which implies that the limits to growth are no longer as restricting as in the past.

The existing literature on energy and development does make clear that energy development is an important component of development more broadly. In this section we have attempted to explore some of the ways in which energy might have influence on the development process. A very simple model of the economy can be used to discuss the possible ways in which increased energy availability might be especially important to economic development. Suppose that

$$Y = F(K_Y, H_Y, E) \tag{3}$$

$$E = E(K_E, H_E) \tag{4}$$

Where Y represents output of final goods and services, and K represents the application of physical capital, H represents human capital services to the production of final good and services, and E represents energy services. Energy services in turn are produced through the application of other physical and human capital services.

A standard assumption from economic growth theory is that the production functions F, E are homogeneous of degree one that if all inputs are increased by some percent, outputs grows at the same percentage.

Energy development – increased availability of energy in quantity and quality terms – is part of enhanced economic development. Energy use per unit of output seem to decline over time in the more advanced stages of industrialization, reflecting the adoption of increasingly more efficient technologies for energy production and utilization as well as changes in the composition of economic activity (Nakicenovic, 1996).

Energy intensity in today's developing countries probably peaks sooner and at a lower level along the development path than was the case during the industrialization of today's developed world. But even with trends toward greater energy efficiency, total energy use and energy use per capita continue to grow in the developed and developing countries. Although, development involves a number of other issues (such as labor markets, financial institutions and provision of infrastructure for water, sanitation, and communications) besides those associated with energy, it is hard to imagine overall economic development succeeding without energy development.

The fact that expanded energy use (in quantity and quality) is associated with economic development still depends on the importance of energy as a factor in economic development, however. Much of the literature on energy and development, focuses mainly on how energy demand is driven by economic development (see, e.g., Barnes and Floor 1996) and on how energy services can be improved for developing countries (Dunkerley et al 1981; Barnes and Floor, 1996). Less is found in the literature on energy and development in the context of the margin of energy advance versus other inputs growth as an agent of economic development. However, there are substantial differences in energy forms and in the nature of economic activities across different stages of development. The linkages among energy, other inputs, and economic activity clearly change significantly as an economy moves through different stages of development. It is referred to "energy ladder" to describe this phenomenon (Barnes and Floor, 1996). At the lowest levels of income, energy sources tend predominantly to be biological sources (wood, dung, sunshine). More processed fuels (charcoal), animal power, and some commercial energy become more prominent in the intermediate stages. Commercial fossil fuels and other energy forms – primary fuels, and ultimately electricity – become predominant in more advanced stages of industrialization and development.

Energy provision requires a variety of different kinds of inputs. Energy utilization also depends on the opportunity costs of other inputs. Finally, the literature makes clear that observed patterns of energy production and utilization reflect a great deal of subtle optimizing behavior, given the constraints faced by the economic actors.

There are different ways in which the economic system might experience some form of increasing returns related to energy services. Therefore, increased energy availability somehow might make a disproportionate contribution to expanded economic activity. Increasing returns in energy services provision would take different forms at different stages of development. The industrial production and distribution of various forms of modern energy show increasing technological returns to scale. Moreover, the transformation of primary energy into deliverable energy (petroleum refining) also exhibits returns to scale. Energy quality is the relative economic usefulness per heat equivalent unit of different fuels. One way for measuring the energy quality is the marginal product of the fuel which means marginal increase in the quantity of a good or service produced by the use of one additional heat unit of fuel (Toman & Jemelkova, 2003).

Over the course of economic development the output mix might change. In the earlier phases of development there is a shift away from agriculture towards heavy industry, while in the later stages of development there is a shift from the more resource intensive extractive and heavy industrial sectors towards services and lighter manufacturing. It is often argued that because of different energy intensities in industries there will be an increase in energy used per unit of output in the early stages of economic development and a reduction in energy used per unit output in the later stages of economic development. Pollution and environmental disruption would be expected to follow a similar path (Panayotou, 1993). This argument can be pursued further to argue that the shift to service industries results in a decoupling of economic growth and energy use. Furthermore, shifts away from the use of coal and particularly towards the use of oil can explain the majority of declining energy intensity. The idea that a shift towards a service-based economy can achieve decoupling is one that is often put forward (Panayotou, 1993). But this notion ignores the large amounts of energy involved in producing services including some service industries such as transport (Stern and Cleveland, 2004). This implies that a shift to a service-based economy cannot achieve a complete decoupling of energy and economic growth.

4. Energy indicators

Energy indicators relating energy to economic issues can be useful tools for policy makers. They provide a way to structure and clarify statistical data to give better insight into the factors that affect energy, environment, economics and social well-being. Indicators can also be used to monitor progress of past policies.

All sectors of an economy – agriculture, manufacturing and mining, and services – require energy. These energy services in turn foster economic and social development at the local level by raising productivity and facilitating local income generation. Energy indicators provide a measure of efficiency and sustainability in economical, social, and environmental programs. Indicators of energy use are usually expressed as normalized quantities of total energy use to facilitate comparison.

The following Energy indicators within the economic dimension are the most commonly used:

- Energy Production and Supply;
- Energy consumption;

- Energy prices;
- Energy intensities of economic sectors;
- Energy mix;
- Energy supply efficiency;
- Energy use per unit GDP;
- Energy consumption per capita;
- Taxation and subsidies;

Energy intensity is a common indicator to measure the relationship between energy use and economic development of a country through time. It means ratios of energy consumption to gross domestic products (GDP) or value added measured in energy units per monetary unit at constant prices. In the same way electricity intensities or carbon intensities can be computed.

This indicator provides an assessment of how much energy intensive is an economy. Energy intensities variations over time indicate trends in "overall economic efficiency" or "energy productivity". The economy is more "energy intensive" when the energy consumption increases more rapidly than the GDP. Energy intensity can be used to indicate the general relationship of energy consumption to economic development and provide a basis for projecting energy consumption and its environmental impacts with economic growth. There are other credible and viable indicators that focus on energy, but they are notable for their flexibility of use and their specific orientation towards sustainability dimensions, such as economic, social and environmental.

Energy consumption indicator is aggregate energy consumption which measured in Tones of Oil Equivalent (TOE). It can be expressed as per capita or per unit GDP (at market prices). This indicator provides an indication of the level and trend of the total annual amount of commercial energy consumed in the country. The indicator can be disaggregated by energy carrier (liquid petroleum fuels, electricity, and coal), then measures the contribution of each commercial energy carrier towards the total national consumption. It can provide warning signals on inefficient and unsustainable utilization of resources and environmental impact.

Energy mix indicator shows the importance of each fuel in the total energy consumption and development scene.

Energy price indicator provides an indication of the efficiency across the different fuels. It also enables prices comparison in one country with other countries. In general fuel consumption tends to be price related. Thus price can be used to influence use of particular fuels.

The energy indicators are useful to track the changes in energy in relation to economic dimensions. From another perspective there are three types of indexes that explain the change in energy use over time:

1. Activity index shows the changes in the level of activity for a sector of an economy.
2. Component-based energy intensity index represents the effect of changing energy intensity for sub-sectors or detailed components of the economy.
3. Structural index shows the effect of changing economic structure.

At the sub-sector or component level, energy intensity is defined as the ratio of energy use per unit of activity. Thus, if Ei is the energy use for component E_i and A_i is the activity for component i, the component-based intensity is defined as:

$$I_i = E_i/A_i \tag{5}$$

When two or more components or sub-sectors are aggregated, the aggregate intensity i is defined as:

$$I = \frac{\sum E_i}{\sum A_i} \tag{6}$$

For some applications, the aggregate energy intensities are useful summary indicators, as they have either a straightforward interpretation expressed in physical units or they can be converted to a time series based index. However, changes in the aggregate intensity over time are influenced not only by changes in the energy intensities of the various components but also the relative shares of each activity components.

Various types of factorization methods have been employed by which structural and compositional effects can be distinguished from the overall change in the energy intensities as represented by the component-based intensity index. A key objective in the system of energy intensity indicators is the development of time series indexes that satisfy a multiplicative relationship of the energy-economy. Decomposition analysis is a tool to quantitative assessment of factors that contributes to changes in energy consumption. It helps in understanding the past trends of energy use for measuring the effectiveness of energy-related policies, and forecasting future energy demand and pollutant emissions. The three main factors that play a significant role in affecting the level of energy consumption in an economy are: the level of overall activity or production, the composition or structure of the economy, and the output or activity per unit of energy consumed (Nooji et al, 2003).

Energy use = activity × structure × intensity

Energy consumption can be expressed as an extended Kaya identity, which is a useful tool to decompose total national energy consumption. It is shown as bellow:

$$EC^t = \sum \frac{E_i^t}{GDP_i^t} \times \frac{GDP_i^t}{GDP^t} \times GDP^t = \sum EI_i^t \times ES_i^t \times G^t \tag{7}$$

The change of energy consumption between a base year 0 and a target year t, denoted by ΔEC, can be decomposed to three effects in additive form: (i) the changes in the energy intensity effect (denoted by EI effect); (ii) the changes in the structural changes effect (denoted by ES effect); and (iii) the growth in the economic activity effect (denoted by G effect), as shown in Eq. (8):

$$\Delta EC = EC^t - EC^0 = EI + ES + G \tag{8}$$

In this regard, energy intensity indicators play a significant role to study the trend and the changes in the output levels. Energy intensity is thought to be inversely related to efficiency. Therefore, declining energy intensities over time may be indicators of improvements in energy efficiencies. The decomposition of the overall change into these three categories can provide policymakers with the information needed to design appropriate strategies for reduction in energy use while helping to mitigate the environmental impacts of industrial energy use. Three factors of structural, activity or technological changes play a significant role in reducing the energy consumption and intensity with respect to the output value.

5. Environmental implications

Would a more efficient use of energy resources reduce the environmental burden of economic activity? This question has become prominent in recent years as governments across the world have implemented energy efficiency program. The environmental impacts of energy use vary depending upon how energy is produced and used, and the related energy regulatory actions and pricing structures. Gaseous emissions and particulates from the burning of fossil fuels pollute the atmosphere and cause poor local air quality and regional acidification.

Energy use has a variety of impacts. Energy extraction and processing always involves some forms of environmental degradation. As all human activities require energy use, in fact all human impacts on the environment could be seen as the consequences of energy use. In this way energy use is sometimes seen as a proxy for environmental impact of human activity. Principal issues related to the environmental dimension include global climate change, air pollution, water pollution, wastes, land degradation and deforestation. The consideration of global environmental issues such as the greenhouse effect and the 'resulting' climate change problem has led to the development of numerous theoretical models and empirical studies that making it hard to distinguish between environment and energy models (Faucheaux and Levarlet, 1999: 1123).

The impacts of climatic problems associated with the increased accumulation of pollution on the world economy have been assessed intensively by researchers since 1990s. Greenhouse gas emissions are directly related to the use of energy, which is an essential factor in the world economy, both for production and consumption. Therefore, the relationship between greenhouse gas emissions and economic growth has important implications for an appropriate joint economic and environmental policy. It is confirmed that in both poor and rich countries, economic development is not in a sufficient condition to reduce emissions. In nearly all industrial countries there is a permanent discussion on policies to reduce greenhouse gas (GHG) emissions and secure energy supply. A shift from lower to higher quality energy sources not only reduces the total energy required to produce a unit of GDP but also may reduce the environmental impact of the energy use. An obvious example would be a shift from coal use to natural gas use. However, we need to consider that kinds of fuels with higher quality necessarily do not have less environmental impacts.

In addition to energy conserving of technological change, it may also change the environmental impact of energy use over time. So that it reduces the emissions of various pollutants or other environmental impacts associated with each energy source. Therefore, despite the strong connections between energy use and economic growth there are several pathways through which the environmental impact of growth can be reduced. However, if there are limits to substitution of other kinds of energy and technological change then the potential reduction in the environmental intensity of economic production is eventually limited.

There is a hypothesis that states an inverted U-shape relation between various indicators of environmental degradation and income per capita. It is named EKC hypothesis in which pollution or other forms of degradation rises in the early stages of economic development and falls in the later stages.

The production of energy (fossil fuel production and power generation) consumes a significant amount of energy and produces much of GHG emissions. These emissions can be reduced, through the adoption of more sustainable forms of energy production, such as community energy systems. Transportation contributes to GHG emissions, partly because the energy used to power vehicles is usually generated from fossil fuels. Regarding the supply side, the identified priority area is to diversify the energy mix, while promoting sustainable development, mainly by increasing the use of renewable energy resources. Comprehensive utilization of energy saving potential can be the main Strategy.

6. Iran case study

This section of the chapter explores energy and economic growth trends in Iran. Taking the example of the Iran economy in the period 1967 to 2007, the final fossil fuel consumption increased by about 617%, and carbon dioxide emissions sharply increased about 610%. This was in accordance with a significant increase in GDP.

The major energy carriers in Iran are liquid petroleum fuels, electricity, coal and biomass. As the growth of petroleum fuels consumption has been higher than population growth, the per capita consumption of petroleum fuels has been increasing. Furthermore, increased petroleum energy use generally indicates air pollution increase, particularly because technologies to control emissions are almost non-existent as in Iran.

Figure .1 shows the GDP trends of Iran during 1994-2007. Empirical evidence on economic growth in Iran over the last decade seems increasing simultaneously as energy increase. National Commercial Energy Consumption indicator provides a broad overview of the energy use situation and makes it possible to compare with other countries. As Fig. 2 illustrates, since the 1994 the amount of energy used has increased in almost all sectors.

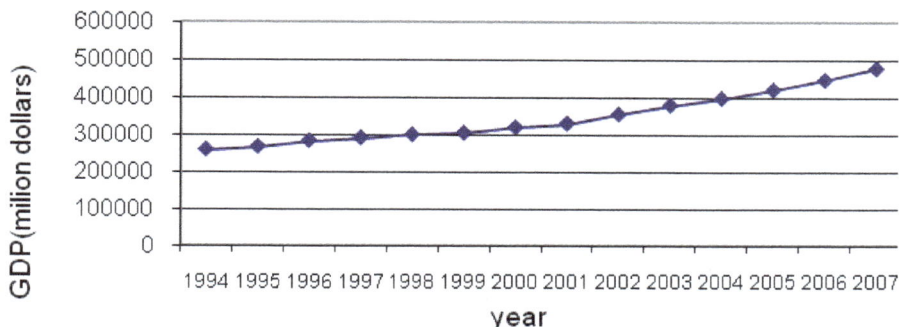

Fig. 1. Iran gross domestic products

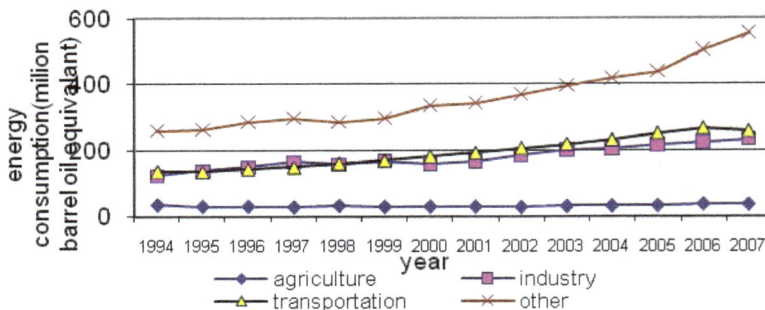

Fig. 2. Iran total energy consumption by sectors

As Fig. 3 demonstrates, once the indicator of energy intensity is used it shows a decreasing trend in some sectors. This has traditionally been assumed to be the result of the application of more energy-efficient technologies within production processes or changes of energy mix inputs. Closer examination of this trend has, however, suggested that this apparent decoupling has in fact been achieved largely by a switch away from the direct use of low-quality fuels to higher quality fuels and energy inputs, electricity in particular.

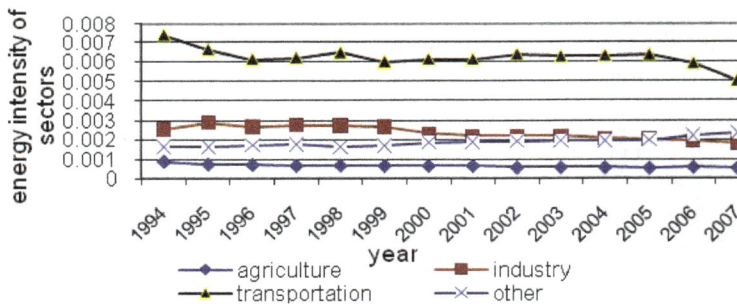

Fig. 3. Iran energy intensity by sectors

Energy intensity is a measure of how much energy is used to produce a unit of economic output. The decoupling of increasing economic activity from increasing energy consumption is a goal for sustainable development. Energy intensity tends to decline over time as a function of underlying efficiency gains and the transition to a more service-based economy. So, it shows less energy needed to generate GDP. It should be considered that Government policies can play a crucial role in how energy intensity changes. Increasing per capita energy consumption is generally associated with development. However, with increasing environmental awareness this is not necessarily a desirable trend.

7. Conclusion

The production, distribution and use of energy create pressures on the environment in the household, workplace and city, and at the national, regional and global levels. Therefore, energy indicators are useful for evaluating impacts of energy systems in all these areas. Trade-offs among three objectives – energy security, environmental protection, and economic growth – has been dominant concerns in energy policy making during recent years. Thus, the main aim of this paper is to present and discuss the use of a particular kind of analytical tool to reinvestigate energy-environment-economy interactions theoretically.

We begin by providing some background that appears to underlie the energy- economy relation which in turn consider environmental issues. Energy poses a challenge to those working to achieve sustainable development goals. We need to use energy to alleviate poverty, promote economic growth and foster social development. But as more energy is consumed, stress is placed on the environment at the local and regional levels.

The fact that expanded energy use is associated with economic development still depends on the importance of energy as a factor in economic development. However, there are substantial differences in energy forms and in the nature of economic activities across different stages of development. The linkages among energy, other inputs, and economic activity clearly change significantly as an economy moves through different stages of development. The energy mix in a developed country can be dominated by non-carbon emitting energy sources, notably hydroelectricity, used in the household and industrial sectors, used in the transport sector. Breaking energy consumption down by sector, by region and for urban and rural areas will be useful in identifying strategies for energy policy implementation. Currently it is possible to disaggregate all data by sector, by region, and between urban and rural.

Internationally, the efficient use of resources has seen a growing role in policy making. In fact, there has been an extensive debate in the energy economics about the impact of improvements in energy efficiency and environmental consequences. If we work together to safeguard the environment without slowing socio-economic development, we look for technological solutions. So, we should change unsustainable patterns of consumption and production toward the least costly ways of achieving sustainable development goals. Analytical tools, such as the energy indicators described before, can be helpful in finding the best solutions in a menu of available options, aimed at achieving these goals.

The cutback in emissions according to the Kyoto Protocol and the maintenance of social well-being require a fragile balance between policies that often have opposite effects. Hence, it is very important to determine the link between economic performance and energy consumption. Emissions of greenhouse gases, especially CO_2 emissions, are closely linked to the energy consumption of primary energy, but the final consumption is crucial for the consumption of primary energy. In fact, energy demand is a derived demand that depends on the productive structure of the economy, the energy content, the sectoral production, the age of the stock of capital, etc. Recently, global and regional environmental problems associated with the utilization of conventional fuels seek to the Renewable Energy Sources (RES) as a competitive participant in the energy scenery. Legislative measures have been adopted in order to reduce dependency from fossil fuels to further integrate non-polluting technologies in the energy mix.

8. References

Aghion, P.& Howitt, P.W. (1998). Endogenous Growth Theory. MIT Press, Cambridge, MA and London.

Barnes, D., & Floor, W.M. (1996). Rural Energy in Developing Countries: A Challenge for Economic Development. Annual Review of Energy and Environment 21: 497–530.

Cleveland C. J., Costanza, R., Hall, C. A. S. & Kaufmann, R. K. (1984). Energy and the U.S. economy: A biophysical perspective, Science 225: 890-897.

Dunkerley, J., Ramsay, W. Gordon, L. & Cecelski, E. (1981). Energy Strategies for Developing Countries. Washington, DC: Resources for the Future.

Faucheaux S., Levarlet F. (1999). Energy-economy-environment models. In: van den Bergh J C J M (eds) Handbook of environmental and resource economics. Edward Elgar, Cheltenham: 123-1145

Hannon B. (1973). The structure of the ecosystem, Journal of Theoretical Biology 41: 535-546.

Kamien M. I. & Schwartz, N. L. (1982). The role of common property resources in optimal planning models with exhaustible resources, in V. K. Smith and J. V. Krutilla (eds.), Explorations in Natural Resource Economics, Johns Hopkins University Press,Baltimore.

Nakicenovic, N. (1996). Freeing Energy from Carbon. Daedalus 125(3): 95–112.

Nooji, M., Kruk, R. & Soest, D. P. (2003). International Comparisons of Domestic Energy Consumption, Energy Economics, 25, PP 259-373.

Ockwell, D.G. (2008). Energy and economic growth: grounding our understanding in physical reality, Energy Policy 36(12): 4600-4604.

Panayotou, T. (1993). Empirical tests and policy analysis of environmental degradation at different stages of economic development", World Employment Programme Research Working Paper WEP2- 22/WP 238

Smulders S. (1999). Endogenous growth theory and the environment, in J. C. J. M. van den Bergh (ed.), Handbook of Environmental and Resource Economics, Edward Elgar,Cheltenham, 89-108.

Solow, R. (1956). A contribution to the theory of economic growth. The Quarterly Journal of Economics 70, 65–94.

Stern D. I. (1997). Limits to substitution and irreversibility in production and consumption: a neoclassical interpretation of ecological economics, Ecological Economics, 21: 197-215.

Stern D. I. (1999). Is energy cost an accurate indicator of natural resource quality? Ecological Economics 31: 381-394.

Stern, D. (2004). Economic Growth and Energy, Encyclopaedia of Energy, vol. 2.

Stern, D.I. & Cleveland, C.J. (2004). Energy and Economic Growth. Rensselaer Working Paper in Economics, no. 0410. Rensselaer Polytechnic Institute, Troy, NY.

Toman M. A., Pezzey, J. & Krautkraemer. J. (1994). Neoclassical economic growth theory and sustainability, in D. Bromley (ed.), Handbook of Environmental Economics , Blackwell, Oxford.

Toman, M. & Jemelkova, B. (2003). Energy and Economic Development: An assessment of the state of knowledge, Volumes 3-13 of Discussion paper, Resources for the Future.

UNDP, UNDESA, WEC, (2000a). World Energy Assessment. United Nations Development Programme, United Nations Department of Economic and Social Affairs, World Energy Council, New York.

12

Carbon Capture and Storage – Technologies and Risk Management

Victor Esteves and Cláudia Morgado
Federal University of Rio de Janeiro
Brazil

1. Introduction

1.1 The greenhouse effect and climate changes

The greenhouse effect (GHE) that allowed the emergence and expansion of life on earth has been growing due to made-man greenhouse gases (GHG) emissions. The increasing use of fossil fuels since the beginning of the industrial revolution has been increasing the GHE and consequently gradually raising the earth's temperature, affecting the conditions for species survival.

GHGs can be subdivided into two groups: those present in the atmosphere since before the industrial revolution and those that are chemical compounds created and produced by humans. The first group includes carbon dioxide (CO2), methane (CH4) and nitrous oxide (N2O), whose concentrations in the atmosphere have been rising as a consequence of intensification of human activity. The second group includes perfluorocarbons (PFCs), chlorofluorocarbons (CFCs), hydrofluorcarbons (HFCs), hydrofluorchlorocarbons (HCFCs) and sulfur hexafluoride (SF_6). Each of these gases has a different potential to absorb infrared radiation.

Table 1 shows the global warming potential (GWP) over a 100-year horizon of some of the main GHGs (IPCC, 1996). The GWP represents the capacity of a gas present in the atmosphere to absorb energy from infrared radiation.

Gas	GWP
CO_2	1
CH_4	21
N_2O	310
CFC-113	4.800
HFC-23	11.700
CF_4	6.500
C_2F_6	9.200
SF_6	23.900

Table 1. Global Warming Potentials (GWP) (100-Year Time Horizon) - Source: IPCC, 1996

The GWP of each gas is the relative warming potential of that gas in relation to CO_2, which has a normalized value of one. For example, N_2O has a GWP of 310, meaning its warming effect is 310 times that of CO_2. Although the GWP indicates in exaggerated form the importance of each GHG over the short term in the atmosphere, particularly for methane, it is the standard defined by the IPCC in its Second Assessment Report (SAR) in 1996 and is utilized by the majority of emissions inventories.

Therefore, although nitrous oxide (N_2O) and methane (CH_4) are present in the atmosphere in much lower concentrations than carbon dioxide and their annual emission levels are far below that of CO_2, their molecules have much greater capacity to absorb infrared energy and hence contribute to increase the earth's temperature on the same order of magnitude as CO_2.

Table 2 shows the anthropic emissions of GHGs of the United States (US EPA, 2011a) in 2008, the 27 countries of the European Union (EEA, 2010) in 2008 and of Brazil (MCT BRASIL, 2010) in 2005. The emissions of all the gases except for CO_2 are expressed by their GWP rather than in absolute mass values. By determination of the United Nations Convention on Climate Change, CFCs and HCFCs are not included in these inventories because they are controlled by the Montreal Protocol, which regulates emissions of gases that destroy the ozone layer. In the case of the United States and European Union (columns 2 to 5 in Table 2), the total emissions are expressed net of the emissions related to changing land use and forestry, which generate negative emissions in these countries. Therefore, changing land use and forestry in these countries cause an increase in the biological capture of CO_2, thus acting as carbon sinks. Just to have an idea of the order of magnitude, changing land use and forestry in the United States in 2008 accounted for negative emission of 1,140.5 $MtCO_2$, representing 16% of the total of 6,961.9 $MtCO_2$. In the European Union this negative emission was 256 $MtCO_2$, representing about 8% of the total of 3318 $MtCO_2$.

	USA	2008	EU	2008	Brazil	2005
	Mt CO_2 eq.		Mt CO_2 eq.		Mt CO_2 eq.	
CO_2	5.921,400	83,9%	3.062,000	82,3%	1.637,905	74,70%
CH_4	676,700	9,6%	302,000	8,1%	380,241	17,34%
N_2O	310,800	4,4%	282,000	7,6%	169,259	7,72%
FCs e HFCs	136,000	1,9%	66,000	1,8%	4,593	0,21%
SF_6	16,100	0,2%	9,000	0,2%	0,602	0,03%
Total	7.061,000	100%	3.721,000	100%	2.192,600	100%

Sources: US EPA (2011a), EEA (2010) and MCT BRASIL (2010)

Table 2. GHG Emissions – USA and EU – Year: 2008 and Brazil – Year: 2005 (using GWP)

The second emissions inventory carried out in Brazil (MCT BRASIL, 2010) presents the emissions for 1990, 1994, 2000 and 2005. Columns 6 and 7 of Table 2 show the GHG emissions of Brazil in 2005. Unlike columns 2 to 5, the figures in columns 6 and 7 include emissions because of changing land use and forestry. The variation in the percentage shares of CO_2 and methane in comparison with those in the United States and European Union is the result of the less intensive industrial activity in Brazil. Besides this, the GWP methodology overstates methane emissions, which have a relatively high value in Brazil due to the importance of farming and stock breeding in comparison with industrial activity.

Until the industrial revolution, natural causes, such as large forest fires caused by lightening and volcanic eruptions, were the main sources of CO_2 release. But since the industrial revolution and the expansion of farming and animal husbandry, human activity has become increasingly relevant. Among the main human activities that contribute to growing CO_2 emissions are the following:

- Thermoelectric plants that burn fossil or other fuels;
- Extraction of fossil fuels;
- Industrial processes that use any type of combustion;
- Land, waterborne and aerial vehicles that used combustion engines;
- Burning to clear land to plant crops or create pastures for animals.

According to the report "CO_2 Emissions from Fuel Combustion", published by the International Energy Agency (IEA), in developed countries the use of energy is by far the human activity that produces the most GHG emissions. Figure 1 depicts the distribution of made-man GHG emissions from developed countries (Annex I of the Kyoto Protocol), excluding those generated by changing land use and forestry, which as mentioned before are negative in these countries. The emissions resulting from production, transformation, manipulation and consumption of all types of energy commodities in Annex I countries account for 83% of all GHG emissions (IEA,2010a).

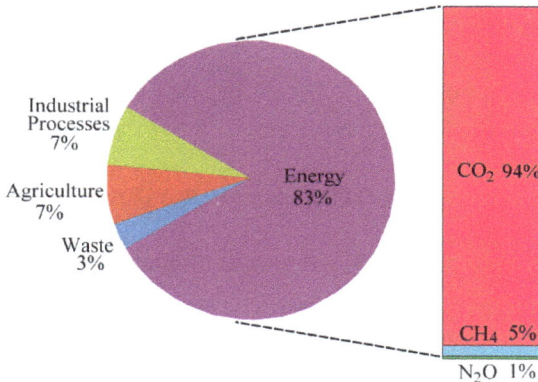

Source: IEA (2010a)

Fig. 1. Shares of made-man GHG emissions in developed countries. Year: 2008.

Figure 2 presents the evolution of the total primary energy supply (TPES) in the world. It can be seen that this doubled between 1971 and 2008. The fact that the share of non-fossil fuels rose from 14% to 19% is due to the increased use of energy from "clean" sources, such as hydroelectric, nuclear and from renewable fuels. Nevertheless, the generation of energy from fossil fuels grew in absolute terms of some 5 gigatonnes of oil equivalent (IEA, 2010a).

The increasing energy demand from the growing consumer markets in emerging economies, of which China and Brazil are leading examples, can only be satisfied over the short term by the use of fossil fuels. In this respect, China is stepping up its use of coal to generate electricity, while Brazil has the option in the medium term of using its immense reserves of natural gas, largely untapped so far.

The use of renewable fuels, such as ethanol from sugarcane, theoretically has the advantage of not adding new carbon to the atmosphere, since the carbon generated by burning it is captured from the atmosphere by the plants from which it is produced. However, it is necessary to perform a complete life cycle analysis of the production of renewable fuels such as ethanol. Practices such as burning off litter in cane fields to facilitate harvesting and the use of farm machinery and trucks that burn fossil fuels diminish the comparative advantage, not to mention social questions. Besides this, the use of renewable energy from biofuels in general competes with land use to produce food, to meet the exploding global demand caused by the inclusion in the consumer market of lower classes from densely populated emerging countries like China and India.

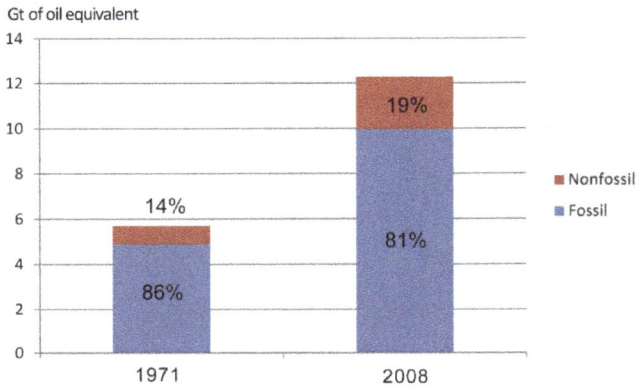

Fig. 2. World Primary Energy Supply (TEPS) - Source: IEA (2010a)

1.2 Carbon capture and storage (CCS) – A way to mitigate climate change

Carbon capture and storage (CCS), also known as carbon capture and geological storage (CCGS), is a process to mitigate climate change by which the CO_2 generated by concentrated industrial activities, such as thermoelectric plants, fossil fuel extraction and refining facilities and other industrial processes that rely on combustion, is captured and stored in geological formations.

One may question the importance of using CCGS to reduce CO_2 emissions since nowadays vehicles are the main contributors to the greenhouse effect. Nevertheless, vehicles are becoming cleaner through better efficiency and the shift to different engines and fuels, such as electric cars. While the electricity used by these cars may be generated by a power plant burning coal, considered a "dirty" source, the CO_2 emitted in concentrated form at this plant can be sequestered while the capture of that emitted in dispersed form by thousands of vehicles with internal combustion engines is economically unfeasible.

The study by the International Energy Agency (IEA, 2010b) shows that the reduction of GHG emissions can only be attained by adopting a series of technological measures. As seen in Figure 3, by the lines traced out to 2050, the IEA believes that if we continue emitting GHGs indiscriminately, global emissions can reach 57 $GtCO_2$ a year over that horizon. But with an intense effort to reduce emissions, through a mixture of CCGS, carbon sequestration by biomass, increased use of renewable energies such as nuclear and enhanced energy efficiency, the world can reduce its emissions to 14 $GtCO_2$ a year.

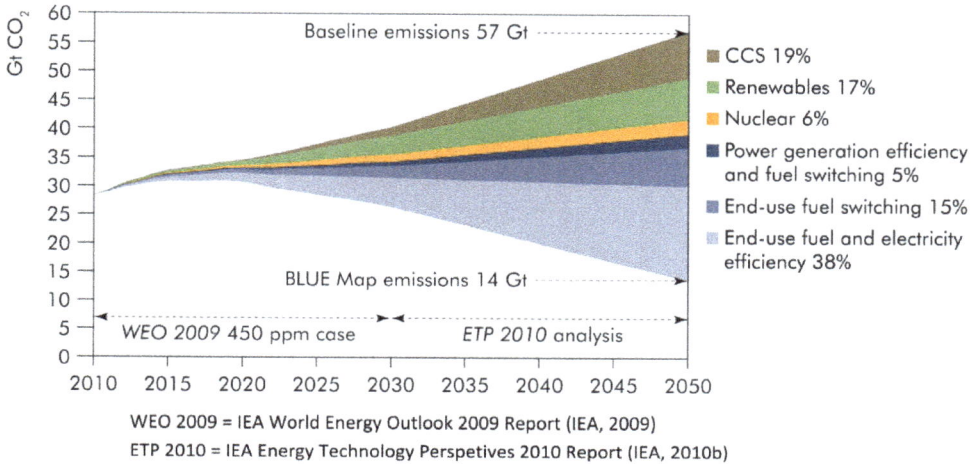

Fig. 3. Technologies for reducing CO_2 emissions - Source: IEA (2010b)

The IEA together with the CSLF (Carbon Sequestration Leadership Forum) prepared a report called "Carbon Capture and Storage – Progress and Next Steps" (IEA & CSLF, 2010) for the G8 summit meeting held in Muskoka, Canada, on June 25-26, 2010. This report lists 80 CCGS projects that fit under a series of criteria, among them the capture of over 500 $MtCO_2$ per year and being in operation between 2015 and 2020. Of these 80 projects, 9 are already in operation and the remaining 71 are in one of the four phases (identification, assessment, definition or execution) that precede operation. Among these 80 projects, 73 are located in developed countries, 4 are in China, 2 in the Middle East and 1 in Africa.

In a graph, shown in Figure 4, the report predicts growth to as many as 3,400 projects in 2050, of which 65% will be located in countries not belonging to the Organization for Economic Cooperation and Development (OECD). These 3,400 projects will be responsible for capturing some 10 $GtCO_2$ annually, representing a yearly average of 3 $MtCO_2$ per project.

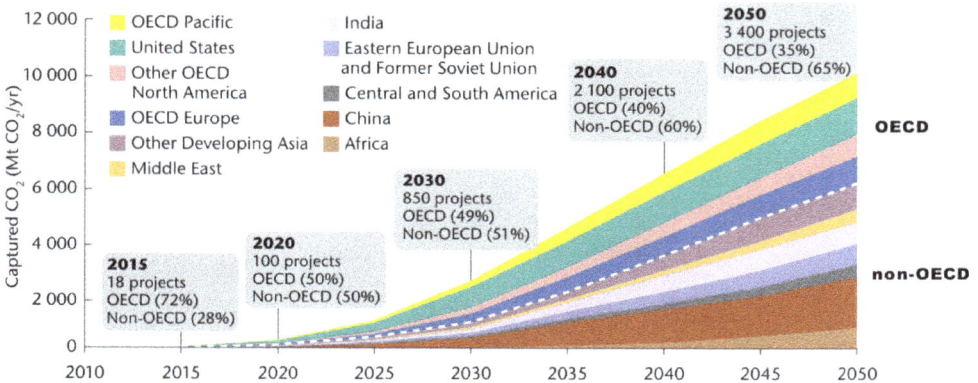

Fig. 4. Global deployment of CCGS 2010-2050 by region – Source: IEA & CSLF (2010)

2. CCS steps – Involved technologies

The CCS process can be divided into six basic steps:

- Separation
- Dehydration;
- Compression
- Transport;
- Injection;
- Storage and monitoring.

2.1 Separation

At present there are basically four cases where the concentration of CO_2 emissions makes its separation for geological sequestration technically and commercially viable. The first is related to the processes of extraction of natural gas, which depending on where and how it is extracted brings with it a varying percentage of CO_2 along with a series of other gases and impurities. The second case the process of gasification of coal, which generates large amounts of CO_2. The third is the generation of hydrogen, in which CO_2 is generated as a byproduct. And the fourth situation, which contributes most to emissions, is the generation of CO_2 from industrial processes involving combustion. Figure 5 presents, as an example of this fourth case, a coal-fired power plant. The coal is burned to heat a boiler to generate steam, which drives the turbines coupled to the generators. The exhaust gases, composed of roughly 15% CO_2, 85% N_2 and under 1% of other compounds such as sulfur oxides (SO_x) and nitrogen oxides (NO_x), pass through a desulfurization system for removal of most of the sulfur-based compounds. The exhaust gases then go to the capture unit, where the CO_2 is separated from the other constituents, which are discharged into the atmosphere. The part discharged is mainly composed of nitrogen (N_2).

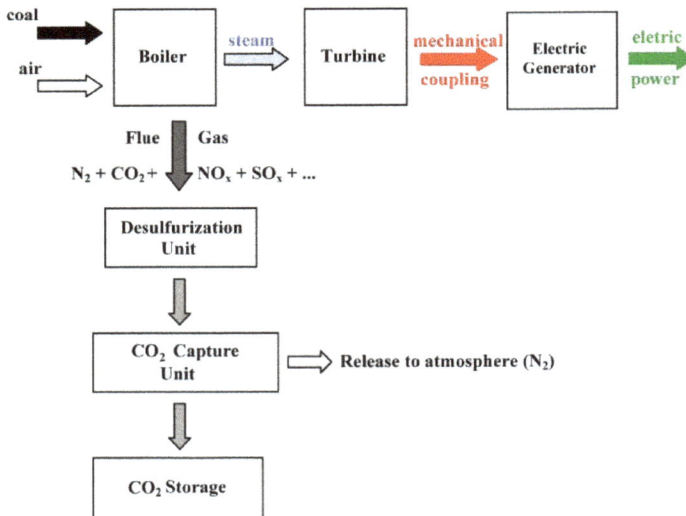

Fig. 5. Coal thermoelectric plant with carbon capture

Today there are a series of CO_2 separation methods already developed or under development, among them the more used are:

- Chemical absorption;
- Physical adsorption;
- Oxy-combustion.

Various factors influence the choice among these separation methods: available space for allocation and consumption of energy by the separation plant, concentration of CO_2 in the gases to be processed, pressure of these gases, level of purity and percentage of CO_2 separation.

2.1.1 Chemical absorption

Chemical absorption is a widely used process with a series of pilot plants distributed around the world. The oldest commercial CCGS plant is located in Sleipner, Norway and has used this process since 1996 (Solomon, 2007).

The process entails using a solvent, normally an amine, which chemically reacts with CO_2, forming a compound. As shown in Figure 6, this reaction occurs in an absorption tower, whose size basically depends on the flow of the flue gas from the industrial process. The compound thus formed is transferred to the regeneration unit where its temperature is raised to release the CO_2. The solvent free of CO_2 then returns to the absorption tower to repeat the cycle.

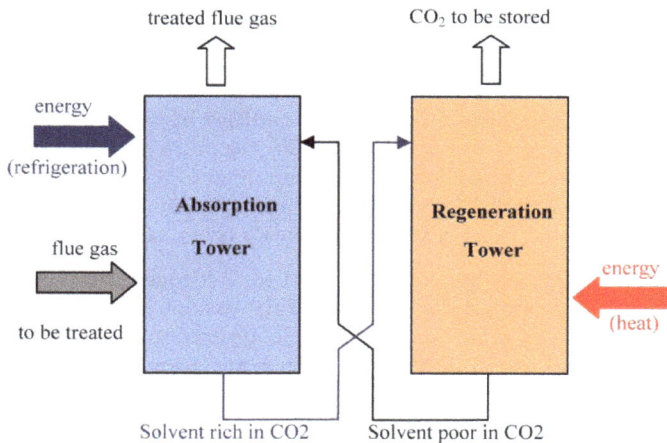

Fig. 6. Absorption and regeneration processes

One example of commercial chemical absorption processes is the chilled ammonia process (CAP), which was developed by Alstom Power and is utilized in pilot plants to capture carbon developed by that company in partnership with American electric utilities. The first pilot plant, with generating capacity of 1.7 Mwatts, was the Pleasant Prairie thermoelectric plant of WE Energies in Wisconsin. The second was the Mountaineer thermoelectric plant, with capacity of 20 Mwatts, owned by American Electric Power in West Virginia (Sherrick et al., 2009). This plant operated from October 2009 to May 2011, for a total of over 6,500 hours, and reached the goal of validating the technology, capturing over 50 $KtCO_2$ in this period and permanently storing over 37 $KtCO_2$ in a saline aquifer located at a depth of 2,400 meters.

The overall chemical reactions associated with the CAP are defined by equations 1 to 4:

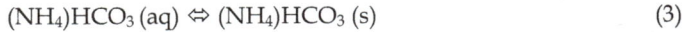

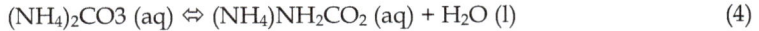

$$CO_2 (g) \Leftrightarrow CO_2 (aq) \tag{1}$$

$$(NH_4)_2CO_3 (aq) + CO_2 (aq) + H_2O (l) \Leftrightarrow 2(NH_4)HCO_3 (aq) \tag{2}$$

$$(NH_4)HCO_3 (aq) \Leftrightarrow (NH_4)HCO_3 (s) \tag{3}$$

$$(NH_4)_2CO3 (aq) \Leftrightarrow (NH_4)NH_2CO_2 (aq) + H_2O (l) \tag{4}$$

These reactions are all reversible and their directions depend on the pressure, temperature and concentration in the system. The equations are exothermic from left to right and endothermic from right to left, requiring the removal or addition of heat.

Besides capturing CO_2, the CAP also removes other residual gases in its cleaning and cooling stages, such as SO_2, SO_3, HCl and HF. Equations 5 and 6 show the overall chemical reactions associated with the removal of SO_2.

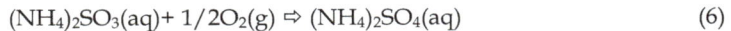

$$SO_2(g) + 2NH_3(g) + H_2O(aq) \Rightarrow (NH_4)_2SO_3(aq) \tag{5}$$

$$(NH_4)_2SO_3(aq) + 1/2O_2(g) \Rightarrow (NH_4)_2SO_4(aq) \tag{6}$$

2.1.2 Physical adsorption

Physical adsorption consists of capturing CO_2 by the surface of a solid material, such as activated charcoal or a zeolite, placed in the path of the flow of the gas targeted for removal of CO_2. The CO_2 adsorbs to the surface of the solid particles by surface forces (non-chemical forces). The adsorption process is facilitated by keeping the process at low temperature or high pressure. Once the adsorbent material reaches a determined CO_2 saturation level, the exhaust gas flow is diverted to another path and the chamber containing the adsorbent material is heated or its pressure is reduced to release the CO_2, in a process called desorption.

An example of the physical adsorption is a project for hydrogen production units in Port Arthur, Texas run by the company Air Products. This was one of the three projects chosen in Phase II of the Industrial Carbon Capture and Sequestration Program (ICCS) of the US Department of Energy (US DOE). The Port Arthur Units 1 and 2, whose block diagrams are shown in Figure 7, work based on the traditional process of reform of natural gas by the action of steam.

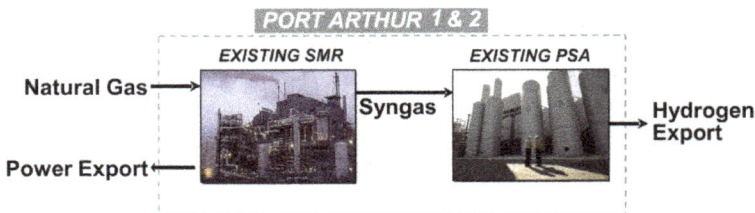

Fig. 7. Port Arthur 1 and 2 –Hydrogen production units - Source: Air Products (2011)

Equations 7 and 8 show the chemical reactions that produce hydrogen from methane.

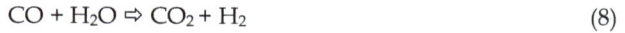

$$CH_4 + H_2O \Rightarrow CO + 3H_2 \tag{7}$$

$$CO + H_2O \Rightarrow CO_2 + H_2 \tag{8}$$

Equation 7 is highly endothermic, consuming a high amount of heat, while equation 8 is slightly endothermic, producing only a small amount of heat. After the reformation process, which is carried out in the steam methane reformer (SMR) unit, the synthetic gas (syngas) generated is composed basically of hydrogen and carbon dioxide associated with some impurities, depending on the composition of the natural gas reformed. The syngas is then sent to the adsorption unit, which works by the principle of pressure swing adsorption (PSA) to separate the hydrogen to be exported.

The project, which received funding of U$ 284 million from the US DOE, will include a CO_2 separation unit and a drying and compression unit in the process (Figure 8), besides interconnection with an existing pipeline to send the CO_2 to the site for geological sequestration. The units are slated to start operating at the end of 2012 and start of 2013 and will capture 1 MtCO_2 per year.

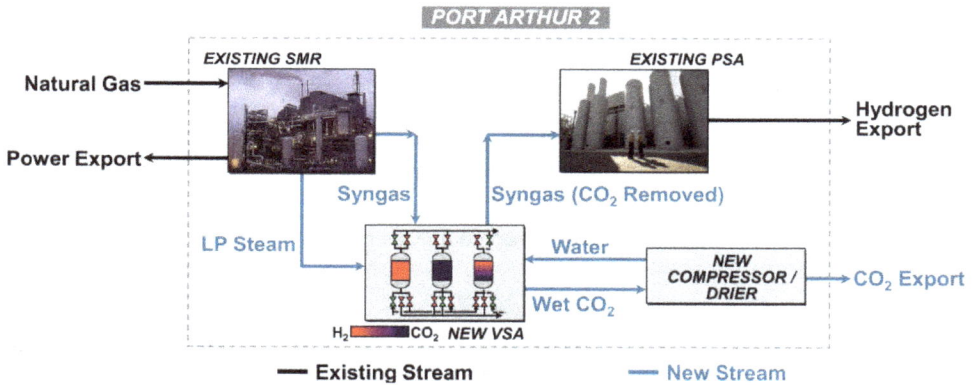

Fig. 8. Port Arthur 2 with CO_2 separation and compressor/drier units

The vacuum swing adsorption (VSA) process is a variation of the PSA process, whereby the adsorption is carried out at a pressure near atmospheric pressure and the desorption occurs by producing a vacuum in the chambers.

2.1.3 Oxy-combustion

In theory the oxy-combustion process involves burning a fuel using O_2 instead of air as the oxidant. In this process, the N_2 is separated in advance, eliminating the presence of nitrous oxide (N_2O) in the exhaust gas. Since the sulfur removal units are already obligatorily included in industrial processes that burn fossil fuels, except for particulates and other impurities the exhaust gas contains a high concentration of CO_2. However, all oxy-combustion systems in practice work with a mixture of O_2 with recirculated exhaust gas. Therefore, the oxy-combustion only increases the CO_2 concentration in the exhaust gas, making its separation more feasible. As a result, the oxy-combustion process must be associated with at least one of the other separation processes. The Figure 9 shows a diagram of oxy-combustion system in a pulverized coal power plant.

The Carbon Capture and Low Emission Coal Research program mandated by the American Recovery and Reinvestment Act (ARRA), signed into law by President Obama in February 2009, calls for investment of US$ 3.4 billion for research aimed to make burning or gasification of coal an activity with low GHG emissions. One of the simplest ways to modernize a coal power plant is to introduce O_2 separation units to feed the burners of the boilers.

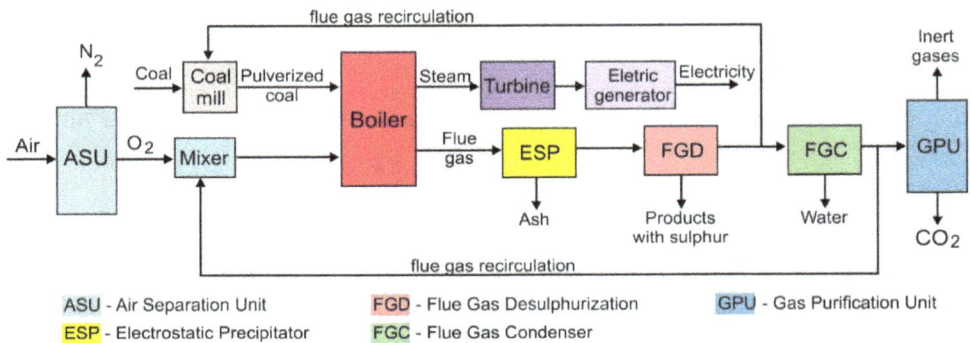

ASU - Air Separation Unit   FGD - Flue Gas Desulphurization   GPU - Gas Purification Unit
ESP - Electrostatic Precipitator   FGC - Flue Gas Condenser

Source: Alstom Power (2011)

Fig. 9. Oxy-combustion system in a pulverized coal power plant

2.2 Dehydration

The objective of dehydration is to reduce the level of moisture of the CO_2 as much as possible so that it will be less prone to cause erosion in the mechanical elements involved in the injection process.

2.3 Compression

To be transported, CO_2 needs to be compressed. The compression range depends on how it will be transported. For pipeline transport, the CO_2 needs to be compressed in the range between 1100 and 3100 psi to assure single phase flow, because above 1100 psi, CO_2 remains in single phase within a broad range of temperatures. Since pipelines are subject to great temperature variations, it is important to avoid the formation of two phases, which can cause pressure spikes that can in turn rupture pipes (Barrie et al., 2004).

The pressure required is much lower for transport in tank trucks, railcars or ships, because the temperature can be kept low through thermal insulation, something that is uneconomic in the case of pipelines. Therefore, pressures of 250 to 400 psi are sufficient to keep the CO_2 in the liquid phase.

2.4 Transport

As indicated above, there are four ways of transporting CO_2 between the emission source and the underground injection site:

- Tank trucks;
- Trains made up of tank cars;
- Tanker ships; and
- Pipelines, which in the case of CO2 are called carbon pipelines.

Of these four transport means, only pipelines are viable for EOR projects, where the distances can run into the hundreds of kilometers and the volumes of CO_2 are in the millions of tonnes per year. This high carrying capacity compensates for the high costs of building, maintaining and operating a carbon pipeline.

Because of the high initial investments and operating expenses of a carbon pipeline and the large damages that could be caused by a rupture, as well as the fact it may cross land held by many owners, special attention must be given to the commercial, legal and insurance aspects to minimize the economic risks. Suppliers and consumers of the CO_2 carried by pipeline along with the line operator must participate in detailed multilateral agreements with well-defined rights and obligations. The other types of transport are feasible for industrial processes that use CO_2 as an input, in cases where the quantity is small and does not compensate the cost of building and operating a pipeline and/or when the production and consumption sites are very close.

2.5 Injection

In this step, the CO_2 is injected through injection wells, basically into three types of geological formations:

- Exhausted or declining oil reservoirs;
- Saline aquifers; and
- Coal beds.

2.5.1 Injection in exhausted or declining oil reservoirs

The option for injection in oilfields where production is waning serves another function besides carbon sequestration: it maximizes oil recovery. This process is called enhanced oil recovery (EOR). The standard production process always involves injection of water to maintain the producing pressure. The EOR process, shown in Figure 10, involves injection of water and CO_2 in alternation. The CO_2 injection increases the oil's fluidity, releasing the oil stuck in the rock pores, while the water, which is by nature not compressible, pushes the oil toward the producing well.

An example of the injection of CO_2 in EOR projects is the Weyburn project, located on the border between Canada and the United States. It has been in operation since 2000. The CO_2, with 95% purity, captured in a coal gasification plant in Beaulah, North Dakota, is carried by a pipeline to an oil production field in Weyburn, Saskatchewan, where it is injected (Zouh et al., 2004). Figure 11 shows the pipeline in yellow that connects Beulah and Weyburn. The red dots show possible derivations for use of the CO_2 in new EOR projects in the region.

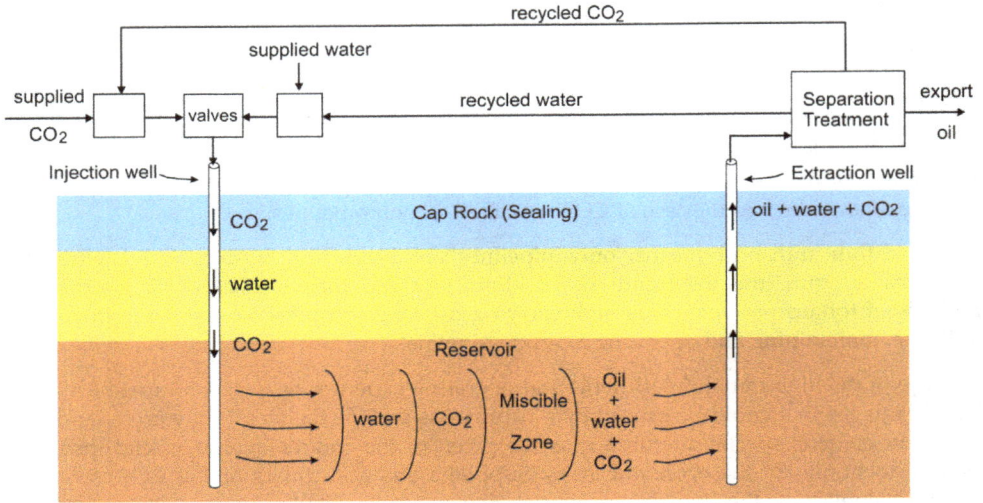

Fig. 10. Enhanced Oil Recovery (EOR) Process

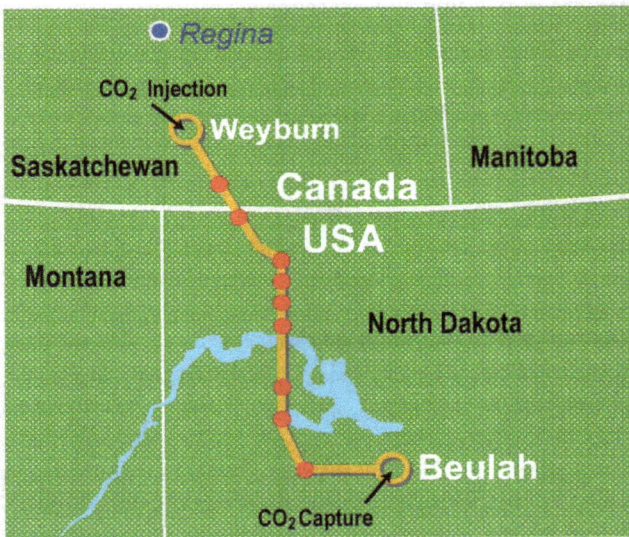

Fig. 11. Carbon pipeline linking Beulah and Weyburn – Source: Cenovus Energy (2011)

Figure 12 presents a graph of oil production in Weyburn since the start of the operation in December 2010. The brown area represents the increase in output because of the EOR process if the process had not begun in 2000, production in December 2010 would have been approximately 10 thousand barrels per day (10 kbopd). The EOR boosted this output in December 2010 to roughly 28 kbopd.

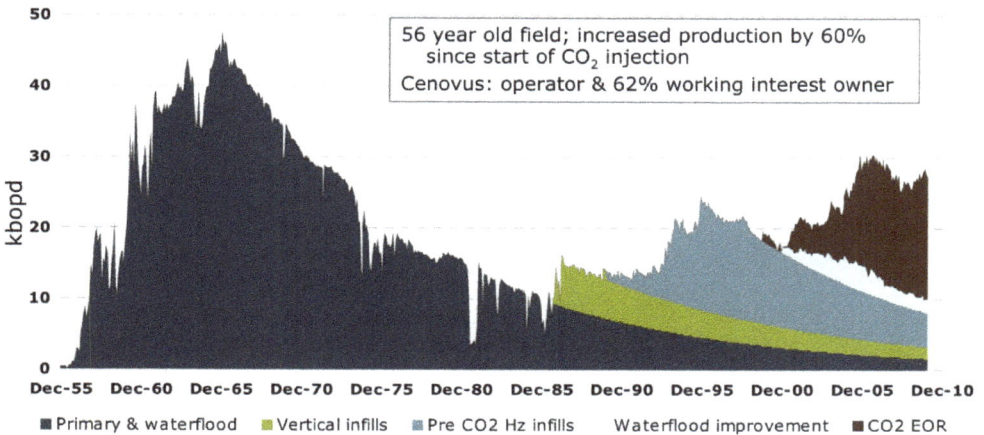

Fig. 12. Weyburn Historical Oil Production - Source: Cenovus Energy (2011)

2.5.2 Saline aquifers

Saline aquifers exist in the great majority of the world's regions. Since this water cannot be used for drinking or farming, the option to store CO_2 in these aquifers appears very promising. The first project to capture carbon of this type was developed by Norway's Statoil in its Sleipner natural gas field in the North Sea.

According to Statoil, the percentage of CO_2 in the natural gas of its Sleipner field is approximately 9% (BGS, 2011), which is above the level tolerated by its consumers. In 1991, the Norwegian government introduced a tax of US$ 50 dollars per tonne of CO_2 emitted. These two aspects combined (standards required by consumers and government taxation) prompted Statoil to develop the geological capture project.

Physically the project is composed of two platforms. On the first one the natural gas rich in CO_2 is extracted. This gas is sent to the second platform where the CO_2 is separated by chemical absorption, then compressed and injected into a saline aquifer located 1000 meters beneath the seabed. According to the projections of a special report of the IPCC (IPCC, 2005), the total storage capacity of the Sleipner project is 20 $MtCO_2$, of which nearly 11 $MtCO_2$ had already been stored by the end of 2008 according to Statoil.

2.5.3 Coal beds

For the storage of CO_2 in coal beds to be feasible, this process must be associated with the production of methane from the bed. The injection of CO_2 enhances the production of methane, hence the name enhanced coal bed methane recovery (ECBM). The process is being studied by, among others, the Swiss Federal Institute of Technology (ETH) and other research organizations funded by European Commission and US Department of Energy (US DOE). These studies aim to obtain the necessary knowledge to apply the technology in large scale.

A pilot ECBM project financed by the US DOE was developed in the San Juan Basin in New México, with the use of 4 CO_2 injection wells and 16 methane production wells, besides an observation well. The methane production started in July 1989 and the CO_2 injection began in April 1995 and continued until August 2001, when the operations were suspended to study the results. Figure 13 shows the results of the variations in methane output as a result of the injection of CO_2 (Reeves & Clarkson, 2003).

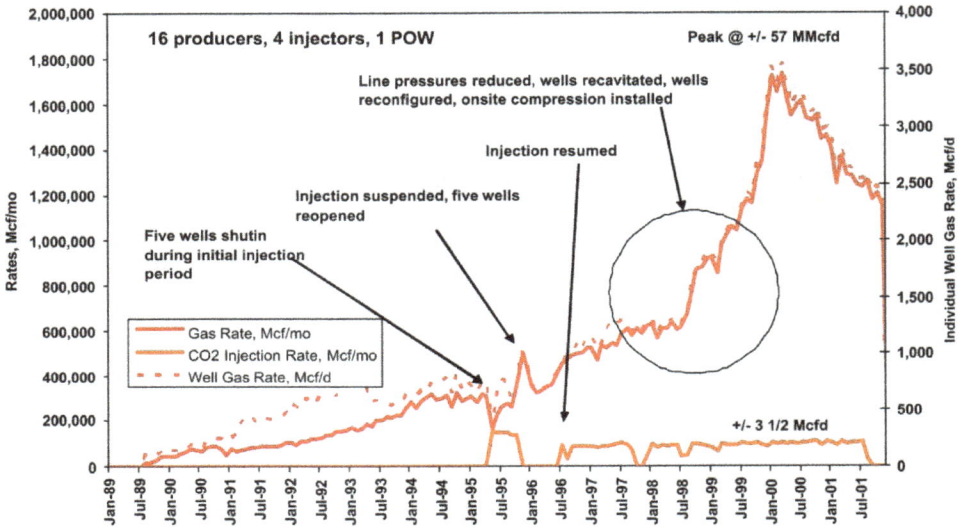

Source: US DOE.

Fig. 13. Evolution of Production/Injection of the UCBM Pilot Project of Alison

2.6 Storage and monitoring

Storage and monitoring are considered to be single step, because monitoring is required to assure that the CO_2 stored will not leak out to the atmosphere. According to the report of the Special Intergovernmental Panel on Climate Change (IPCC, 2005), this monitoring aims to verify possible leaks or other aspects indicating deterioration of the storage over the long term, to assure there are no risks to the environment. Various technologies can be used to perform different types of monitoring:

- Monitoring of the injection flow and pressure;
- Monitoring of the underground CO2 distribution;
- Monitoring of the integrity of the injection wells;
- Monitoring of the local environmental effects; and
- Monitoring by a network of sensors placed at points distant from the injection sites.

All the data gathered by these monitoring efforts are fed into computer systems equipped with "intelligent" software as part of a risk management system, which besides indicating tendencies that can foretell risky situations and determine operational changes, also indicates mitigation routes in case of leaks or malfunctions of the system.

3. Risk assessments

Risk is the product of the probability of a negative event's occurrence and the magnitude of the consequences. Risk management is a tool used to make decisions to help manage adverse events. For proper assessment of risks, it is necessary to identify all the possible causes of risk and their consequences. This can be done by preparing a chart showing the series of risk-posing events that can lead to a catastrophe, as shown in Figure 14.

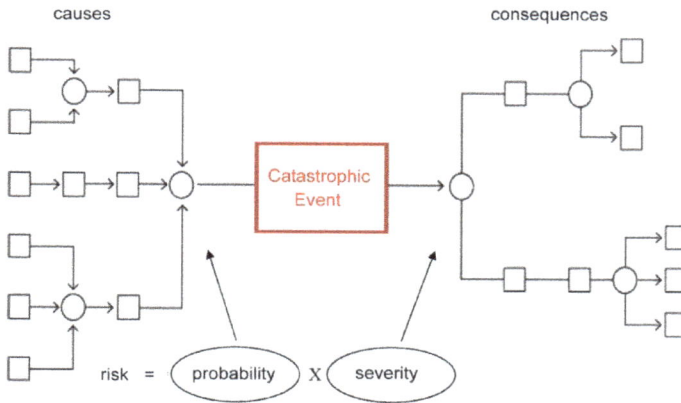

Fig. 14. Series of Risks

Normally in industrial undertakings, the causes of events with large adverse effects are treated by managing the technology, that is, by specifying the equipment and materials, preparing rules and procedures, training programs, etc. The effort to reduce risk is concentrated in diminishing the probability of the occurrence of the causes that can trigger a series of events that lead to catastrophe and to assess the consequences. These consequences are analyzed by using the data on the area surrounding the project, its population and natural resources. Therefore, contingency plans are drawn up for mitigation of the catastrophic events if they occur. However, the focus is on the causes.

The risks of CCS projects are hybrid in nature, meaning they are a combination of technological and natural risks, because the possibility of leaks and other problems does not depend on the technology alone. The size of the reservoir, demographic changes, seismic behavior of the region, micro-climate and many other factors can modify the characteristics of the process and thus its complexity. Hence, there is less control over the causes that can lead to a catastrophic event, and it is important to monitor and identify anomalies in the process that can require taking action to control the emergency, by application of contingency plans prepared in advance.

The magnitude and complexity of the events involved in CCS projects prevent the application of traditional risk management based on administrative procedures and operational controls. Unlike an industrial plant, the CCS process is part of a natural formation that is responsible for its final function. The activities of the people in the surrounding area and the possibility of seismic events that trigger natural geophysical and geochemical changes in the reservoir are just some of the aspects that must be considered to manage the risk of a CCS project. This imposes the need for an adaptive intelligence able to accompany this dynamic interplay of factors.

The complexity of managing the risks of a CCS process depends on a series of aspects inherent to each project, among them the following:

- Separation technology;
- Separation and injection flow;
- Distance between the separation and injection sites;
- Injection purpose (besides carbon sequestration);
- Characteristics of the storage reservoir;
- Monitoring technology; and
- Substances that form the gas to be injected.

The varying combination of these aspects will determine the analyses that must be undertaken.

3.1 Risks in the separation, dehydration and compression steps

Due to their individual character, the risks of these steps are similar to those involved in the industrial process that will be the source of the CO_2. In the case of Weyburn, the CO_2 comes from the coal gasification plant in Beaulah, while in the Sleipner project it comes directly from the natural gas production well. In any event, the addition of CO_2 separation, dehydration and compression units increases the complexity of the endeavor, raising the cross-risks and consequently changing the risk analysis drastically.

3.2 Risk during transportation

According to a study by the Pipeline & Hazardous Materials Safety Administration (PHMSA) of the U.S. Department of Transportation (US DOT), in that country in 2008 there were 5,580 Km of carbon pipelines in operation, mainly involving enhanced oil recovery (EOR) projects. These pipelines are located in North Dakota (ND), Wyoming (WY), Utah (UT), Colorado (CO), New Mexico (NM), Texas (TX), Oklahoma (OK), Mississippi (MS) and Louisiana (LA). Most of these lines cross sparsely populated regions, a characteristic that reduces the severity factor of the risk associated with transporting the CO_2. This is clearly intended, since the severity reflects the direct effects of possible accidents on people. Nevertheless, while effects on natural biomass may not directly affect local populations, they can cause secondary effects on more distant population centers. If these effects are neglected for not being direct, the losses can be greater and broader in scope, ceasing to be local and becoming regional.

In densely populated and highly industrialized regions such as Central and Northern Europe, carbon pipelines linking CO_2 sources with storage sites will have to traverse populated areas, potentially prompting public opposition. The current risk perception places the risks of onshore storage above those of onshore transportation. This is understandable because people have lived for decades with oil and gas pipelines but are not accustomed to the idea of having geological formations beneath their feet containing millions of tonnes of CO_2 "ready to escape". But while onshore storage projects may face low acceptance, offshore projects require a much greater investment in constructing the necessary pipelines.

Failures of carbon pipelines can be caused by holes or complete ruptures. In both cases the failure can be the result of:

- Corrosion;
- Construction defects;
- Materials defects;
- Soil movement;
- Operational errors; and
- Human activities in surrounding areas.

The climatic and geological aspects of the area where a carbon pipeline is or will be installed directly influence the effects suffered by the materials used in their construction. Besides this, these aspects also influence the choice between a buried or aboveground pipeline. In the case of failure of a high-pressure underground pipeline that causes a large leak, the pressure will fall rapidly, releasing a large quantity of energy. This energy will cause the soil above to be ejected, potentially resulting in large damages to structures and loss of lives.

Accidents in densely populated areas represent a greater risk both in terms of probability and severity. This fact requires a larger investment in security and ongoing monitoring of urban expansion in the areas through which the pipeline passes.

The main aspects that influence the amount of CO_2 that can escape during an accident are: internal diameter of the pipeline, size of the hole, operating temperature and pressure and distance between shut-off valves.

Because CO_2 is heavier than air, when released in large quantities it behaves differently than gases that are lighter than air. The release of CO_2 occurs in the form of a cloud that moves near the ground and its progress depends closely on the local topography and weather.

The most important aspect to be analyzed is the impact of CO_2 leaks on human health. In this respect, the concentration and exposure time are the two factors that must be assessed. A CO_2 concentration of 150,000 parts per million (ppm), or 15% by volume, can cause a person to lose consciousness in less than one minute. Exposure for one hour to concentrations between 100,000 and 150,000 ppm can cause mortality ranging from 20% to 90% (Koornneef et al., 2010).

3.3 Risk of leakage to the atmosphere

When injected, the CO_2 is less dense than the saline fluids of the reservoirs, so it can migrate to other geological formations or to the surface. The escape to the atmosphere, besides causing risks to human health and the environment in nearby areas, also obviously reduces the effectiveness of the effort to control GHG emissions intended by the CCS project in the first place. The leakage of high concentrations to the atmosphere can have catastrophic effects on the local biota.

- CO_2 leakage to the surface can occur because of:
- Pre-existing geological fractures or faults;

- New geological fractures caused by seismic movements;
- Abandoned production or injection wells; and
- Long-term changes in the properties of the reservoir's rock formations.

In an EOR project, the drilling of new injection wells continues until no longer economically feasible. The abandoned wells, although sealed with cement, can provide paths for CO_2 to escape. This can happen due to degradation of the sealing materials. Contact with CO_2 in brackish water increases the attack on cement by around tenfold in comparison with freshwater (Barlet-Goue´dard et al., 2009). The Weyburn project currently has over 1,000 wells along its extension. One of the assumptions of the studies conducted there is an increase in 100 years in the permeability of the sealing cement from an initial level of 0.001 md to 1 md (Zhou et al., 2004).

Changes in the porosity and permeability of the reservoir's rocks can be caused by the effect of the chemical interactions between the carbonic acid and the minerals forming the rocks. Carbonic acid is generated directly by the reaction of CO_2 with the water present in the reservoir. This effect is stronger in storage projects that use saline aquifers, such as the Sleipner project, but is also occurs on a lesser scale in EOR projects, such as Weyburn.

A study carried out at the University of Nottingham (Patil et al., 2009) to assess the possible effects of CO_2 employed injection at a controlled rate. The study utilized two types of ground: a pasture and a fallow plowed field awaiting planting. The results showed that the concentration of CO_2 displaced the O_2 from the soil and reduced its pH. The consequences of these alterations were impairment of the action of earthworms and reduced grass growth in the pasture and crop germination after planting, with consequent diminished productivity of both the pasture and planted field.

3.4 Risk of underground movements

One of the most important aspects that must be analyzed regarding injected CO_2 is its capacity to carry metals in the underground that can contaminate groundwater.

A comprehensive risk assessment must consider the main composition of the storage reservoir's rock formation. There are basically two types of formations:

- Carbonate rocks (calcite, argonite, dolomite. etc.); and
- Silicate rocks (quartz, feldspar, etc.).

The presence of saltwater, as in storage in saline aquifers, is important because it promotes the formation of carbonic acid, which reacts with the surrounding minerals and can carry the metals present in them. This transport can contaminate nearby potable water aquifers.

In the case of silicate rocks, the carbonic acid reacts very slowly with the rock so there is practically no change in the porosity and permeability. In contrast, carbonate rocks react more quickly with the CO_2, altering the porosity and permeability. This effect, however, is damped by the rapid increase of the pH of salt water, which leads to a decrease of acid action on the rocks (Wilson et al., 2007).

An example where the risk of underground movement is present is the project developed by In Salah Gas (ISG), a joint venture among British Petroleum (33%), Statoil (32%) and

Sonatrach, the Algerian national oil company (35%). The gas produced by the production wells in the Sahara Desert region has an average CO_2 concentration of 7%, a level that needs to be lowered to under 0.3% for the gas to be exported to Europe. Therefore, a purification plant was built at the Krechba Oásis, 700 Km from Algiers (Iding & Ringrose, 2009). The purified methane is sent northward in a pipeline that connects to the Algerian gas exportation network, while the captured CO_2 is pressurized, carried by pipeline and injected in a saline aquifer located below the gas field. The main risk of this undertaking is the possibility of migration of the CO_2 toward a drinking water aquifer that lies above the gas reservoir. Investigations demonstrated that the upper part of the reservoir where the CO_2 is being injected has a thick layer of schist that seals this reservoir. However this risk of groundwater contamination should be given priority attention, in this desert region, where there have historically been violent conflicts involving water rights.

Another risk associated with underground movement is the possible generation of seismic events due to the alteration of the underground geophysical characteristics. Such seismic events, besides potentially generating geological fractures capable of releasing large amounts of stored CO_2, can also unleash other catastrophic events that damage structures and endanger lives.

3.5 Risk of using hydrocarbon reservoirs for sequestration

The analysis of the risks of using depleted hydrocarbon reservoirs for geological sequestration of carbon or the employment of CO_2 injection for enhanced oil recovery (EOR) is a complex process that must consider constant changes in the risk factors over time and the various types of wells. In a given reservoir, there can be five basic well types:

- Production well;
- Injection well;
- Sealed wells without monitoring instrumentation;
- Sealed wells with monitoring instrumentation; and
- Monitoring wells.

The status of a well can change, altering the set of instrumentation necessary and the ranking of the importance of the data necessary for risk management. Additionally, the change in the status of a determined well alters the entire system and affects the ability to monitor the system. Therefore, the risk management system must be adaptable to accompany the system's evolution.

4. Policy and regulation

Governments play an essential role in CCS, by setting safety standards and other requirements for operation and obtaining public support. The deployment of CCS projects relies on the approval of civil society, who must believe that the injected CO2 will stay stored in the reservoir for thousands of years. To this end, the analysis of possible risks associated with the escape of CO2 is an essential stage in the life cycle of the storage system and aims to promote and ensure the safety of the activity to the environment and to human health, contributing to the technology's acceptance.

One of the main sticking points for the expanded use of carbon sequestration, mainly in densely populated areas, is the acceptance of the people living above or nearby the reservoir that will be used. The same situation exists for the location of sanitary landfills, prisons, power plants or any other large project with potentially negative impacts. While society at large agrees on the need for such undertakings, those most closely affected generally feel otherwise, often because of a lack of knowledge of the real risks involved. This is the well-known "not in my backyard" conundrum. In the case of carbon sequestration, the benefits accrue to the population of the entire planet, not just a region or state, making this contrast between the general welfare and local concerns as stark as it possibly can be. Winning public support thus requires a major effort to educate the public about the real risks of geological storage of carbon. A real example of the public acceptable importance is the project of Shell in Barendrecht, Holland. This project planned to store some 10 MtCO$_2$ over a period of 25 years, captured from Shell's hydrogen gasification plant at the Pernis refinery near Rotterdam. The CO$_2$ would be transported by a pipeline about 20 km and injected in two depleted natural gas fields over a mile deep under the city of Barendrecht. Despite many public hearings held by the city council and strong support of the central government, through approval of by the Dutch Senate, Ministry of Economic Affairs and Ministry of Housing, Spatial Planning & the Environment, the project faced strong opposition from the citizens of Barendrecht and it finally had to be canceled.

The geological storage in saline aquifers and other formations located on continental shelves is the best option. The study carried out by Dutch researchers (Broek et al., 2009 shows the technical and economic feasibility of a network of carbon pipelines linking power plants and industries that emit high amounts of CO$_2$ to the Utsira aquifer. This aquifer is located below the North Sea, between Great Britain and Norway. Testing has been conducted there since 1996, including through the Sleipner CCS pilot plant run by Statoil. The study took into consideration the perspectives for growth of emissions due to increased energy demand and for growth of taxation on emissions from €25/tCO$_2$ in 2010 to €60/tCO$_2$ in 2030.

In most countries the regulation of CCS is the responsibility of the central (federal) government. In the United States, Australia and Canada there is shared responsibility among the federal, state (provincial) and local spheres. The specific legislation to regulate the activities involved in CCS should start from existing laws on extraction and processing of fossil fuels. Countries like Norway, Canada and Spain are involved in this process of formulating the CCS regulation based on the regulatory powers under existing legislation on exploitation of oil and gas or through amendment of those laws to extend their scope.

Another consideration is the fact that many likely places for CO$_2$ injection lie in international waters and many such schemes involve emissions from multiple countries. Hence, international agreements come into play. Maritime treaties such as the London Protocol limit the exportation of trash or other materials and also the dumping or incineration of such materials on the high seas. Because the Protocol had been interpreted as prohibiting the export of CO$_2$ from one contracting state to another for injection into sub-seabed geological formations, it was amended in 2009 specifically to permit this. To take effect, this amendment must be ratified by at least two-thirds of the contracting states. Without this ratification, densely populated countries not located on coastlines, such as those in Central Europe, are prevented from using the option of sending CO$_2$ for offshore storage in geological formations beneath the continental shelf, even though the populations of the emitting countries and the coastal ones that can provide this service may agree with it. However, only a few countries are involved in the development of CCS schemes and fewer still in offshore storage with cross-border transport of CO_2. Therefore, ratification is far from assured in the short term.

Another important international maritime accord is the OSPAR (Oil Spill Preparedness and Response) Convention for the Protection of the Marine Environment of the North-East Atlantic. It has also been amended to permit injection of CO_2 in sub-seabed formations, an amendment that is also awaiting ratification. Since this agreement only has 15 member states, only two more ratifications are necessary for it to take effect. The greater ease of ratification is also due to the fact that the Sleipner project – the largest offshore CCS project – is located in this region.

There are proposals, accepted even by the World Trade Organization (WTO), to create differentiated import taxes for products from countries with policies and commitments to reduce various emissions. The aim of this policy is to level the competitive playing field for products whose costs include environmental taxation in the country of origin (WTO&UNEP, 2009).

International accords and mechanism such as the United Nations Framework Convention on Climate Change (UNFCCC), which was created in 1992 at the United Nations Conference on the Environment and Development (Rio 92), have an important role in fostering CCS. Among the Kyoto Protocol's features is the Clean Development Mechanism (CDM), which permits developing countries, which are not required to have emission reduction targets, to develop projects to reduce GHG emissions and in return acquire Certified Emission Reduction (CER) certificates. These CERs can be traded with developed countries to enable them to meet their emission reduction goals. But due to the still-existing doubts about the capacity to guarantee the effectiveness of geological sequestration of carbon, CCS projects are not yet eligible to receive CERs. Another reason for this lack of eligibility is the political and economic dispute between consolidated fossil fuel industries and environmentalists and researchers. The first group advocates the use of CCS as a viable way to reduce emissions while the second believe this will just prolong the use of fossil fuels, thus discouraging investments to develop renewable energy sources with smaller carbon footprints.

At the Sixteenth Conference of the Parties (COP 16), in Cancun, Mexico, it was determined that CCS should be included as eligible under the CDM, and the Subsidiary Body for Scientific and Technical Advice, which had proposed the decision, was tasked with preparing the procedures for inclusion of CCS in the CDM, to be decided upon at the COP 17 in Durban, South Africa (in December 2011). The final report enumerates a series of issues about CCS that must be considered before final approval of its inclusion under the CDM, among these are:

* Robust and rigorous criteria for selecting the storage site;
* Strict plans for monitoring, aiming at adequate risk management;
* Study of migration routes; and
* Inclusion of the possibility of dissolving CO2 in groundwater.

The items that follow illustrate the current state of CCS legislation in some countries.

4.1 Australia

In Australia the federal government shares jurisdiction with the state and territorial governments for both onshore and offshore geological sequestration (out to the 3-nautical mile limit). The federal government has sole jurisdiction on the continental shelf beyond this limit.

In June 2011, the Australian government approved the "Offshore Petroleum and Greenhouse Gas Storage (Greenhouse Gas Injection and Storage) Regulations 2011". These regulations, issued under the authority granted by the Offshore Petroleum and Greenhouse Gas Storage Act, approved by Parliament in 2006, basically cover the following interrelated elements:

- Testing the risk of a significant adverse impacts;
- Information necessary for a declaration of a geological formation as adequate for storage;
- Local injection and storage plans;
- Incident reporting;
- Decommissioning; and
- Discharge of securities.

In July 2011, the Australian government presented its Clean Energy Future Plan, which calls for a tax of Au\$ 23 (about US \$25) per tonne of CO_2 starting in July 2012. This taxation includes all activities that emit more than 25,000 tonnes of CO_2 per year. It does not include emissions from light vehicles and farming activities. To maintain the country's industrial competitiveness, steelmakers, coal miners and electricity generators will receive compensations. An energy security plan will assure sufficient electricity generation in the face of possible problems, since 75% of the country power is generated by coal-fired plants. Tax cuts for consumers are also planned to offset possible increases in the cost of living due to the CO_2 emission tax. The adoption of this tax was the fruit of suggestions by companies from the coal mining sector, which in 2010 proposed that the government adopts a CO_2 emission tax along with a requirement that part of the revenue be allocated to develop clean technologies to permit the companies to remain competitive in the global market.

4.2 Canada

In Canada, the central government shares jurisdiction over CCS with the provincial governments. The latter governments' jurisdiction covers natural resources within the borders of each province, including exploration and development of non-renewable natural resources and management of power plants. This means the provinces have authority over certain aspects of CCS while others fall under federal jurisdiction, such as international and inter-provincial commerce, taxation and criminal legislation.

4.3 Norway

Norway established a CO_2 emission tax in 1991, mainly applying to offshore oil and gas extraction. Other sectors in the country with a large carbon footprint were exempted, such as fishing, metallurgy, cement making, aviation and others. The power generation sector was not affected because 98% of the country's electricity comes from hydroelectric plants. Because of this policy of taxation centered on exploitation and consumption of fossil fuels, the pump prices of gasoline and diesel in Norway are among the highest in Europe (equivalent in July 2011 to US$ 2.30). But because of the many exceptions, Norway's carbon tax has not managed to reduce emissions as much as envisioned.

Regarding specific regulations on CCS, the Ministries of Petroleum and Energy, Labor, and the Environment as of May 2011 were still working on new regulations on the transport and storage of CO_2 in subsea reservoirs under the country's continental shelf. The work was being delayed due to the conflicts of interest within and among the ministries, and no draft regulations had been put out for public consultation as of that date.

4.4 European Union

Both the European Commission and the governments of the state members are involved in regulating the geological sequestration of carbon. The member states are required to put into practice the directives and regulations issued by the European Union, including the Emission Trading System (RTS) and the CCS Directive, which function as framework legislation. The CCS Directive has to be transposed to the law of each member state by June 2011. This process permits each country to develop its own legislation on CCS to fit the particular circumstances of each one, within the overall European Union framework.

4.5 United States

In the United States, the Clean Air Act of 1970, which was substantially amended by Congress in November 1990, with further small alterations since then, entrusts responsibility for CO2 emissions to the Environmental Protection Agency (US EPA).

Specifically regarding GHG emissions, in December 2009 the US EPA issued a note indicating it had concluded that the current and projected atmospheric concentrations of GHGs jeopardized current and future public health (US EPA, 2009).

Due to some projections made by the US EPA and the United States Energy Information Administration (US EIA), such as slow growth of electricity demand, low natural gas prices and strong gas supply, the only projects for new coal power plants other than those already under construction are a small number of medium-sized plants subsidized by federal programs for carbon capture and storage. As seen in Figure 15, the projections indicate that the growth of electricity demand of approximately 700 TWh (tera watt hours) between 2015 and 2030 will be almost all met by the entrance into operation of combined cycle natural gas plants, utilizing gas and steam turbines in the same cycle.

Another aspect that can be observed from Figure 15 is that nearly half of the 4.1 million GWh (giga watt hours) forecast for 2015 will be generated by traditional coal-fired plants with turbines driven by steam. As can be observed in Table 3, of the current 1,266 coal power plants in the United States, more than one-third are classified as large, with average capacity of 532 MW (mega watts), which together account for 76% of the energy generated by coal-fired plants.

Based on these projections, in December 2010 the US EPA announced the preparation of rules to cover GHG emissions from power plants that burn fossil fuels and that generate more than 25 MW. The rules will establish performance standards for new emission sources (New Source Performance Standards - NSPS), applicable both to new plants and revamped ones.

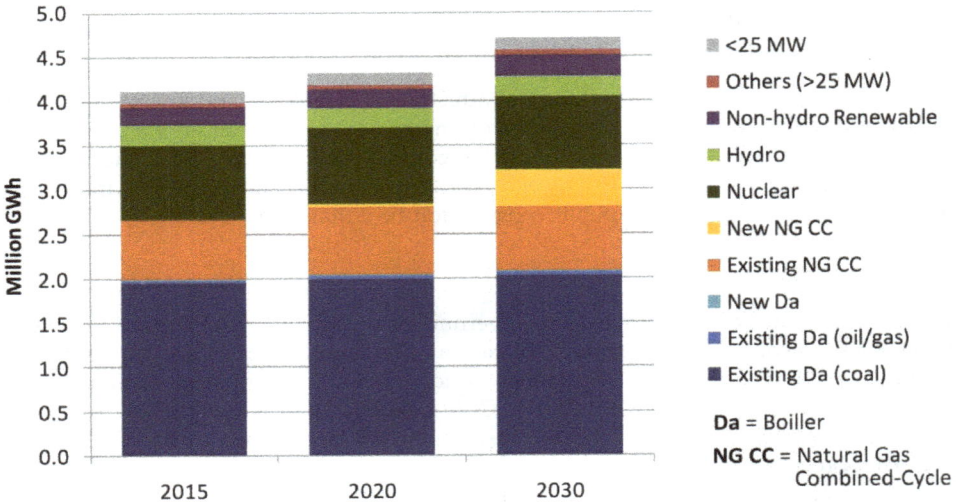

Source: US EPA (2011b)

Fig. 15. Projections for distribution of power plants in the United States

Capacity (MW)	Number of units	Avg. age (years)	Avg. capacity (MW)	Total capacity (MW)	Share	Avg. thermal efficiency (Btu/KWh)
< 25	193	45	15	2,849	1%	11.154
25 - 49	108	42	38	4,081	1%	11.722
50 - 99	162	47	75	12,132	4%	11.328
100 - 149	269	49	141	38,051	12%	10.641
150 - 249	81	43	224	18,184	6%	10.303
> 250	453	34	532	241,184	76%	10.193
Totals	1,266			316,480		

Table 3. Coal-fired power plants in the United States - Source: US EPA (2011b)

Besides this, the rules will establish an emissions guide for existing units. The Agency is slated to present the proposed regulations in September 2011, and discussion is scheduled to last until May 2012, when the final versions will be presented.

4.6 Brazil

The current Brazilian Constitution, promulgated in 1988, deals with the environment specifically in its Article 225. Among other aspects, this article refers to the concept of sustainability, in line with what was presented by the United Nations World Commission on the Environment and Development (WCED/UN, 1987), according to which all people have the right to an ecologically balanced environment, essential to a healthy quality of life, and the government and the community have the duty to defend and preserve it for present and future generations. It is also determined that public authorities must require the preparation and publicity of environmental impact studies for all activities that can potentially cause degradation of the environment. CCS can fit under this because although the aim is to benefit the environment by reducing the concentration of CO2 in the atmosphere, it can also bring possible negative impacts that must be dealt with through preventive and/or mitigating measures.

Law 6,938/81 established the National Environmental Policy and created the National Environmental System (*Sistema Nacional do Meio Ambiente* - SISNAMA). Within the SISNAMA structure, the National Environmental Council (*Conselho Nacional do Meio Ambiente* - CONAMA) was created as the consultative and deliberative entity of the SISNAMA. CONAMA issues resolutions that create general guidelines, rules and standards.

CONAMA Resolution 01/86 contains the definitions, responsibilities and basic criteria for the use and implementation of environmental impact assessment. To build and operate any project involving an activity considered potentially polluting, it is mandatory to prepare an environmental impact study (*Estudo de Impacto Ambiental* - EIA) and accompanying environmental impact report (*Relatório de Impacto Ambiental* – RIMA). The activities listed as potentially polluting that are related to an CCS project are: (a) gas pipelines; (b) extraction of fossil fuel, which would apply in the case of using the CO_2 captured for enhanced oil recovery (EOR); (c) power plants, applicable in case of capture of exhaust gases from these plants; and (d) industrial plants, which would apply to a wide range of industrial activities, both in the petroleum industry (refineries, fertilizer plants, coal gasification plants) and others (steel mills, cement factories, chemical plants, etc.). Presentation of the EIA/RIMA set is a mandatory step of the licensing by the environmental agency (federal, state or municipal) and besides setting out the magnitude of probable impacts (positive and negative), muse define the mitigating measures of the negative ones.

Annex I of CONAMA Resolution 237/97 lists which activities need to be licensed at the federal, state or municipal level. Projects whose "environmental impacts exceed the territorial limits of the country or of one or more of its states" fall under the remit of the Brazilian Institute of the Environment and Renewable Natural Resources (*Instituto Brasileiro do Meio Ambiente e dos Recursos Naturais Renováveis* - IBAMA), the federal environmental agency. Therefore, except for very small CCS projects, federal licensing is required.

Law 9,605/98, known as the "Environmental Crimes Law", defines these crimes and the penalties that can be imposed on companies and individuals to deter commission of acts harmful to the environment. For individuals found liable, the penalties include imprisonment or other restriction of rights and fines. Companies that commit environmental crimes can receive fines and temporary or even permanent interdiction of activities. As far as the duty to pay compensation for damages, the general rule is strict liability, under the polluter pays principle, whereby the polluter is obligated to repair the damage caused to other parities and the public at large, regardless of blame or intention – it is enough to have caused the damage. So, if a company commits an environmental crime and it can be established that the owners, officers/managers acted negligently or with willful misconduct in the commission of that crime, they can be held personally liable.

5. Conclusions

An increasing supply of energy is an essential factor for economic growth and to improve living standards and quality, especially in developing countries. However, the current global energy mix, which is heavily reliant on burning fossil fuels, is responsible for the majority of GHG emissions. The search for new technologies that can reduce these emissions must be approached as a long-range policy. In the short and medium terms, due to this intense use of fossil fuels, CCS is the only technologically feasible option to mitigate GHG emissions on a large scale in a process of transition to a global energy system dominated by carbon-free sources.

The future of the CCS industry unquestionably depends on public acceptance and government support and encouragement, positively through subsidies and/or tax breaks and negatively through prohibitions on certain activities and setting of emissions limits. In this respect, public policymakers and legislators will play a defining role. Various projects whose pilot phases have been technically approved and have public and/or private funding committed are still waiting for definition of the applicable regulations so they can be scaled up. Because the implementation of a CCS project raises operating costs, there need to be general rules and public mechanisms (tax breaks, subsidies and/or carbon trading schemes) to defray these costs.

6. References

Air Products (2011). Port Arthur Project Update: Demonstration of CO_2 Capture and Sequestration for Steam Methane Reforming Process for Large Scale Hydrogen Production, presentation at *NETL CO2 Capture Technology Meeting*, Pittsburgh, PA, USA, August 22 - 26, 2011

Alstom Power (2011). Recovery Act: Oxy-Combustion Technology Development for Industrial-scale Boiler, presentation at *NETL CO2 Capture Technology Meeting*, Pittsburgh, PA, USA, August 22 - 26, 2011

Barlet-Goue´dard, V.; Rimmele, G.; Porcherie, O.; Quisel, N. & Desroches, J. (2009). A solution against well cement degradation under CO_2 geological storage environment. *International Journal of Greenhouse Gas Control*, 3, p. 206–216.

Barrie, J.; Brown, K.; Hatcher, P.R. & Schellhase, H.U. (2004). Carbon dioxide pipelines: A preliminary review of design and risk. *Proceedings of VII International Conference on GHG Control and Technologies (GHGT)*

BGS (2011). Removal of CO_2 from natural gas processing plant, Available from www.bgs. ac.uk/discoveringGeology/climateChange/CCS/RemovalCO2NaturalGas.html

Boe, R.; Magnus, C.; Osmundsen, P.T. & Rindstad, B.I. (2002). Geological storage of CO_2 from combustion of fossil fuel. Summary Report of the GESTCO Project. Geological Survey of Norway, Trondheim

Broek, M.; Ramírez, A.; Groenenberg, H.; Neele, F.; Viebahn, P.; Turkenburg, W. & Faaij, A. (2009). Feasibility of storing CO_2 in the Utsira formation as part of a long term Dutch CCS strategy. An evaluation based on a GIS/MARKAL toolbox. *International Journal of Greenhouse Gas Control*

Cenovus Energy (2011). Weyburn CCS Overview, presentation at *Petrobras CCS Mission to Canada*, Regina, SK, Canada, March 21, 2011.

EEA (European Environment Agency) (2010). Annual European Union greenhouse gas inventory 1990-2008

Iding, M. & Ringrose, P. (2009). Evaluating the impact of fractures on the long-term performance of the In Salah CO_2 storage site. *Energy Procedia I* p. 2021-2028

IEA (International Energy Agency) (2010a). CO_2 Emissions from Fuel Combustion

IEA (International Energy Agency) (2010b). Energy Technology Perspectives 2010, Scenarios and Strategies for 2050

IEA&CSLF (International Energy Agency & Carbon Sequestration Leadership Forum) (2010). Carbon Capture and Storage – Progress and Next Steps

IPCC (Intergovernmental Panel on Climate Change) (1996). Climate Change 1995: The Science of Climate Change. Intergovernmental Panel on Climate Change, J.T. Houghton, L.G. Meira Filho, B.A. Callander, N. Harris, A. Kattenberg, and K. Maskell. (eds.). Cambridge University Press. Cambridge, United Kingdom

IPCC (Intergovernmental Panel on Climate Change) (2005). IPCC Special Report on Carbon Dioxide Capture and Storage, *prepared by Working Group III of the IPCC*

Koornneef, J.; Spruijt, M.; Molag, M.; Ramírez, A.; Turkengurg, W. & Faaij, A. (2010). Quantitative risk assessment of CO_2 transport by pipelines — A review of uncertainties and their impacts. *Journal of Hazardous Materials* 177, p. 12-27

MCT Brazil (Brazilian Science and Technology Ministry) (2010). Brazilian Inventory of made-man emissions by sources and removals by sinks of GHG, Brazil.

Patil, R.H.; Colls, J.J. & Steven, M.D. (2009). Effects of CO_2 gas as leaks from geological storage sites on agro-ecosystems, *Proceedings of 3rd International Conference on Sustainable Energy and Environmental Protection*

Reeves, S. & Clarkson, C. (2003). The Allison Unit CO_2-ECBM Pilot: A Reservoir Modeling Study. *Topical Report US DOE*

Sherrick, B.; Hammond, M.; Spitznogle, G.; Muraskin, D.; Black, S.; Cage, M. & Kozak, F. (2009). CCS with Alstom's Chilled Ammonia Process at AEP's Mountaineer Plant. Alstom Power, Inc.

Solomon, S. (2007). Carbon Dioxide Storage: Geological Security and Environmental Issues - Case Study on the Sleipner Gas Field in Norway. *Bellona Foundation*

US EPA (United States Environmental Protection Agency) (2009). Endangerment and Cause or Contribute Findings for Greenhouse Gases Under Section 202(a) of the Clean Air Act; Final Rule

US EPA (United States Environmental Protection Agency) (2011a). Inventory of U.S. Greenhouse Gas Emissions and Sinks: 1990-2009

US EPA (United States Environmental Protection Agency) (2011b). Rulemaking of Greenhouse Gas Emissions from Electric Utility Generating Units, presentation at *NETL CO2 Capture Technology Meeting*, Pittsburgh, PA, USA, August 22 - 26, 2011

Wilson, E.J.; Friedmann, S.J. & Pollak, M.F. (2007). Research for deployment: Incorporating risk, regulation, and liability for carbon capture and sequestration. *Environmental Science Technology*. 41, p. 5945–5952

WTO&UNEP (World Trade Organization & United Nations Environment Program) (2009). Trade and Climate Change WTO-UNEP Report

Zhou, W.; Stenhouse, M.J.; Arthur, R.; Whittaker, S.; Law, D.H.S.; Chalaturnyk, R. & Jzrawi, W. (2004). The IEA Weyburn CO_2 Monitoring and Storage Project – Modeling of the Long-Term Migration of CO_2 from Weyburn. *Proceedings of the 7th International Conference on Greenhouse Gas Control Technologies (GHGT-7)*

13

Fossil Fuel and Food Security

Richard E. White
Smith College
USA

1. Introduction

The ongoing growth of global population, projected at about 9 billion by mid-century, has prompted increasing attention to the challenge of adequate nutrition. Food production outpaced population growth during the late 20th century, owing to increases in land devoted to food production, large increases in fertilizer use and irrigation, and notably the introduction of high yielding strains of major grain crops. Even so, roughly 1 billion people, primarily in developing countries, remain undernourished, while comparable numbers, mostly in rich nations, have become obese.

Fossil fuels play critical roles in the contemporary global food system, yet potential limitations in fossil fuel supplies receive scant attention in current discussions of food security. This chapter reviews elements of food security that depend on fossil fuels, highlights the potential instability of fossil fuel supplies, and considers the corresponding impact on food security over the coming decades as the human population increases.

The narrative proceeds as follows. Many authorities acknowledge food security as a global challenge for the coming decades (Section 2). These narratives highlight the enormous success of the Green Revolution in expanding food supplies in the late 20th century, but underplay the reliance of this success on the widespread availabilty of inexpensive fossil fuels (Section 3). Climate change, which is driven primarily by fossil fuel burning and by deforestation to expand agricultural lands, increases the food security challenge (Section 4). Finite supplies and increasingly difficult access to fossil fuel resources already have impacted fuel and food prices; their impact is virtually certain to grow in the coming four decades that form the primary focus of food security discussions (Section 5). Sustainable agricultural methods, particularly including reduced dependence on fossil fuels, are essential to meet growing human nutritional needs in a stable way, and sustainability in the food system also requires attention to other dimensions of food security (Section 6). Increasing fossil fuel costs will prompt evolutionary changes in the movement of food from farm to fork in different parts of the world (Section 7). Addressing the food security challenge in the face of fossil fuel scarcity is a critical element in the transition to a sustainable human economy based on renewable energy resources (Section 8).

2. Food security

The United Nations Food and Agriculture Organization (FAO) provides a definition of food security that was formulated at the 1996 World Food Summit (FAO, 2008):

Food security exists when all people at all times have access to sufficient, safe, nutritious food to maintain a healthy and active life.

The FAO document identifies four dimensions of food security: (1) availability—food supply from production, stocks, or trade; (2) access—the ability of individuals to purchase food at local market prices; (3) utilization—household processing and physiological intake of adequate nutrition; and (4) stability—continuity over time of availability, access, and utilization. More recent discussions of food security at the local level have added a fifth dimension: cultural appropriateness (Community Food Security Coalition, 2011).

Since the 1980s, food production has only kept pace with population growth, as shown in Figure 1. Consequently, recent discussions of food security (e.g., Worldwatch, 2011; *Science*, 2011) focus on expanded food production. They identify multiple challenges.

- Limited availability of additional lands for cultivation
- Low productivity in regions, especially sub-Saharan Africa, largely untouched by the Green Revolution
- Declining soil fertility, particularly owing to population pressure undermining traditional practices of fertilization with manure and regular fallowing of fields
- Soil erosion under intensive cultivation
- Declining water resources as surface flows become fully appropriated and aquifers are mined unsustainably
- Losses of both agricultural and wild biodiversity as intensive production of a limited number of crops and application of agricultural chemicals affect natural ecosystems
- Substantial losses of crops to pests or spoilage, as well as consumer waste
- Land use competition between food production and other uses, especially biofuel production
- Climate change, with projected increases in temperature and variations in water availability

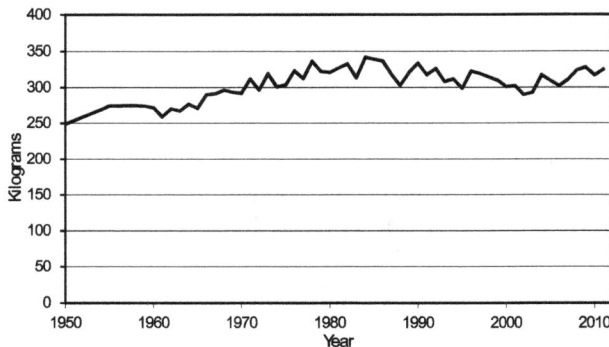

Fig. 1. Grain production per capita, 1950-2011 (Earth Policy Institute [EPI], 2008, 2011a; U.S. Census Bureau, 2011).

Although these reviews include stress on agricultural methods that can be sustained over long time periods, they scarcely mention the dependence of global agriculture on fossil fuels. As detailed in Section 5, a high probability exists that in the coming decades, even apart from policies to mitigate climate change, energy supply limitations will lead to much higher costs for these non-renewable resources. This will impose an additional challenge to meeting global food security needs.

Fossil fuel limitations primarily impact the availability and stability dimensions of food security. The Worldwatch (2011) and *Science* (2011) reviews also address food access as a critical dimension of food security, a point revisited in Section 6.

3. The green revolution

Increasing food supplies, together with improved medical care, made possible the explosive growth of world population from 2.5 billion in 1950 to 7 billion today.

Much of the spectacular improvement in agricultural production is rightly attributed to development of high-yield strains of the major cereal crops, rice, maize, and wheat (Evenson & Gollin, 2003). Dwarf varieties allowed plants to bear more fruit without collapsing, enabling mechanized application of fertilizer (Figure 2), as well as pesticides and herbicides, greatly enhancing the productivity of land while reducing farm labor (Freebairn, 1995). A great increase in irrigation (Figure 3) further improved productivity and brought addidtional, mostly marginal, lands under cultivation (Figure 4). Notably, however, irrigated area per person has been constant within a few percent since 1960 (EPI, 2011b).

The level of fossil fuel dependence differs significantly between developed and developing countries. Although total primary fossil energy input into farm production is comparable between developed countries and developing countries, as illustrated in Figure 5, developed countries use more than four times the energy per capita (8.0 gigajoules/capita/year) than developing countries (1.7 GJ/capita/year). Moreover, Figure 5 further reveals very different distribution of energy use across agricultural inputs. For developing countries, nitrogen fertilizer accounts for more than half the energy inputs, with fuel and irrigation forming the next largest inputs. By contrast, in developed countries, fuel and machinery account for more than half the inputs, with nitrogen accounting for about one quarter.

Primarily as a consequence of this disparity, the farm inputs alone in developed countries average about 2 units of fossil fuel energy inputs for every unit of food energy, whereas in developing countries, the ratio is less than 1 to 2 (Giampietro, 2004, Table 10.7). These figures illustrate the heavy fossil fuel dependence of industrial agriculture.

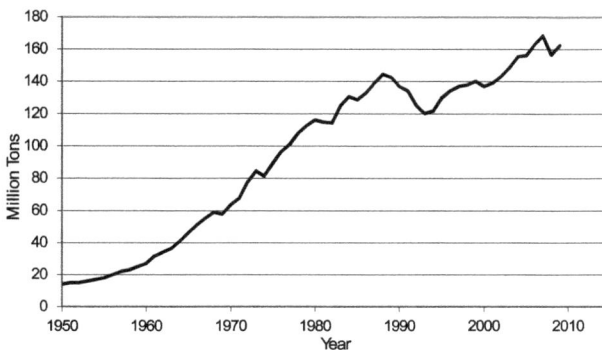

Fig. 2. Global fertilizer use, 2009 (EPI, 20011c).

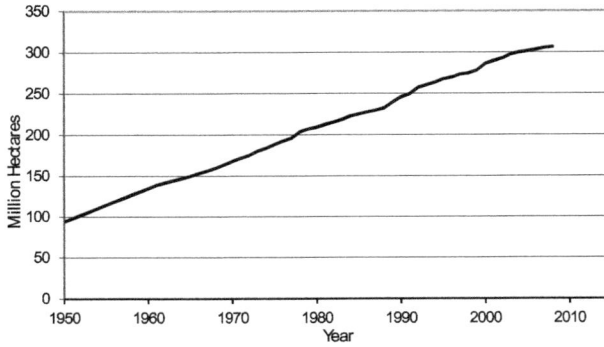

Fig. 3. Global irrigated area, 1950-2008 (EPI, 2011b).

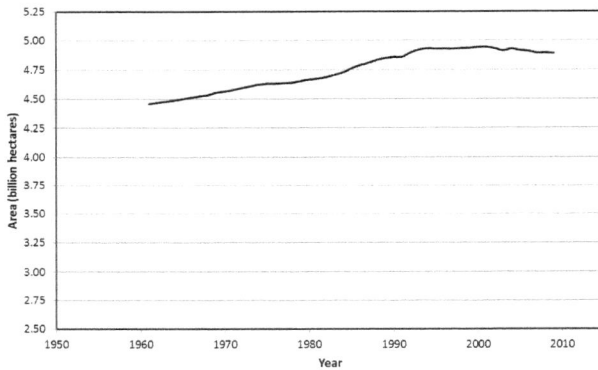

Fig. 4. Global agricultural area, 1961-2009 (FAO, 2011a).

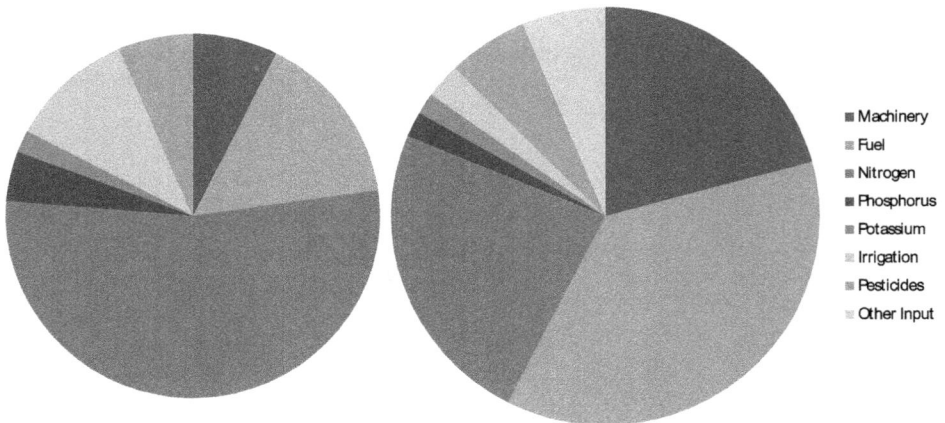

Fig. 5. Distribution of farm energy inputs in developing countries (left) and in developed countries (right) (Giampietro, 2004, Table 10.2). The areas are proportional to total energy use: 8 EJ/year and 10 EJ/year, respectively. Successive input sectors appear clockwise from the top.

Pimentel et al. (2008) have assembled case studies of cultivation of a variety of crops in both developed and developing countries. In general, their data indicate that developed countries achieve much higher yields with much less human labor, with the difference coming from fossil energy. Consequently, developed country production is more expensive and less energy efficient. In most cases the output energy is less than the input, whereas the opposite is true in developing countries (with the few exceptions probably coming from animal energy — which is provided by forage unavailable for human consumption). To cite an extreme example, apple production in the United States achieves 9 times the yield per hectare, but with overall energy efficiency of 40% compared to India and nearly 100 times the economic cost.

The fossil fuel dependence of the food system in developed countries, however, is much larger still, because farm production accounts only for about 20% of the total system inputs. Indeed, the largest single contribution comes from food preservation and preparation in the home, with additional substantial contributions from transportation and food processing (Figure 6).

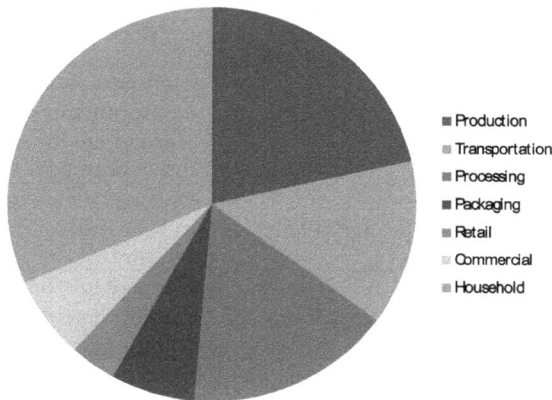

Fig. 6. Distribution of energy inputs in the U.S. food system (Heller & Keoleian, 2000, p. 41). Successive input sectors appear clockwise from the top.

4. Climate change

The Intergovernmental Panel on Climate Change (IPCC, 2007) has summarized the evidence for climate change, its likely impacts, and possible mitigation and adaptation measures. Heat-trapping by so-called greenhouse gases, most importantly carbon dioxide (CO_2), is warming the global climate (IPCC, 2007, p. 2). Most emissions (57%) of these gases come from fossil fuel burning, with an additional 17% contribution from deforestation, decay of organic matter, and peat (IPCC, 2007, p. 5). Deforestation is largely driven by expanding populations bringing additional land under cultivation. A breakdown of emissions by sector attributes nearly 14% to agriculture and another 17% to forestry, although these figures do not include other post-farm contributions from the food system. Agriculture accounts for roughly 50% of methane emissions (mostly from rice paddies and ruminant animals) and 70% of nitrous oxide emissions (mostly associated with nitrogen fertilizer) (IPCC, 1996, pp. 49-53).

Climate change will have significant impact on agricultural productivity and consequently on food security (Table 1).

Agricultural systems have long adapted to slow variations in climate and it is likely that they will do so under the projected warming of the 21st century. Nevertheless, climate change impacts on global agriculture already have had demonstrable negative effects on agricultural output (Lobell, Schlenker, & Costa-Roberts, 2011). Beyond such impacts, a growing rate of extreme events promises to be especially disruptive. The correlated events of the 2010 drought in Russia and flooding in Pakistan (Lau & Kim, 2011) had global consequences through their impact on food prices. In particular, high food prices in the Middle East, which resulted in part from unavailability of Russian grain, along with other obvious social and political factors, contributed to the political unrest that unseated several governments (Lagi, Bertrand, & Bar-Yam, 2011). Although no individual weather event, much less individual civil events, can be attributed solely to climate change, models consistently predict increasing frequency of extreme weather events (IPCC, 2007, p. 13). Together, both cumulative slow impacts and severe local events will exacerbate the challenge of achieving global food security.

Phenonomenon & Trend	Estimated Likelihood	Agricultural/forestry/ecosystem Impact
Over most land areas, warmer and fewer cold days and nights, warmer and more frequent hot days and nights	Virtually certain	Increased yield in colder environments; decreased yields in warmer environments; increased insect outbreaks
Warm spells/heat waves. Frequency increases over most land areas	Very likely	Reduced yields in warmer areas due to heat stress; increased danger of wildfire
Heavy precipitation events. Frequency increases over most areas	Very likely	Damage to crops; soil erosion, inability to cultivate land due to waterlogging of soils
Area affected by drought increases	Likely	Land degradation; lower yields/crop damage and failure; increased livestock deaths; increased risk of wildfire
Intense tropical cyclone activity increases	Likely	Damage to crops; windthrow (uprooting) of trees; damage to coral reefs
Increased incidence of extreme high sea levels (excludes tsunamis)	Likely	Salinisation of irrigation water, estuaries, and fresh water systems

Table 1. Projected impacts of climate change on agriculture, forestry, and ecosystems (IPCC, 2007, p. 13)

Global diplomatic action to mitigate climate change has stalled. Most nations that ratified the 1997 Kyoto Protocol have made significant progress toward reducing greenhouse gas emissions, as have individual states in the U.S. and many cities around the world. With the two largest emitters, China and the United States, at loggerheads, current negotiations seem likely to remain stalled (Bodansky, 2011). In the absence of a wide-ranging international agreement, business as usual prevails, with emissions rising even faster than most earlier projections by the IPCC (Figure 7).

Climate change clearly will afflict agriculture and overall human well-being in the 21st century. Along with other challenges to food security, climate impacts on agriculture will exert upward pressure on food prices. Policy action to mitigate climate change by putting a price on greenhouse gas emissions, especially from fossil fuel combustion, will increase fossil fuel prices and further impact food security. Yet another complication is diversion of cropland to biofuel production, which has forged a tight link between the prices of oil and of biofuels (Figure 8). As shown and discussed in Section 6 (Figure 11), this linkage extends to food prices. Further discussion of the impact of biofuels appears in Section 6.4.

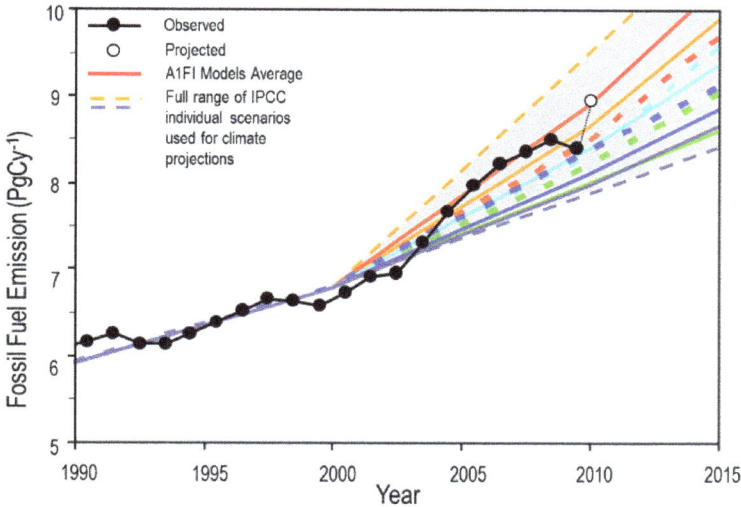

Fig. 7. Comparison of actual and projected emissions with multiple emission scenarios from the IPCC (2000, solid lines). The outer dashed lines show the full range of projected emissions. The highest solid line represents the fossil-fuel intensive scenario (Raupach, 2007, updated by Canadell, 2011).

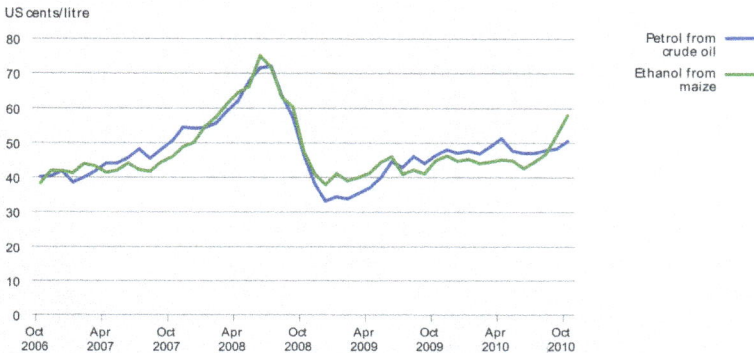

Fig. 8. Linkage between gasoline and ethanol prices (FAO, 2011b, p. 80).

5. Peak hydrocarbons

The great abundance of easily accessed fossil fuels and the correspondingly low price of energy that fueled the Green Revolution were temporary phenomena. Consequently, impending maxima in production of oil and coal, and eventually natural gas, combined with the dependence of global food production on fossil fuels and novel demands for agricultural production of biofuels, pose a daunting challenge for food security.

Agriculture around the world depends to varying degrees on gasoline and diesel for mechanized farm machinery, for transportation of supplies to farm and ranch and deliveries of products to market, and for off-grid energy to power irrigation pumps. It depends critically on natural gas to create ammonia-based nitrogen fertilizers. In many areas, especially in developed countries, agriculture depends on coal-fired electricity for irrigation, food processing, preservation, and cooking (see sector for "Other Input" in developed countries in Figure 5). In addition, fossil fuel inputs also contribute to energy-intensive mining of phosphorus, a critical nutrient with limited mineable resources (Elser & Bennett, 2011). The fossil oil, natural gas, and coal that power modern society, in particular contemporary global agriculture, were deposited by geological processes over millions of years. Large-scale human exploitation of coal has occurred for little more than 200 years; for oil and gas the timescale is little more than 100 years. Evidence is accumulating that production of these non-renewable resources will reach a maximum in the coming decades during which humankind must meet the challenge of food security. Peak production could come much sooner.

5.1 Oil

Global production of oil reached a plateau in about 2005 and did not increase even during the extreme price rise in 2007 and 2008. Hamilton (2009) has analyzed the price spike and concluded that it resulted primarily from growing demand and inelasticity in oil supplies. Some analysts and even oil executives have asserted that the production plateau indicates the arrival of Peak Oil, the time when the global production of this finite resource reaches its maximum value (Post Carbon, 2011). Although roughly half of the global recoverable resource remains available, ability to accommodate increased demand is limited at best. Consequently, continuously escalating demand from the rapidly developing economies, especially those of China and India, which represent roughly 1/3 of humanity, promises higher prices. The oil price rise early in 2011 surely had political roots in "Arab Spring," especially the revolution and civil war in Libya, but constrained supplies already had caused rising prices before the political events unfolded (Hargreaves, 2011).

Strong differences of opinion exist about the probable timing of Peak Oil. Prominent oil analysts such as Daniel Yergin (Smil, 2011) and Vaclav Smil (2003) remain very skeptical that Peak Oil will occur in the near future. On the other hand, Fatih Birol, chief economist for the International Energy Agency (IEA) has stated, "[I]t will be very challenging to see an increase in the production to meet the growth in the demand, and as a result of that one of the major conclusions we have from our recent work in the energy outlook is that the age of cheap oil is over. We all have to prepare ourselves, as governments, as industry, or as a private car driver, for higher oil prices" (Williams, 2011). Also, military establishments have begun to include Peak Oil in their contingency planning (e.g., Bundeswehr, 2010).

The 2010 IEA World Energy Outlook projects slowly increasing oil production (less than 1% per year) to about 96 million barrels per day (Mb/d) in 2035, with falling production in producing fields compensated by oil fields yet to be developed and others yet to be found (Figure 9). The projected growth is insufficient to meet escalating demand from developing countries, especially China and India. Furthermore, successive IEA projections have been declining: 121 Mb/d in 2004, 116 Mb/d in 2006, 102 Mb/d in 2008, and 96 Mb/d in 2010. Moreover, IEA projections have been criticized by the Association for the Study of Peak Oil and Gas. For example, Aleklett et al. (2010) project just 76 Mb/d in 2030, 26 Mb/d less than the 2008 IEA projection for 2030.

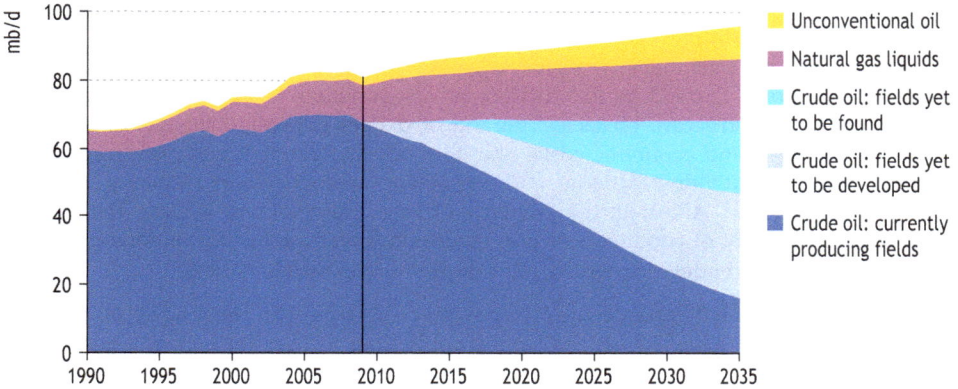

Fig. 9. Oil production by type in the IEA preferred New Policies scenario (IEA, 2010).

A major issue here is the lack of transparency about oil reserves by national oil companies, especially in Saudi Arabia. To what extent does excess capacity exist to stabilize oil prices in the coming decades? Beyond that, to what extent can unconventional sources, such as deep water fields, tar sands and shale oil raise future production and defer sharp price increases and limit volatility? One characteristic of these resources is that they yield less energy returned in sale product for the energy invested in discovery and production (Guilford et al., 2011). Consequently, they require greater energy investment and therefore greater capital investment to produce. As a result, the future — even the relatively near future — will hold higher oil prices. Besides direct impact on the cost of agricultural production, higher oil prices will increase pressure to divert cropland from food to biofuel production, raising food prices even more. Both trends will exacerbate difficulties for the poor to maintain food access.

5.2 Coal

Coal produces 48% of electricity in the United States (Energy Information Administration [EIA], 2011a) and approximately 42% in the world (EIA, 2011b, Table 66) Many utilities promote coal-burning as a low-cost alternative, but recent studies suggest that the often quoted 200 years of coal supply, both globally and in the US, is a serious overestimate.

Heinberg (2009) reviewed several available investigations and concluded that exhaustion of high quality coal reserves and infrastructure limitations on development and marketing of lower quality resources will advance the timing of Peak Coal into the relatively near future (one to two decades, not one to two centuries). Independently, Rutledge (2011) has analyzed the patterns of coal development and concluded that actual developed reserves typically are about ¼ of the early reserve estimates of the geological resource. He concludes that the world will consume 90% of producible coal by 2070. Although he refrains from discussing peak production, his analysis again points to a time no more than a few decades into the future. Patzek & Croft (2010) project a peak already in 2011, with a production decline to 50% of the peak by 2037.

Glustrom (2009) provides a bottom-up analysis of coal reserves, focusing primarily on the western United States. Analyzing the production potential of individual mines, especially in the Powder River Basin of Wyoming and Montana, which accounts for about 40% of U.S. coal production, she finds that extant surface mines on the basin perimieter have 10-20 year production horizons. Expanding production by development of new surface mines faces regulatory and infrastructure obstacles; mining of deeper deposits faces these, as well as additional energetic and economic costs. She does not analyze the coal resources of other countries in detail, but citing Rutledge and one of the same studies as Heinberg, she infers that the issue is global. Although the existence of vast coal resources is clear, the energetic and economic viability of production from lower and lower quality formations in less and less accessible places renders increasing rates of production problematical.

Consequently, expanded reliance on coal-powered electricity to meet agricultural needs faces economic challenges. These challenges apply also to post-farm components of the food system, implying an overall rise in household food expenditures, especially in developed countries where these components account for a much larger share of food system energy.

5.3 Natural gas

The situation for natural gas, the primary energy and hydrogen source for creation of synthetic nitrogen fertilizer, is more promising but still uncertain. Technological developments of hydraulic fracturing (fracking) and horizontal drilling have begun to unlock oil and natural gas from extensive shale formations. This has led to sharp increases in estimates of U.S. natural gas reserves (EIA, 2011c), as well as large increases in drilling activity in formations such as the Barnett Shale in Texas and the Marcellus Shale in Pennsylvania. The broad distribution of comparable shale formations around the world suggests that similar initiatives will lead to an abundance of natural gas potentially lasting many decades into the future. However, doubt has been cast on the most optimistic projections.

First, fracking is controversial because of its potential environmental impacts. The process involves injection into the target formation at high pressure of a large volume of water mixed with sand and a brew of chemicals, some of them toxic. The major concern is the potential for contamination of surface or groundwater. This could occur through failure of well linings intended to isolate deep wells from shallower geological strata, though spills of concentrated fracking fluids, or through inadequate treatment of fracking fluids released into local streams. An additional concern is the potential for resource conflict associated with the sheer volume of water required for fracking. As a result of such concerns, France has banned the technology (Patel, 2011). In the United States, the potential impact on drinking water prompted a Congressional mandate to the Environmental Protection Agency (EPA) to study the issue. Preliminary results are due in 2012 and a final report in 2014 (EPA, 2011, p. x). Fracking has been widely used for production of coalbed methane in the western U.S. and the potential energy resource is exceedingly valuable, so that the practice will continue in most countries, possibly under greater regulatory scrutiny.

Second, questions have surfaced concerning the magnitude and potential cost of shale gas production, as reviewed by Hughes (2011a). Evidence exists that shale gas wells deplete rapidly, so that the ulimtate resource is smaller than conventional projections (Figure 10). Morever, the technology-intensive drilling process (even apart from environmental concerns) requires elevated prices to be profitable—higher than present prices and higher than EIA projections for a decade or more. Consequently, either shale gas resources will prove to be smaller than early optimistic estimates or prices will rise so that shale gas can profitably accommodate growing demand.

Finally, doubt has arisen over the life-cycle carbon emissions of shale gas, particularly compared to coal, and therefore over the potential of natural gas to reduce emissions by displacing coal and thus to serve as bridge fuel in a transition to a renewable energy economy. Hone (2011) reviews this issue and Hughes (2011a) also compares two discordant shale gas-coal comparison studies. Shale gas probably would reduce total emissions, especially as best-practices evolve to minimize fugitive emissions (methane, the principal component of natural gas has roughly 20 times the heat-trapping effect as CO_2), but perhaps not by the 50% projected by the most optimistic estimates.

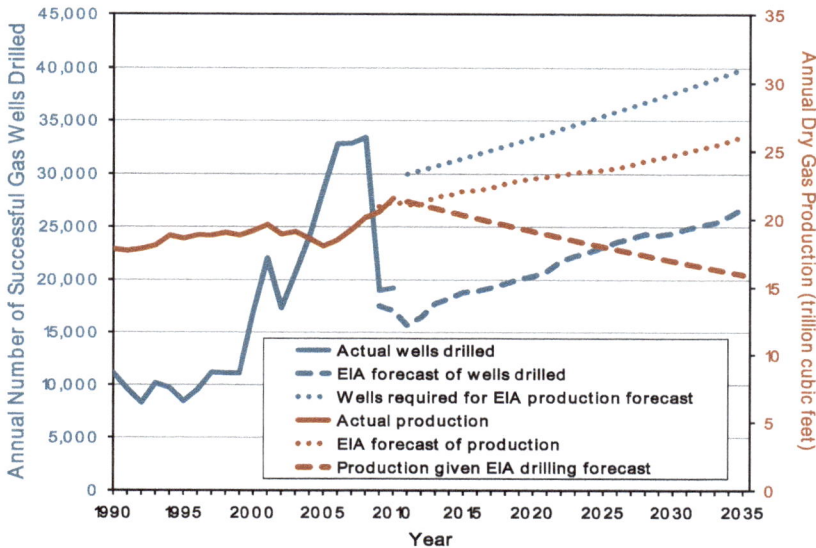

Fig. 10. Shale gas drilling rates (blue) and production (red)(Hughes 2011a). Solid lines represent historical data. Red dashed line and blue dotted line represent Hughes' alternatives to EIA projections.

No one doubts the existence of a vast geological shale gas resource, but, as for coal, conversion to producible reserves depends as much on energetics and economics as on technology. Such considerations about shale gas production will determine the ultimate magnitude of global natural gas production. Given the importance of natural gas for synthesis of nitrogen fertilizer, the evolution of shale gas production in the coming decades will have a direct impact on the cost of conventional efforts to maintain soil fertility.

In this regard and as mentioned above, the dependence of global agriculture on mined phosphorus also is a relevant concern. To the extent that fossil fuel availability contributes to the cost of mining, it will impact the price of this other critical soil nutrient.

5.4 Fossil fuels and climate change

On a marginally hopeful note, Rutledge (2010) concludes that actual fossil fuel consumption will be less than projected in *any* of the emissions scenarios considered by the IPCC (2000), yielding a peak atmospheric CO_2 concentration of 455 parts per million. If he is correct, the world has already experienced roughly half of the maximum temperature rise that will occur from fossil fuel burning—although impacts such as rising sea levels and vanishing glaciers will continue to unfold beyond 2150.

It is unlikely that the possible stabilization of long-term climate will greatly relieve stress on agricultural systems by mid-century—especially compared to the associated direct challenge of rising fossil fuel prices.

6. Sustainable agriculture

The vulnerability of the world food system and hence global food security to fossil fuel prices, as illustrated in Figure 11, renders critical the need for a transition to sustainable agricultural systems.

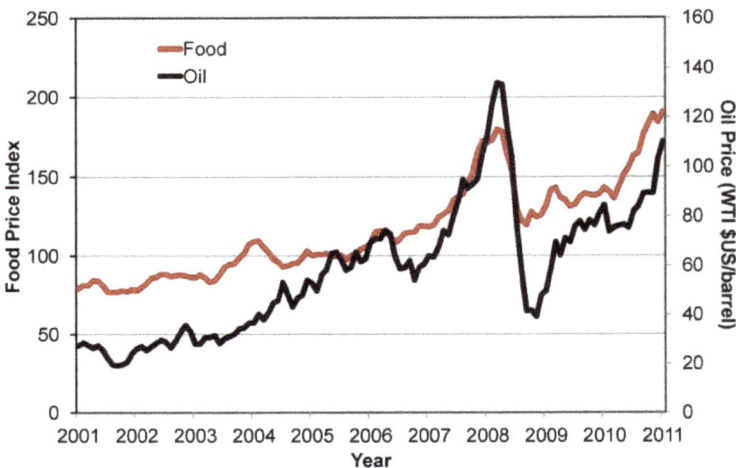

Fig. 11. Coupling of oil and food prices (Hughes, 2011b). The red line represents the FAO food price index; the black line represents the cost of West Texas Intermediate crude oil.

The great success of the Green Revolution in expanding food production faster than population during the second half of the 20th century depended on multiple developments in plant genetics, expanded use of synthetic fertilizer, increased irrigation, mechanization, petroleum-based herbicides and pesticides, and policies at the national and international levels (Smedshaug, 2010, pp. 219-222). Many of the innovations depended on low energy costs, especially for oil, that are unrepeatable. Particularly in developed countries, cheap energy has led to widespread intensification, indeed, industrialization of agriculture, with capital and fossil fuel inputs producing very high yields (e.g., in kg/ha) with very low inputs of human labor (Pimentel & Pimentel, 2008, Chapter 10). A social consequence of this has been disruption of agricultural communities and migration to cities where unemployment has been a common outcome (Berry, 2010).

For all its success, industrial agriculture is unsustainable (Tilman, 1998; Kimbrell, 2002), owing to its diverse negative effects on the environment and on social systems, which include the following.

- Reduction of agricultural biodiversity through monoculture plantings of a small number of crop cultivars
- Reduction of wild biodiversity through habitat destruction and pesticide poisoning
- Contamination of groundwater with pesticide runoff
- Eutrophication of waterways from runoff of excessive nutrients
- Reduction of soil fertility through loss of soil organic matter
- Soil erosion far in excess of natural replenishment
- Increased incidence of crop and animal diseases
- Pollution from concentrated animal wastes
- Release of greenhouse gases
- Disruption of agricultural communities
- Health impacts of agricultural chemicals, antibiotic residues in human food, and poor diets
- Opportunity costs of public agricultural subsidies

All of these impacts threaten the stability of global food production, and hence threaten food security. In addition, for all its productivity, industrial agriculture has failed to provide adequate food access to roughly 15% of the global population. Consequently, discussions of food security increasingly stress the need for agricultural systems to move to methods that can be sustained over generations (Pimentel & Pimentel, 2008, Chapter 23; Science, 2011; Smedshaug, 2010, pp. 222-225; Smil, 2010; Worldwatch, 2011). The coupling of food prices to rising prices of fossil fuels compounds this need.

The following subsections highlight proposed approaches to sustainable agriculture. Owing to the complexity of the global agricultural system, including huge differences between developed and developing countries, as well as linkages to economic and social policy, it can only provide a sampling of available information. Topics include agroecology; organic cultivation; crop breeding, including both genetically modified organisms (GMOs) and perennial crops; competition with biofuels; and proposed broad strategies.

6.1 Agroecology

The central theme of evolving global agriculture in the 21st century is "sustainable intensification," which FAO (2011c, Chapter 1) has defined as "producing more from the same area of land while reducing negative environmental impacts and increasing contributions to natural capital and the flow of environmental services." A key element of this is agroecology, particularly practiced by small producers in developing countries, as advocated by the International Assessment of Agricultural Knowledge, Science and Technology for Development (IAASTD, 2009). De Schutter & Vanloqueren (2011) provide a brief for agroecology with the following definition. "Agroecology is the application of ecological science to the study, design, and management of sustainable agriculture. It seeks to mimic natural ecological processes, and it emphasizes the importance of improving the entire agricultural system, not just the plant." Following ecological principles, agroecology seeks to recycle biomass and nutrients; enhance organic matter deposition to build soil; carefully manage resources of sun, water, and nutrients; enhance biological and genetic diversity; and encourage beneficial biological synergies while minimizing pesticides. De Schutter and Vanloqueren provide sample cases of successful implementation in the developing world, identify obstacles to wider implementation (including marginalization of the targeted small-scale farmers by past policies), and articulate policies for scaling up these innovations. By following ecological principles, agroecology seeks to minimize inputs and recycle nutrients, thus greatly reducing fossil fuel inputs and improving sustainability.

6.2 Organic agriculture

In the developed world, which already practices intensive agriculture, a popular alternative to conventional methods is organic cultivation. In a major comparison between organic and conventional cultivation, Gomiero, Pimentel, & Paoletti (2011) provide the following definition. "Organic agriculture refers to a farming system that enhances soil fertility through maximizing the efficient use of local resources, while foregoing the use of agrochemicals, the use of GMOs, as well as that of many synthetic compounds used as food additives. Organic agriculture relies on a number of farming practices based on ecological cycles, and aims at minimizing the environmental impact of the food industry, preserving the long term sustainability of soil and reducing to a minimum the use of non-renewable resources."

Some commentators flatly state that organic agriculture cannot feed the world. Gomiero et al. (2011), however, offer a more complex picture, which is framed by the difficulty in making apt comparisons between conventional and organic agriculture. Studies differ in how they define system boundaries, e.g., including or excluding indirect energy costs, leading to wide divergence in resulting estimates. Further, most such studies focus on single crops and analyze data for just a few years. Whereas conventional agriculture increasingly relies on monocultures, organic agriculture flourishes by rotating and varying crops over multi-year cycles. Moreover, long-term trends, especially on soil fertility—in which the two systems exhibit opposite effects—do not emerge in short-duration investigations. The authors document that organic agriculture is superior on virtually every aspect of environmental performance, particularly energy efficiency. Productivity data are mixed, with yields generally higher for conventional agriculture by perhaps 20%, but with differences between developed countries, where organic yields tend to be lower, and developing countries, where they tend to be higher. Given what is now decades of research and investment in conventional agriculture, the likelihood that organic yields could become equal or better given comparable investment of resources deserves serious consideration.

The Rodale Institute, long a leader in research on organic agriculture, recently issued a report on its 30-year study comparing conventional and organic cultivation (Rodale, 2011). The report documents comparable yields between the two systems, with fewer inputs, lower carbon emissions, and higher profitability from the organic fields. While some of the higher financial returns depend on the market premium paid for organic crops, the large profit disparity (organic: \$224/ha/year; conventional: \$60/ha/year) implies that much of the difference comes from the much lower cost of inputs. Although the contribution of organic agriculture is growing rapidly, critics point out that organic cultivation still accounts for only about 1% of U.S. production. Moreover, as pointed out by Pollan (2006, p. 184), "Big Organic" cultivation (eschewing synthetic fertilizer, pesticides, and GMOs, but not industrialized production methods) is not necessarily sustainable or free of concerns about fossil fuel scarcity: "As in so many other realms, nature's logic has proven no match for the logic of capitalism, one in which cheap energy has always been a given. And so, today, the organic food industry finds itself in a most unexpected, uncomfortable, and, yes, unsustainable position: floating on a sinking sea of petroleum."

A case study in agriculture after Peak Oil comes from the Cuban experience in the 1990s following the collapse of the Soviet Union, which eliminated both the source of almost all of its oil imports and also markets for Cuban agriculture. Wright (2009) has studied this example and documented the dramatic shift to organic methods. The example may be imperfect, because Cuba's isolation from global markets and industrial inputs are unique historically, but it does indicate the ability of organic agriculture and ample labor to produce an adequate food supply with minimal fossil fuel inputs.

6.3 Crop breeding

An unquestioned need exists for continued advances in crop breeding to produce cultivars adapted for specific habitats and circumstances, including climate change. Whether these techniques should include genetic engineering is controversial. Acknowledging the challenges of sustainable agriculture, many food security experts, such as Fedoroff et al. (2010), strongly advocate GMOs as necessary to the solution. Goals include breeding grain crops that would fix nitrogen, eliminating the need for synthetic nitrogen fertilizer, the most fossil fuel-dependent agricultural input in the developing world. Others, such as Benbrook (2011) take a more skeptical view, especially for developing countries. To the extent that genetically engineered cultivars depend on fossil fuel-dependent technologies, they will fail to meet the coming challenge of increasing fossil fuel costs. Likewise, the profit-driven choice of cultivars, with restrictions on seed saving, local experimentation, and innovation, appears inadequate to address the essential needs of small-scale agriculture that feeds 80% of the world's people.

Regardless of the resolution of the GMO debate, advances in molecular biology have provided powerful tools for advancing conventional crop breeding. One advocate of this approach is the Kansas-based Land Institute. Consistent with the agroecology approach, Land Institute founder Wes Jackson and his colleagues advocate a sustainable "next synthesis" based on cultivation of perennial grains (Jackson, Cox, & Crews, 2011; Glover et al., 2010). Jackson et al. argue that these crops can reconcile ecological sustainability with the productivity needed to meet human needs, in the process providing both a model and metaphor for the material economy.

6.4 Competition with biofuels

Demand for alternative liquid fuels has driven diversion of cropland to biofuel production, resulting in close coupling between the prices of oil and agricultural commodities illustrated in Figures 8 & 11. Advocates such as Collins & Duffield (2005) are optimistic about the ability of conventional U.S. agriculture to meet world food needs, as well as to make a significant contribution to biofuel production. Sustainability-minded analysts Giampietro & Mayumi (2009), however, argue that biofuels (at least those produced from agricultural crops) reduce food supply, increase CO_2 emissions, and retard rural development. Addressing the broader question of meeting a large fraction of human energy needs with biomass, Smil (2010, p. 721) dismisses the idea as an insufferable intrusion on the necessary functioning of the biosphere. Smil does not address specifically either crop residues or grasses as possible biofuel sources, but his general energetic analysis underscores concerns that extensive exploitation these non-crop biological resources would undermine necessary nutrient recycling. Acknowledging the impact of energy prices on food prices and the volatility of food markets, Koning & Mol (2009) call for new institutions to balance food and energy markets.

6.5 Proposed strategies

Two recent reviews (Godfray et. al, 2010; Foley, et al., 2011) identify broad strategic approaches to sustainable food security that incorporate to some extent all sides of the issues addressed in Sections 6.1 to 6.4. The suggestions of the two articles overlap to some degree and include the following.

- Stop expanding agriculture. The environmental benefits of preserving sensitive ecosystems would outweigh the marginal loss of increasing production, especially in the tropics.
- Close yield gaps. Improving realized productivity toward what is achievable with locally available genetic material, technology, and management could improve yields in many regions by tens of percent. This is the thrust of the sustainable intensification efforts discussed in Section 6.1, though there are many complexities and necessary local variations.
- Increase agricultural resource efficiency. Careful management of both water and nutrient inputs can avoid both deficiencies and excess applications that produce environmental degradation. Great scope exists for more precise applications in time and space.
- Increase production limits. Greater yields would result from optimizing cultivars for specific conditions, especially those in developing countries, by conventional plant breeding or genetic engineering. Preservation of agricultural biodiversity, especially of locally adapted crops and livestock, is an important component of this effort.
- Increase food delivery by shifting diets and reducing waste. Limiting diversion of crops to uses such as animal feed and biofuel production would increase the amount available for direct human consumption. Likewise, the 30% or more of harvested food lost to pests, degradation, and discard could feed many more people.
- Expand aquaculture. This initiative would continue current trends, but also focus on minimizing environmental impacts.

Godfray et al. (2010, p. 817) conclude by stating, "The goal is no longer simply to maximize productivity, but to optimize across a far more complex landscape of production, environmental, and social justice outcomes." Figure 12 vividly illustrates the challenge.

Notably, neither of these reviews acknowledges possible fossil fuel scarcity and high costs as challenges to food security.

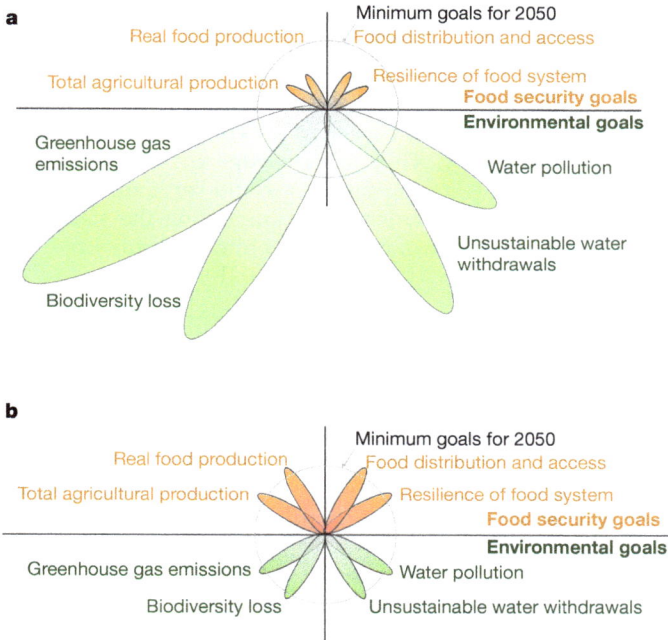

Fig. 12. Qualitative comparison of (a) the present state of global agriculture and (b) projection for meeting food security and environmental goals for 2050 (Foley et al., 2011.)

6.6 Other commentaries

Smil has written extensively on both food (2000) and energy (2003). He is optimistic that fossil fuels will remain abundant for several decades and that the agricultural system has sufficient inefficiencies to accommodate the needed productivity growth while undergoing transition to a renewable energy base. In *Energy at the Crossroads* (2003), he stresses the fact that an energy transition on the societal scale projected in the coming decades will itself require decades, owing to the massive capital investment required and the time needed for these investments to bear fruit. The same consideration clearly applies to the agricultural system as well. As an optimist about energy supply, Smil does not consider the possibility, as does Heinberg (2009), that the upfront energy investment required for renewable energy technologies and potential limits in absolute energy supply could prevent the needed investments. He does, however, stress the need for more equitable distribution of both energy and nutritional resources at levels intermediate between the current consumption levels of developing nations and of the highest consuming nations, particularly the United States.

Bottom-up analyses such as Smil's, typically take little account of institutional inertia that creates significant obstacles to widespread achievement of possible system efficiencies. One such potential obstacle is the disparate influence on agricultural systems and food trade between rich nations and transnational corporations on the one hand, and poor nations and small farmers on the other (FAO, 2003). The new movement for food sovereignty (Wikipedia, 2011) is a grass roots attempt to redress this imbalance, asserting as a human right access to healthy, culturally appropriate, sustainably produced food.

Pimentel and Pimentel (2008, Chapter 23) are less sanguine than Smil about the energy future. They review future food needs, energy requirements for producing the food, constraints on land and water, climate change, and environmental pollution. They stress, as do none of the other sources discussed here, that lowering birthrates is an additional food security and sustainability strategy. They acknowledge the social challenge involved in conveying to parents that having smaller families would serve their interests and those of their children, but do not mention evidence that improving the status and education of women yields multiple societal benefits, including voluntary fertility reduction. The gender gap in world agriculture is highlighted by the FAO (2011b).

Smedshaug (2010) provides a long view of the history of agriculture and the role of national and international policy in regulating food production. He takes as given the constriction of energy resources in this century, but provides no detailed projections of its impact. He does, however, stress the critical role of policy in moderating the fluctuating effects of markets on production and on farm incomes. The history of intermittent overproduction that he documents suggests that providing adequate food supplies will be possible—even as it is today. The question is whether policy makers can achieve a state that meets the broad definition of food security, particularly including not just food availability, but universal access to food on a stable basis.

7. Fossil fuels in the food system

Section 6 focused primarily on the farm and the interaction of the farm system with the environment. This section seeks to articulate likely impacts of rising fossil fuel costs and conceivable adaptions in developed and developing countries, both on the farm and in post-farm segments of the food system.

Because of its annual cycle of production, the food system adapts relatively quickly to altered prices of inputs, leading to optimism that the system will evolve relatively smoothly in a changing cost environment. Price spikes, however, can be more disruptive then gradual increases, although they may also induce long-term adjustment. An example is the radical improvement in energy efficiency of American refrigerators since the oil price shocks of the 1970s. Since then, energy efficiency has improved by a factor of about 3.5, while average sizes have grown about 10%, and real prices have decreased by 2/3 (Appliance Standards Awareness Project, 2011).

7.1 Developed countries

Figures 5 and 6 show the heavy fossil fuel dependence of the agricultural system in the U.S., which should be broadly typical of other developed countries. Increasing energy costs will prompt evolution toward the more sustainable cultivation systems described in

Section 6, in particular, the reduced material and energy inputs and greater labor intensity illustrated by Rodale (2011). Technological inputs from improved crop strains, precision application of water and nutrients, and other innovations will also contribute. Higher energy prices will accelerate the longstanding evolution of agriculture and the larger economy toward higher energy efficiency; they may also reverse the corresponding trend toward greater overall energy use. Because of the significant food system costs to individual households, food and energy price increases will directly impact household budgets, leading to choices of more efficient appliances and possibly dietary changes. Owing to the long life of home appliances, however, evolution will be slow. Fuel costs will directly impact the transport sector, while electricity costs will impact food processing. Possible responses are greater reliance on biofuels, which exacerbate food price increases, reliance on renewable energy resources, and localization of processing and distribution facilities. The evolution probably will include all three, in a variable mix depending on national and regional market forces.

Smil's (2000, 2003) optimism about the adaptability of both the food and energy systems is likely well-placed, unless severe economic shocks of the more pessimistic Peak Oil forecasts disrupt the economy on a large scale and undermine capital investment. Then, the more drastic localization scenario represented by the Cuban experience (Wright, 2009) may be relevant.

Agricultural and energy policy initiatives could either accelerate or inhibit adaptation to an environment of gradually increasing energy prices; they also could reduce or increase vulnerability to price shocks. Unfortunately, politically powerful vested interests in conventional agriculture and fossil fuel production are likely to oppose policy innovations to promote more sustainable systems.

7.2 Developing countries

Figure 5 also displays the distribution of farm energy inputs in developing countries, showing that fertilizer embodies the largest energy cost. Rising prices will make this input increasingly expensive, and probably unreachable for most small farmers, unless subsidized by government policy. The cost of fertilizer will exert pressure to accelerate agroecological innovations, although soil fertility already compromised by reduction in animal nutrients and rotation cyles (Bunch, 2011) may warrant targeted use of synthetic fertilzer as part of a long range plan for land restoration. Scarcity of other fossil fuel inputs will confirm existing patterns of labor-intensive farming. Success in global efforts to raise crop yields on small farms in the developing world will require sustained policy commitments from national and international agencies.

Little information is available about post-farm energy inputs in developing country agricultural systems (Ziesemer, 2007). Organic systems in developed countries may provide a partial model, although these data also are few. Transportation costs will largely confine distribution to local and regional markets, although success in intensifying production may increasingly satisfy local needs and require expansion of regional marketing opportunities. Urban farming should have a role to play (Karanja & Njenga, 2011), as should innovative approaches that simultaneously minimize waste, generate income, and provide food, e.g., cultivating mushrooms on invasive water hyacinths in Africa (Pauli, 1998, Chapter 11). Because the largest population growth rates also occur in countries where agricultural productivity is low, systemically addressing the role of women in agriculture, along with provision of women's health services and educational opportunities, could advance food security both by improving productivity and by reducing human fertility, in addition to providing broader societal benefits.

8. Conclusion

Creating food security for the projected mid-century global population of 9 billion is an enormous challenge. Meeting the challenge will require efficient use of arable land without continued deforestation, efficient use of water, mindful choices between use of agriculture for food and for energy, altering diets for optimum balance of animal and plant protein, supplementing diets with unconventional foods such as mushrooms and algae, attention to global nutrient cycles, restoration of soils, protection of biodiversity, improved crop varieties, and adaptation to climate change.

The goal of this chapter is to demonstrate that over the coming decades this multi-faceted challenge must be met in an economy with increasingly limited supplies of oil and coal, and probably natural gas, with correspondingly rising energy prices. In particular, continuation and expansion of conventional energy- and chemical-intensive agriculture will become uneconomic, even apart from its contribution to climate change and other negative ecological impacts. In the arena of food security, "business as ususal" is unsustainable. Even if conventional agriculture can meet short-term productivity demands, it already fails to provide food access to a billion people, and it simply cannot provide a stable long-term solution to the food security challenge.

The challenge of food security represents just one dimension of the growing conflict between the dominant economic paradigm of unending growth and the finite capacity of planet Earth to supply resources and absorb wastes. Crucial elements of this conflict have been identified in *Limits to Growth* (Meadows, Meadows, & Randers, 2004, p. 178): (1) the cultural acceptance of growth as desirable, (2) the existence of physical limits, such as the land area available for agriculture, which may be erodable by overexploitation, and (3) the existence of delays in the system between signals, such as declining crop yields, and responses, such as altered land management.

To project possible trajectories of human welfare, *Limits to Growth* broadly represents key components of the economy: population, food production, industrial output, pollution, and resource depletion. It further identifies interactions that provide positive or negative feedbacks among them. For example, increasing food production positively impacts population, wheres pollution has a negative impact. Conceptual scenarios explore the possible evolution of the coupled systems. Most of the scenarios exhibit peaks in population and industrial output by mid-century, followed by collapse; only a few scenarios that embody conscious choices acknowledging ecological limits show a relatively smooth transition to sustainability.

Many of global challenges of recent years, such as increases in oil prices, the growing climate crisis, and the continuing economic crisis, can be interpreted as signals of a global economy that has overshot the carrying capacity of the planet. Unfortunately, owing to slow responses in natural systems, if policy makers wait for an unequivocal signal, such as prolonged economic depression caused by declining oil supplies, their belated responses will fail to forestall collapse. Moreover, even with courageous leadership focused on long-term outcomes, slow responses in political systems will delay decisive policy action.

Thus, leaders at every level of government and society face a critical challenge to identify and implement wise policies to build long-term sustainability into the interlinked food, resource, environment, and economic systems. *Limits to Growth* (pp. 259-260) offers a list of guidelines for this effort. Other sources, such as Brown (2011), provide book-length treatments.

Food security requires sustainable agriculture, which in turn requires farsighted leadership to guide evolution away from heavy dependence on fossil fuel inputs and toward the alternatives discussed in Section 6. Failure to achieve food security will take a tragic toll of human suffering through famine and social chaos. The need to act is urgent.

9. Acknowledgements

The author thanks original authors or publishers for permission to reproduce their figures or data, particularly Dr. Josep Canadell for Figure 7. He thanks Dr. James Boyce and especially Dr. Tina Evans for comments on drafts of the manuscript; to Dr. Wes Jackson for a copy of his manuscript in advance of publication; to Dr. Richard Grossman for prompting his initial foray into this subject (White & Grossman 2010); and to Smith College for funds toward the cost of publication. He particularly thanks Dr. Angeline Schrater for astute editorial comments, careful proofreading, and patient support.

10. References

Aleklett, K., Höök, M., Jakobsson, K. Lardelli, M., Snowden, S., & Söderbergh, B. (2010). The Peak of the Oil Age, *Energy Policy*, Vol. 38, No. 3, (March), pp. 1398-1414, ISSN 0301-4215

Appliance Standards Awareness Project (2011). Average Household Refrigerator Energy Use, Volume, and Price Over Time, accessed at http://www.appliance-standards.org/sites/default/files/Refrigerator%20Graph_July_2011.PDF

Benbrook, C. (2011). Innovations in Evaluating Agricultural Development Projects, In *State of the World, 2011: Innovations that Nourish the Planet*, pp. 169-172, W. W. Norton, ISBN 978-0-393-33880-5, New York

Berry, W. (2010). What Are People For?, In *What Matters? Economics for a Renewed Commonwealth*, pp. 105-107, Counterpoint, ISBN 978-1-58243-606-7, Berkeley, CA

Bodansky, D. (2011). W[h]ither the Kyoto Protocol? Durban and Beyond, *Viewpoints*, Harvard Project on Climate Agreements, (August), Cambridge, MA, accessed at http://www.pewclimate.org/docUploads/whither-kyoto-protocol-durban-and-beyond.pdf

Brown, L. R. (2011). *World on the Edge: How to Prevent Environmental and Economic Collapse*, W. W. Norton, ISBN 978-0-393-33949-9, New York

Bunch, R.(2011). Africa's Fertility Crisis and the Coming Famine, In *State of the World, 2011: Innovations that Nourish the Planet*, pp. 59-70, W. W. Norton, ISBN 978-0-393-33880-5, New York

Bundeswehr (2010). *Peak Oil: Security policy implications of scarce resources*, Bundeswehr Transformation Centre, Future Analysis Branch; Armed Forces, Capabilities and

Technologies in the 21st Century: Environmental Dimensions of Security, Sub-study 1, (November), Strausberg, Germany, accessed via http://baobab2050.org/2011/09/04/bundeswehr-peak-oil-report-now-officially-translated-in-english/

Canadell, J. G. (2011). Private communication

Collins, K. J., & Duffield, J. A. (2005). Energy and Agriculture at the Crossroads of a New Future, In *Agriculture as a Producer and Consumer of Energy*, J. L. Outlaw, K. J. Collins, & J. A. Duffield (eds.), pp. 1-29, CAB International, ISBN 0-85199-018-5, Cambridge, MA

Community Food Security Coalition (2011). Community Food Security Programs: What Do They Look Like?, accessed at http://www.foodsecurity.org/CFS_projects.pdf

De Schutter O., & Vanloqueren, G. (2011). The New Green Revolution: How Twenty-First Century Science Can Feed the World, *Solutions*, Vol. 2, No. 4, pp. 33-44, (August 2011)

EIA (2011a). *Annual Energy Review, 2010*, U. S. Energy Information Administration, Report Number DOE/EIA-0484 (2010), (October), Washington, DC, accessed at http://www.eia.gov/totalenergy/

EIA (2011b). *International Energy Outlook, 2011*, U. S. Energy Information Administration, Report Number DOE/EIA-0484, (September), Washington, DC, accessed at http://www.eia.gov/forecasts/ieo/

EIA (2011c). Review of Emerging Resources: U.S. Shale Gas and Shale Oil Plays, U. S. Energy Information Administration, (July), accessed at http://205.254.135.24/analysis/studies/usshalegas/

Elser, J., & Bennett, E. (2011). A Broken Geochemical Cycle, *Nature*, Vol. 478, No. 7367, pp. 29-31, (6 October), ISSN 0028-0836

EPA (2011). *Plan to Study the Potential Impacts of Hydraulic Fracturing on Drinking Water Resources*, U. S. Environmental Protection Agency, Office of Research and Development, (November), Washington, DC, accessed at http://www.epa.gov/hfstudy/HF_Study_Plan_1102_FINAL_508.pdf

EPI (2008). World Grain Production per Person, 1950-2007, Earth Policy Institute, accessed at http://www.earth-policy.org/data_center/xls/update72_11.xls

EPI (2011a). World Grain Production and Annual Gain or Loss, 1980-2010; Grain Production in Selected Countries and the World, 2010 and Estimates for 2011, Earth Policy Institute, accessed at http://www.earth-policy.org/data_center/xls/update92_2.xls & /update92_5.xls

EPI (2011b). World Irrigated Area and Irrigated Area Per Thousand People, 1950-2008, Earth Policy Institute, accessed at http://www.earth-policy.org/data_center/book_wote_ch2_4.xls

EPI (2011c). *Plan B 4.0*, Chapter 9 Data: Feeding Eight Billion People Well, Earth Policy Institute, accessed at http://www.earth-policy.org/data_center/xls/book_wote_Ch12_10.xls

Evenson, R. E. & Gollin, D. (2003). Assessing the Impact of the Green Revolution, 1960-2000, *Science,* Vol. 300, No. 5620, (2 May), pp. 758-762, ISSN 0036-8075

FAO (2003). The Role of Transnational Corporations, Chapter 9 In *Trade Reforms and Food Security*, UN Food and Agriculture Organization, Rome, accessed at http://www.fao.org/docrep/005/y4671e/y4671e0e.htm#bm14

FAO (2008). *An Introduction to the Basic Concepts of Food Security*, UN Food and Agriculture Organization, Rome accessed at http://www.fao.org/docrep/013/al936e/al936e00.pdf

FAO (2011a). FAOSTAT database, UN Food and Agriculture Organization, Rome, accessed at http://faostat.fao.org/site/377/default.aspx#ancor

FAO (2011b). *The State of Food and Agriculture 2010-2011*, UN Food and Agriculture Organization, ISBN 978-92-5-106768-0, Rome, accessed at http://www.fao.org/docrep/013/i2050e/i2050e00.htm

FAO (2011c). *Save and Grow: A policymaker's guide to the sustainable intensification of smallholder crop production*, UN Food and Agriculture Organization, Rome, accessed at http://www.fao.org/ag/save-and-grow/index_en.html

Fedoroff, N. V., Battisti, D. S., Beachy, R. N., Cooper, P. J. M., Fischhoff, D. A., Hodges, C. N., Knauf, V. C., Lobell, D., Mazur, B. J., Molden, D., Reynolds, M. P., Ronald, P. C., Rosegrant, M. W., Sanchez, P. A., Vonshak, A., & Zhu, J.-K. (2010). Radically Rethinking Agriculture for the 21st Century, *Science*, Vol 327, No. 5967, (12 February), pp. 833-834, ISSN 0036-8075

Foley, J. A., Ramankutty, N., Brauman, K. A., Cassidy, E. S., Gerber, J. S., Johnston, M., Mueller, N. D., O'Connell, C., Ray, D. K., West, P. C., Balzer, C., Bennett, E. M., Carpenter, S. R., Hill, J., Monfreda, C., Polansky, S., Rockström, J., Sheehan, J., Siebert, S., Tlman, D., & Zaks, D. P. M. (2011). Solutions for a cultivated planet, *Nature*, Vol. 478, No. 7369, (20 October), pp. 337-342, ISSN 0028-0836

Freebairn, D. K. (1995). Did the Green Revolution concentrate incomes, *World Development*, Vol. 23, No. 2, (February), pp. 265-279, ISSN 0305-750X

Giampietro, M. (2004). *Multi-scale Integrated Analysis of Agroecosystems*, CRC Press, ISBN 0-8493-1067-9, Boca Raton, FL

Giampietro, M, & Mayumi, K. (2009). *The Biofuel Delusion: The Fallacy of Large-Scale Agro-biofuel Production*, Earthscan, ISBN 978-1-84407-681-9, London

Glover, J. D., Reganold, J. P., Bell, L. W., Borevitz, J., Brummer, E. C., Buckler, E. S., Cox, C. M., Cox., T. S., Crews, T. E., Culman, S. W., DeHaan, L. R., Eriksson, D., Gill, B. S., Holland, J., Hu, F., Hulke, B. S., Ibrahim, A. M. H., Jackson, W., Jones, S. S., Murray, S. C., Paterson, A. H., Ploschuk, E., Sacks, E. J., Snapp, S., Tao, D., Van Tassel, D. L., Wade, L. J., Wyse, D. L., & Xu, Y. (2010). Increased Food and Ecosystem Security via Perennial Grains, *Science*, Vol. 328, No. 5986, (25 June), pp. 1638-1639, ISSN 0036-8075

Glustrom, L. (2009). Coal: Cheap and Abundant ... or Is It?, Clean Energy Action, (February), accessed at http://www.cleanenergyaction.org/sites/default/files/Coal_Supply_Constraints_CEA_021209.pdf

Godfray, H. C. J., Beddington, J. R., Crute, I. R., Haddad, Lawrence, D., Muir, J. F., Pretty, J., Robinson, S., Thomas, S., & Toulmin, C. (2010). Food Security: The Challenge of Feeding 9 Billion People, *Science*, Vol. 327, No. 5967, (12 February), pp. 812-818, ISSN 0036-8075

Gomiero, T., Pimentel, D., & Paoletti, M. G. (2011). Environmental Impact of Different Agricultural Management Practices: Conventional vs. Organic Agriculture, *Critical Reviews in Plant Sciences*, Vol. 30, No. 1, pp. 95–124, ISSN 1549-7836

Guilford., M. C., Hall, C. A. S., O'Connor, P., & Cleveland, C. J. (2011). A New Long Term Assessment of Energy Return on Investment (EROI) for U.S. Oil and Gas Discovery and Production, *Sustainability*, Vol. 3, No. 10, (October), pp. 1866-1887, ISSN 2071-1050

Hamilton, J. D. (2009). Causes and Consequences of the Oil Shock of 2007-08, National Bureau of Economic Research, NBER Working Paper No. 15002

Hargreaves, S. (2011). Gas Prices High and Might Get Higher, CNN Money, (17 January), accessed at
http://money.cnn.com/2011/01/21/markets/gasoline_prices_rising/index.htm

Heinberg, R. (2009). *Blackout: Coal, Climate and the Last Energy Crisis*, New Society, ISBN 978-0-86571-656-8, Gabriola Island, BC, Canada

Heller, M. C., & Keoleian, G. A. (2000). *Life Cycle-Based Sustainability Indicators for Assessment of the U.S. Food System*, Center for Sustainable Systems Report CSS00-04, University of Michigan, Ann Arbor, accessed at http://css.snre.umich.edu/css_doc/CSS00-04.pdf

Hone, D. (2011). Natural Gas, CO2 Emissions and Climate Change, The Energy Collective, (19 September), accessed at
http://theenergycollective.com/davidhone/65490/natural-gas-co2-emissions-and-climate-change

Hughes, J. D. (2011a). *Will Natural Gas Fuel America in the 21st Century?*, Post Carbon Institute, (August), Santa Rosa, CA, accessed at
http://www.postcarbon.org/reports/PCI-report-nat-gas-future.pdf

Hughes, J. D., (2011b), Global Energy: A View from 50,000 feet, presentation at Adhesive and Sealant Council Chemical Conference, Indianapolis, Indiana, (17 October)

IAASTD (2009). *Agriculture at the Crossroads*, International Assessment of Agricultural Knowledge, Science and Technology for Development, Synthesis Report, B. D. McIntyre, H. R. Herren, J. Wakhungu, & R. T. Watson (eds.), Island Press, ISBN 978-1-59726-550-8, Washington, DC

IEA (2010). Oil Production Becomes Less Crude, *World Energy Outlook*, 2010 ©OECD/International Energy Agency 2010, Key Graph #7, accessed at http://www.worldenergyoutlook.org/docs/weo2010/key_graphs.pdf

IPCC (1996). *Technologies, Politics and Strategies for Mitigating Climate Change*, Intergovernmental Panel on Climate Change, IPCC Technical Paper I, W. T. Watson, M. C. Zinyowera, & R. H. Moss (eds.), ISBN: 92-9169-100-3

IPCC (2000). *Emissions Scenarios: Summary for Policy Makers*, Intergovernmental Panel on Climate Change, IPCC Special Report, ISBN: 92-9169-113-5

IPCC (2007). *Climate Change: 2007, Synthesis Report Summary for Policy Makers. Contributions of Working Groups I, II, and III to the Fourth Assessment Report of the Intergovernmental Panel on Climate Change*, Intergovernmental Panel on Climate Change, Core Writing Team, R. K. Pachauri, and A. Reisinger (eds.), Geneva

Jackson, W., Cox., S., & Crews, T. (2011). The Next Synthesis, In *Nature as Measure: The Selected Essays of Wes Jackson*, Counterpoint, ISBN 978-1-58243-700-2, pp. 199-232, Berkeley, CA

Karanja, N., & Njenga, M. (2011). Feeding the Cities, In *State of the World, 2011: Innovations that Nourish the Planet*, pp. 109-120, W. W. Norton, ISBN 978-0-393-33880-5, New York

Kimbrell, A. (ed.). (2002). Seven Deadly Myths of Industrial Agriculture, In *Fatal Harvest: The Tragedy of Industrial Agriculture*, pp. 49-63, Island Press, ISBN 1-55963-940-7, Washington, DC

Koning, M., & Mol, A. P. J. (2009). Wanted: institutions for balancing global food and energy markets, *Energy Security*, Vol. 1, No. 3, pp. 291-303, ISSN:1876-4525

Lagi, M., Bertrand, K. Z., & Bar-Yam, Y. (2011). The Food Crises and Political Instability in North Africa and the Middle East, arXiv:1108.2455v1 [physics.soc-ph], preprint, accessed at http://arxiv.org/abs/1108.2455

Lau, W. K. M., & Kim, K-M. (2011). The 2010 Pakistan Flood and Russian Heat Wave: Teleconnection of Hydrometeorologic Extremes, *Journal of Hydrometeorology*, in press, ISSN 1525-7541

Lobell, D. B., Schlenker, W., & Costa-Roberts, J. (2011). Climate Trends and Global Crop Production Since 1980, *Science*, Vol. 233, No. 6042, pp. 616-620, ISSN 0036-8075

Meadows, D., Meadows, D., & Randers, J. (2004). *Limits to Growth: The 30-Year Update*, Chelsea Green, ISBN1-931498-58-X, White River, VT

Patel (2011). France Vote Outlaws 'Fracking' Shale for Natural Gas, Oil Extraction, Bloomberg, (1 July), accessed at http://www.bloomberg.com/news/2011-07-01/france-vote-outlaws-fracking-shale-for-natural-gas-oil-extraction.html

Pauli, G. (1998). *Upsizing: The Road to Zero Emissions: More Jobs, More Income, and No Pollution*, Greenleaf, ISBN 1 874719 18 7, Sheffield, UK

Patzek, T. W., & Croft, G. D. (2010). A global coal production forecast with multi-Hubbert cycle analysis, *Energy*, Vol. 35, No. 8, (August), pp. 3109-3122, ISSN 0360-5442

Pimentel, D., & Pimentel, M. H. (eds.). (2008). *Food, Energy, and Society (3rd. Ed)*, CRC Press, ISBN 978-1-4200-4667-0, Boca Raton, FL

Pimentel, D., Doughty, R., Carothers, C., Lamberson, S., Bora, N., & Lee, K. (2008). Energy Inputs in Crop Production in Developing and Developed Countries, In *Food, Energy, and Society (3rd. Ed)*, D. Pimentel & M. H. Pimentel (eds.), pp.137-159, CRC Press, ISBN 978-1-4200-4667-0, Boca Raton, FL

Pollan, M. (2006). *The Omnivore's Dilemma*, Penguin, ISBN 1-59420-082-3, New York

Post Carbon (2011). Peak Oil Primer, Post Carbon Institute Energy Bulletin, accessed at http://www.energybulletin.net/primer.php

Raupach, M. R., Marland, G., Ciais, P., Le Quéré, Corinne, Canadell, J. G., Klepper, G., & Field, C. B. (2007). Global and regional drivers of accelerating CO_2 emissions, *P.N.A.S.*, Vol. 104, No. 24, (June 12), pp. 10288-10293

Rodale (2011). *The Farming Systems Trial*, Rodale Institute, accessed at http://www.rodaleinstitute.org/files/FSTbookletFINAL.pdf

Rutledge, D. (2010). Coal, Climate Change, and Peak Oil, ASPO-USA meeting presentation, (10 October), accessed at http://www.aspousa.org/2010presentationfiles/10-8-2010_aspousa_CoalQuestion_Rutledge_D.pdf

Rutledge, D. (2011). Estimating long-term world coal production with logit and probit transforms, *International Journal of Coal Geology*, Vol. 85, No. 1, (January), pp. 23-33, ISSN 0166-5162

Science (2011). Food Security (Special Section), *Science*, Vol. 327, No. 5967, (12 February), pp. 797-834, ISSN 0036-8075

Smedshaug, C. A. (2010). *Feeding the World in the 21st Century*, Anthem, ISBN 978-1-84331-867-5 or 1-84331-867-9, London

Smil, V. (2000). *Feeding the World: A Challenge for the Twenty-First Century,* MIT Press, ISBN 0-262-19432-5, Cambridge, MA

Smil, V. (2003). *Energy at the Crossroads,* MIT Press, ISBN 0-262-19492-9, Cambridge, MA

Smil, V. (2010). Science, energy, ethics, and civilization, In: *Visions of Discovery: New Light on Physics, Cosmology, and Consciousness,* R.Y. Chiao, M. L. Cohen, A. J. Leggett, W. D. Phillips, & A. L. Harper, Jr. (eds.), pp. 709-729, Cambridge University Press, (ISBN-13: 9780521882392, Cambridge, UK

Smil, V. (2011). Burning Desires, *Nature,* Vol. 477, No.7365, p. 403, (22 September), ISSN 0028-0836, review of D. Yergin (2011), *The Quest : Energy, Security, and the Remaking of the Modern World,* Penguin, ISBN 978-159-4202-83-4, New York

Tilman, D. (1998). The Greening of the Green Revolution, *Nature,* 396, No. 6708, (19 November), pp. 211-212, ISSN 0028-0836

U.S. Census Bureau (2011). Total Midyear Population of the World: 1950-2050, accessed at http://www.census.gov/population/international/data/idb/worldpoptotal.php

White, R. E., & Grossman, R. (2010). Food Security: Fossil Fuels, *Science,* Vol. 328, No. 5975, (9 April), p. 173, ISSN 0036-8075

Wikipedia (2011). Food Sovereignty, Wikipedia, accessed at http://en.wikipedia.org/wiki/Food_sovereignty

Williams, R. (2011). Peak oil: just around the corner, The Science Show, (23 April), ABC Radio National, Australia, accessed at http://www.abc.net.au/rn/scienceshow/stories/2011/3198227.htm

Worldwatch (2011). *State of the World, 2011: Innovations that Nourish the Planet.* W. W. Norton, ISBN 978-0-393-33880-5, New York

Wright, J. (2009). *Sustainable agriculture and food security in an era of oil scarcity – lessons from Cuba,* Earthscan, ISBN 978-1-84407-572-0, London

Ziesemer, J. (2007). Energy Use in Organic Food Systems, FAO Natural Resources Management and Environment Department, Rome, accessed at http://www.fao.org/docs/eims/upload/233069/energy-use-oa.pdf

Permissions

The contributors of this book come from diverse backgrounds, making this book a truly international effort. This book will bring forth new frontiers with its revolutionizing research information and detailed analysis of the nascent developments around the world.

We would like to thank Shahriar Khan Associate Professor, for lending his expertise to make the book truly unique. He has played a crucial role in the development of this book. Without his invaluable contribution this book wouldn't have been possible. He has made vital efforts to compile up to date information on the varied aspects of this subject to make this book a valuable addition to the collection of many professionals and students.

This book was conceptualized with the vision of imparting up-to-date information and advanced data in this field. To ensure the same, a matchless editorial board was set up. Every individual on the board went through rigorous rounds of assessment to prove their worth. After which they invested a large part of their time researching and compiling the most relevant data for our readers. Conferences and sessions were held from time to time between the editorial board and the contributing authors to present the data in the most comprehensible form. The editorial team has worked tirelessly to provide valuable and valid information to help people across the globe.

Every chapter published in this book has been scrutinized by our experts. Their significance has been extensively debated. The topics covered herein carry significant findings which will fuel the growth of the discipline. They may even be implemented as practical applications or may be referred to as a beginning point for another development. Chapters in this book were first published by InTech; hereby published with permission under the Creative Commons Attribution License or equivalent.

The editorial board has been involved in producing this book since its inception. They have spent rigorous hours researching and exploring the diverse topics which have resulted in the successful publishing of this book. They have passed on their knowledge of decades through this book. To expedite this challenging task, the publisher supported the team at every step. A small team of assistant editors was also appointed to further simplify the editing procedure and attain best results for the readers.

Our editorial team has been hand-picked from every corner of the world. Their multi-ethnicity adds dynamic inputs to the discussions which result in innovative outcomes. These outcomes are then further discussed with the researchers and contributors who give their valuable feedback and opinion regarding the same. The feedback is then collaborated with the researches and they are edited in a comprehensive manner to aid the understanding of the subject.

Apart from the editorial board, the designing team has also invested a significant amount of their time in understanding the subject and creating the most relevant covers. They scrutinized every image to scout for the most suitable representation of the subject and create an appropriate cover for the book.

The publishing team has been involved in this book since its early stages. They were actively engaged in every process, be it collecting the data, connecting with the contributors or procuring relevant information. The team has been an ardent support to the editorial, designing and production team. Their endless efforts to recruit the best for this project, has resulted in the accomplishment of this book. They are a veteran in the field of academics and their pool of knowledge is as vast as their experience in printing. Their expertise and guidance has proved useful at every step. Their uncompromising quality standards have made this book an exceptional effort. Their encouragement from time to time has been an inspiration for everyone.

The publisher and the editorial board hope that this book will prove to be a valuable piece of knowledge for researchers, students, practitioners and scholars across the globe.

List of Contributors

Antonio Mariani, Biagio Morrone and Andrea Unich
Dept. of Aerospace and Mechanical Engineering - Seconda Universitá degli Studi di Napoli, Italy

Kazuhiro Hayashida
Department of Mechanical Engineering, Kitami Institute of Technology, Japan

Katsuhiko Haji
Advanced Technology and Research Institute, Petroleum Energy Center (PEC) (Present affiliation: Research & Development Division, JX Nippon Oil & Energy Corporation), Japan

Stanisław Gil
Silesian University of Technology, Poland

José Tavira-Mondragón, Guillermo Romero-Jiménez and Luis Jiménez-Fraustro
Electric Research Institute, Mexico

Pavel Kolat and Zdeněk Kadlec
VŠB-Technical University Ostrava, Czech Republic

R. W. Wies, R. A. Johnson and A. N. Agrawal
University of Alaska Fairbanks, USA

Nuno Luis Madureira
ISCTE-IUL, CEHC, Portugal

Shahriar Khan
Independent University, Bashundhara R/A, Dhaka, Bangladesh

Bharat Raj Singh
SMS Institute of Technology, Lucknow, India

Onkar Singh
Harcourt Butler Technological Institute, Kanpur, India

Salvador Vega, Rutilio Ortiz, Rey Gutiérrez and Beatriz Schettino
Laboratorio de Análisis Instrumental, Departamento de Producción Agrícola y Animal Universidad Autónoma Metropolitana Unidad Xochimilco, Colonia, Coyoacán, México

Richard Gibson
Institute of Agri-Food and Land Use, School of Biological Sciences Queen's University Belfast, Ireland

Mohammad Reza Lotfalipour
Department of Economics, Ferdowsi University of Mashhad, Iran

Malihe Ashena
Department of Management and Economics, Tarbiat Mdares University, Tahran, Iran

Victor Esteves and Cláudia Morgado
Federal University of Rio de Janeiro, Brazil

Richard E. White
Smith College, USA